DEEP WATER

DEEP WATER

THE EPIC STRUGGLE OVER DAMS, DISPLACED PEOPLE, AND THE ENVIRONMENT

JACQUES LESLIE

FARRAR, STRAUS AND GIROUX   NEW YORK

Farrar, Straus and Giroux
19 Union Square West, New York 10003

Library of Congress Cataloging-in-Publication Data
Leslie, Jacques, 1947–
    Deep water : the epic struggle over dams, displaced people,
and the environment / Jacques Leslie. — 1st ed.
        p.   cm.
    ISBN-13: 978-0-374-28172-4 (hardcover : alk. paper)
    ISBN-10: 0-374-28172-6 (hardcover : alk. paper)
    1. Dams.   2. Dams—Environmental aspects.   3. Dams—
Social aspects.   I. Title.

TC540.L495 2005
333.91'62—dc22

                                                2005013379

Designed by Jonathan D. Lippincott
Maps designed by Jeffrey L. Ward

www.fsgbooks.com

1   2   3   4   5   6   7   8   9   10

To Leslie,

who inspired, guided, and abided,

and Sarah,

an emerging writer

Whoever wishes to take over the world
will not succeed.
The world is a sacred vessel
and nothing should be done to it.
Whoever tries to tamper with it
will mar it.
Whoever tries to grab it
will lose it.

—Lao Tzu

CONTENTS

DEEP WATER

PROLOGUE

Start with the primal dam, Hoover. The first dam of the modern era is America's Great Pyramid, whose face was designed without adornment to emphasize its power, to focus the eye on its smooth, arcing, awe-inspiring bulk. Yet the dam nods to beauty, with a grace that grows more precious year by year: its suave Art Deco railings, fluted brass fixtures, and a three-mile-long sidewalk's worth of polished terrazzo-granite floors are the sort of features missing from the purely utilitarian public works projects of more recent decades. Hoover is a miraculous giant thumbnail that happens to have transformed the West. Take it away, and you take away water and power from twenty-five million people. Take it away, and you remove a slice of American history, including a piece of the recovery from the Depression, when news of each step in the dam's construction—the drilling of the diversion tunnels, the building of the earth-and-rock cofferdams, the digging to bedrock, the first pour of foundation, the accretion of five-foot-high cement terraces that eventually formed the face—heartened hungry and dejected people across the country. And you take away the jobs the dam provided ten or fifteen thousand workers, whose desperation compelled them to accept risky, exhausting labor for $4 a day—more than two hundred workers died during Hoover's construction.

The dam and Las Vegas more or less vivified each other; if Hoover evokes glory, Las Vegas, only thirty miles away, is its malignant twin. Even now, Hoover provides 90 percent of Las Vegas's water, turning a desert outpost into the fastest-growing metropolis in the country—by all means, take away Las Vegas. Take away Hoover, and you might also have to take away the Allied victory in World War II, which partly depended on warplanes and ships built in Southern California with its hydroelectric current. And take away modern Los Angeles, San Diego, Phoenix: you reverse the twentieth-century shift of American economic power from East Coast to West. Take away Hoover and the dams it spawned on the Colorado—Glen Canyon, Davis, Parker, Headgate Rock, Palo

Verde, all the way to Morelos across the Mexican border—and you re-store much of the American Southwest's landscape, including a portion of its abundant agricultural land, to shrub and cactus desert. Above all, take away Hoover, and you take away the American belief in technology, the extraordinary assumption that it above all will redeem our sins. At Hoover's September 30, 1935, dedication, Interior Secretary Harold Ickes exactly reflected the common understanding when he declared, "Pridefully, man acclaims his conquest of nature."

Hoover's image became one of the nation's most popular exports: after it, every country wanted dams, and every major country, regardless of ideology, built them. Between Hoover and the end of the century, more than forty-five thousand large dams—dams at least five stories tall—were built in 140 countries. By now the planet has expended $2 trillion on dams, the equivalent of the entire 2003 U.S. government budget. The world's dams have shifted so much weight that geophysicists believe they have slightly altered the speed of the earth's rotation, the tilt of its axis, and the shape of its gravitational field. They adorn 60 percent of the world's two hundred–plus major river basins, and the water behind them blots out a terrain bigger than California. Their turbines generate a fifth of the world's electricity supply, and the water they store makes possible as much as a sixth of the earth's food production. Take away Hoover Dam, and you take away a bearing, a confidence, a sense of what nations are for.

Yet in a sense, that's what's happening. Even if Hoover lasts another eleven hundred years (by which time Bureau of Reclamation officials say Lake Mead will be filled with sediment, turning the dam into an expensive waterfall), its teleological edifice has already begun to crumble. In seven decades we have learned that if you take away Hoover, you also take away millions of tons of salt that the Colorado once carried to the sea but that have instead been strewn across the irrigated landscape, slowly poisoning the soil. Take away the Colorado River dams, and you return the silt gathering behind them to a free-flowing river, allowing it again to enrich the downstream wetlands and the once fantastically abundant, now often caked, arid, and refuse-fouled delta. Take away the dams, and the Cocopa Indians, whose ancestors fished and farmed the delta for more than a millennium, might have a chance of avoiding cultural extinction. Take away the dams, and the Colorado would again bring its nutrients to the Gulf of California, helping that depleted fishery to recover the status it held a half century ago as an unparalleled repository of marine life. Take away the dams, finally, and the Colorado River returns to its virgin state: tempestuous, fickle, in some stretches astonishing.

From the peak of dam construction in the early 1970s, when large dams rose at the rate of nearly a thousand a year, the pace has dramatically slowed. One part of the explanation is simple topography: particularly in the United States and Europe, the best dam sites have been used. The other part reflects a gradual appreciation of dams' monumental destructiveness. Dam-planning processes, once the province of bureaucrats, engineers, and economists, have expanded to include environmentalists and anthropologists charged with limiting dams' harm. And environmental and human rights activists in the United States and Europe have allied with groups in poor countries whose members are threatened with displacement. Though limited by their tiny budgets (and, typically, police intimidation), the groups discovered that if they could tie up projects in long delays, investors might withdraw.

The battle over dams now is at the core of conflicts throughout the world involving water scarcity, environmental degradation, biodiversity loss, development and globalization, social justice, the survival of indigenous peoples, and the growing gap between rich and poor. As water has grown scarce in one river basin after another, some people have predicted water wars, but the mortal struggle involving dams is already a couple of decades old. How it is resolved will determine the fate of countless river basins and the life—human, animal, and plant—that they support.

Despite their paltry resources, dam opponents in recent years have won the more telling victories. Under pressure from its critics, the world's largest dam financier, the World Bank, established policies to protect indigenous peoples and tightened its regulations to improve resettlement and limit environmental harm—but the Bank often ignored its own policies. In 1993, it established an appeals mechanism, the Inspection Panel, which allowed people adversely affected by Bank development projects to file claims—and dams became by far the likeliest Bank projects to elicit complaints. These constraints constricted dam construction. Between 1970 and 1985, the Bank supported an average of twenty-six dams a year, but as the projects grew politically charged, the number dropped to four a year over the next decade. And dams became the Bank's most problem-ridden projects; as Bank senior water adviser John Briscoe put it, a major dam project "will often account for a small proportion of a country director's portfolio but a major proportion of his headaches."

By the mid-1990s, the Bank was staggering from one dam-related embarrassment to another. For the first time in its history, it was forced to withdraw from a project it had begun funding—naturally, a dam. And when the Inspection Panel responded to its first claim—also involving a dam—by questioning the project's value, the Bank canceled it. Led by the tiny but effective International Rivers Network of Berkeley, California, dam opponents campaigned for the creation of an independent commission that could arrive at an honest assessment of Bank dams' performance. On the defensive, the Bank agreed, with the proviso that the commission study not just the Bank's dams, but *all* large dams—an apparent attempt to divert attention from the Bank's problem-ridden dams. The result was the formation of the World Commission on Dams, an independent body of twelve commissioners, charged with assessing dams' impacts, positive and negative, and providing guidelines for future construction. "Truce called in battle of the dams," said a 1997 *Financial Times* headline over a story about the commission's creation. "The end result," the story said, "may be the development of pathbreaking international guidelines for building and operating dams which balance the competing demands of the economy and the surrounding environment."

In pursuit of independence and across-the-spectrum representation, the commission's organizers drew commissioners equally from three categories of nominees: "prodam," "mixed," and "antidam." Among the commissioners were Göran Lindahl, president of ABB Ltd., then the world's largest supplier of hydropower generators, and Medha Patkar, the world's foremost antidam activist, an Indian firebrand whose protests against dams repeatedly involved courting her own death. The commissioners were so diverse that few people who followed the commission thought they could achieve consensus. Yet in November 2000, two and a half years after its formation, the commission unveiled its final report in London, accompanied by an enthusiastic keynote address by Nelson Mandela. The commission's success seemed to herald the growing role of nongovernmental organizations, and speculation spread that the Bank would use a process similar to the formation of the commission in constituting a newly announced review of Bank participation in the international oil, gas, and mining industries.

The finished report, titled *Dams and Development: A New Framework for Decision-Making*, was more than four hundred pages long. Its first part, based on the findings of the most thorough study of dams' impacts ever conducted, seemed to confirm many dam opponents' claims. It said large dams showed a "marked tendency" toward schedule delays and significant cost overruns; that irrigation dams typically did not recover their costs, did not produce the expected volume of water, and were less profitable than forecast; that their environmental impacts "are more negative than positive and, in many cases, have led to irreversible loss of species and ecosystems"; that large dams' social impacts "have led to the impoverishment and suffering of millions, giving rise to growing opposition to dams by affected communities worldwide"; and that since the environmental and social costs of large dams have never been adequately measured, the "true profitability" of large dam schemes "remains elusive." The commission even challenged the conventional wisdom that a major advantage of dams over fossil-burning energy sources is that they don't contribute to global warming. On the contrary, the commission said, dams, particularly shallow, tropical ones, emit greenhouse gases released by vegetation rotting in reservoirs and carbon inflows from watersheds.

For all that, the document's second part is more important than its first, for it provides a framework for building dams in the future. Most controversially, it lists twenty-six guidelines meant to replace the existing arbitrary and politically weighted process of dam decision making. It calls for examining cheaper and less damaging alternatives before deciding on dams, obtaining the "free, prior and informed consent" of indigenous people threatened by dams, and planning water releases from dams that can mitigate environmental damage by mimicking rivers' predam flow. Given the many disasters that dam projects have produced, the recommendations' guiding concept was to identify problems before dams are built instead of afterward, in the wake of tragedy.

But the report drew only scattered endorsements, and many of those were hedged. Most significantly, the World Bank turned its back on its own creation. The Bank had done something like this once before, but on a smaller scale. In 1991 it had funded an independent panel to review a hugely controversial dam in India, then tried to ignore the panel when it recommended quitting the project. (This dam, Sardar Sarovar, is the subject of Part I.) Now Bank officials said the guidelines were too numerous and cumbersome, and would cause long project delays. The Bank took thirteen months to issue an official response to the report, by which time it was anticlimactic. The statement diplomatically praises the report, calling it "a carefully prepared and well-written" document that "makes a substantial contribution" and "presents innovative ideas"—but none were so substantial or innovative as to cause the Bank to change any of its policies. Indeed, the statement ends by touting the Bank's policies, not the report's. Briscoe, who'd been instrumental in selecting commissioners, charged that dam opponents "hijacked" the commission. A more convincing explanation of the commission's findings is that they arose from dams' historical record. Complacently counting on triumphal conclusions, the Bank gambled on a favorable report and lost.

International commissions typically fade quickly from the collective consciousness, but despite the Bank's disregard, the report so far has escaped this fate. A few countries and regional groups—South Africa, Vietnam, Thailand, Nepal, and the Southern African Development Community—have launched initiatives to consider the report, and many others have taken fledgling steps. Even without the World Bank's endorsement, the commission report has become a standard, a compilation of best practices against which less rigorous approaches are measured. Unheeded but not forgotten, the report hovers over dam projects as an admonition to dam builders in the name of human decency and environmental sanity. Meanwhile, the battle between dam builders and opponents has been rejoined.

Water is one of the great looming subjects of the twenty-first century; the collision of the burgeoning human population and the planet's unchanging supply of freshwater has already started and will grow worse. Five years ago, I wrote a piece for *Harper's Magazine* that laid out the evidence for this claim, but the effort left me, shall we say, thirsty: now that I'd identified the enormous conflicts that water was triggering, I wanted to portray some of them in narrative form, to bring them to life. It seemed obvious that my focus should be on dams, the largest structures built by humans and the sites of so many different sorts of drama, where development's tentacles reach into remote valleys and upend the existing cultural and environmental regimes. Accordingly, I followed the proceedings of the World Commission on Dams until I realized that, for a writer's purposes, it offered an invaluable frame. Just as the commission's organizers sorted commissioner nominees into three categories and picked an equal number from each one, for roughly equivalent reasons I chose as subjects for this book one commissioner from each group—an Indian activist, an American anthropologist, and an Australian water resources manager. Then I depicted them at work. My intent has been to see dams whole, and in doing so to glimpse the fate of the earth.

PART I: INDIA

TO STRUGGLE IS TO LIVE

1 DOMKHEDI

To get to Domkhedi, Medha Patkar's monsoon headquarters, you fly to Bombay, then to Baroda, the second-largest city in the state of Gujarat. You call up Joe Athialy at the Baroda office of the Narmada Bachao Andolan, the "Save the Narmada Movement" that Patkar leads, and Joe invites you over. Note that "office" is misleading to the extent that it implies accoutrements such as desks and chairs—of the Andolan's two offices, in Baroda and Badwani, Baroda's is the nicer one, and it consists of a couple of grimy, largely empty rooms on the second floor of a moldering residential apartment building. You and Joe sit cross-legged on the rug-covered part of the floor and work out the details of the last leg of the trip. In my case, the only rough spot was when I told Joe I wanted to take some bottled water to Domkhedi, and he said the Andolan opposes bottled water because it is an instrument of globalization. I was willing to concede the point on ideological grounds, but I couldn't get over the medical ones. Given the abundance of disease-producing bugs in Indian drinking water, consuming it struck me as highly imprudent. It's a measure of Joe's grace—and the Andolan's—that despite my political lapse, he arranged for Ajit, the Mario Andretti of Andolan drivers, to take me to Domkhedi in a four-wheel-drive jeep and told Ajit to stop at a shop along the way where I could buy a few cases of bottled water.

The drive took five hours. After five minutes, we all smelled of diesel, the odor of modern India, and wore a layer of diesel grime. My friend Bob Dawson, an environmental photographer, and I sat in the backseat; in the front sat Ajit, the race-car driver, and Champalal, an endearingly cheerful young activist. Ajit and Champalal spoke animatedly to one another in Gujarati and occasionally tossed a word or two of broken English in our direction. "Narmada," they'd say, and point off into the distance toward our destination, reputedly India's most beautiful river. "Canal," they said once, but no explanation was needed—at that moment we were on a bridge over the world's largest canal, itself an expensive adornment to one of the planet's largest dams, Sardar Sarovar, which already loomed 295 feet over the Narmada yet was only two-thirds completed. "Dam" was the one word nobody uttered, but we all knew its approximate location, thirty or forty miles downstream from our destination. In a fundamental way, the dam was the reason for our journey.

Sardar Sarovar is a wall, a monument, a modern ziggurat—it's as voluminous as two or three Great Pyramids, and it attracts equivalent adoration. It's a block so massive that its construction would be noteworthy even if it weren't bisecting a riverbed, holding back a seasonally torrential river. Large dams come in two basic shapes: arches, such as Hoover, which are elegant, and use the massive force of reservoir water to fortify their walls; and the blunt instruments of gravity and embankment dams, which fling themselves in a straight line across riverbeds too broad for arches, and rely on raw mass to stay there. Sardar Sarovar is a spectacular specimen of gravity dam: it's a straight cement salient across a bed nearly a mile wide, transforming the Narmada behind it from sinuous river to jagged, elongated lake, and seeming to promise in its bulk and geometric regularity an end to nature. The overlapping sine-wave-shaped hills that arise from both shores still turn throbbingly viridescent during the rainy season, but the dam will be the final indignity that the hills' inhabitants endure. Over the last few centuries, tribal people fled to them to avoid the depredations of dominant Hindus and learned how to survive in the forest. By one anthropologist's count, they found uses for sixty-four tree species, from edible fruit to bows and arrows, from medicine to fuel, from goat fodder to a paste for stupefying fish. Then, over the last century, the state logged the forest, until much of it was degraded. Now Sardar Sarovar will force many tribal people out of the forest entirely. The presumably luckier ones will receive infertile lowland plots; the others will end up on city streets. Either way, the tribal cultures will shatter. Sardar Sarovar is a wailing wall.

The V-shaped canal, the dam's upside-down, scooped-out converse, flows northward, perpendicular to the river, as rigid as a robot's arm. The canal is the chief justification for the dam, the reason the dam's planners felt no compunction about displacing two or three hundred thousand people from the Narmada Valley and a hundred thousand more from the canal's path. At its head, the 270-mile-long canal is as wide as two football fields are long, and will bring water to the "drought-prone" areas of Kutch and Saurashtra in northern Gujarat, turning a desert into an oasis for twenty million people—according to the planners. There are many reasons to doubt that the water will ever get there, including the water-guzzling "development projects" accumulating at the head of the canal to intercept the water: sugarcane plantations, chemical plants, golf courses, hotels, and a water park (a *water park*!). The government itself has conceded that the water won't reach Kutch and Saurashtra until 2025—a date, as Dilip D'Souza wrote in *The Narmada Dammed: An Inquiry into the Politics of Development*, "so remote as to be meaningless." Now, however, as we crossed over the canal, the debate remained hypothetical—the dam hadn't yet been built high enough to divert water into the canal, and if Medha had her way, it never would. We looked down into the canal's symmetrical, brick-lined V and saw near the bottom a pane of nearly motionless steel-blue water—the canal was nearly empty. In a certain way, the dam was a technological marvel, but the women at the water's edge were cleaning clothes in the traditional fashion, pounding their fabrics into submission against the canal wall.

There is no gracious way to say this: we'd come to see Medha try to drown. True, we had plenty of other reasons for making the trip, and we understood that it probably wouldn't be a failure if she didn't drown, but the drama of the spectacle pulled us in. Drowning was the reason we chose to visit Medha at a remote malarial hamlet on the lip of a Narmada tributary, and it was why we were arriving in late July, in the heart of the 2001 monsoon season, even though the heat and humidity would be prodigious, bordering on preposterous. This summer, as in the previous several, she'd announced her intention to drown in the Sardar Sarovar reservoir's rising waters, in hopes that the furor her death might induce would force the government to reconsider its zeal for dams. One sign of the depth of her commitment was that she stationed an Andolan outpost on a low-lying patch of Domkhedi (pronounced dom-KED-ee) ground. Now that the dam was high enough, the rains of a strong monsoon would lift the reservoir's water level above the Andolan's huts, thereby affording Medha and her squad of volunteers the chance to drown. But the monsoon often failed, and when it didn't, the police intervened, to save the Indian government the embarrassment of her death. In September 1999, Medha and several other activists stood in bone-chilling water for thirty hours, until they were neck-deep; then the police used a barge to break down the hut wall, yanked Medha from her comrades' desperate grasp, and detained her until the water level declined. This did not deter Medha from trying again. As our trip approached, Bob and I kept reading announcements on the Friends of the Narmada Web site that made us worry we'd reach Domkhedi too late. One headline said, "Major floods in Narmada Valley in next 24 hours feared; Satyagrahis"—protesters—"to face submergence." A few days later, another one said, "Water level stabilises at two feet below Satyagraha house." As soon as we reached Bombay, I called Joe to find out about the water level, and I asked him again when we got to Baroda. His answer both times was that it had risen no farther. My relief was momentary: I began to worry that we'd arrive in Domkhedi too long *before* the inundation.

We shared the pitted road with goats, dogs, donkeys, rooting pigs, the occasional camel, yoked water buffalo with horns painted in rainbow colors, and, of course, cows. Cows stood or sat contentedly on the road, sometimes stationing themselves perpendicular to the traffic flow, unruffled by the speeding vehicles navigating around them. The nonanimal occupants of the road were just as varied; Ajit maneuvered adroitly among cars, buses, bicycles, motorcycles, auto-rickshaws, motor scooters (sometimes with sari-clad passengers riding sidesaddle on the back while clutching up to three small children), women carrying large water vessels and other containers on their heads (men didn't perform this sort of heavy lifting), and people *sitting* on the highway, for God's sake, shooting the breeze. Plus, at irregular intervals we encountered speed bumps. Among vehicles on the road, Ajit's jeep was relatively unusual in that it went fast enough to have to slow down for the bumps: on more than one occasion I watched with interest as the jeep's odometer hit 75 miles per hour. Ajit handled the car with a hot-rodder's confidence. He was forever venturing out into the incoming lane to pass one or more slower vehicles, then darting back into an emerging opening an instant before, say, an oncoming truck pulverized us, or a lumbering bullock didn't quite get out of our way. As any Indian driver will tell you, a car is only as good as its horn; Ajit used his as justification for assuming that prospective targets of our jeep—see under "goats, dogs . . . ," etc., above—would move away in time. They always did, but I wondered when Ajit would encounter, say, a deaf water buffalo, and envisioned us flying through the windshield of the seat belt–less car.

Every town looked a grimy brown, and the flanks of every road in every town were encrusted with trash. The town where we stopped to buy water was indistinguishable from the towns before and after it where the Indians stopped for "tea" (a euphemism, I learned, for "meal"). Champalal had little trouble finding a merchant selling cartons of Pepsi-Cola-produced Aquafina, one of two brands of water an Indian doctor had told us were certifiably free of bacteria. We bought the water, bread, packaged cookies, and, at a fruit stall on the corner, bananas. In our voluminous bags, now stuffed into the back of the jeep, we also carried energy bars. Though Champalal would accompany us to Domkhedi, he carried nothing.

At the outset the road was bad, and it gradually got worse. For the last few miles, it all but disappeared—we were on the only road toward Domkhedi that the modest monsoon rains had not fully dismantled, and even it was washed out in parts. From flat farmland stretching as far as the horizon, the terrain changed to lush, wild foothills. The weather turned overcast, so weighted with humidity that I thought I could make out balls of gaseous moisture as big as gnats. For the first time on our trip, we saw land that was not densely populated. We crossed a gurgling seasonal stream over and back, and drove straight down its bed a few times. Ajit was spinning the steering wheel like a sea captain, earning his pay. I was beginning to contemplate *floating* to the river just as he found the last stretch of road. It led us to a Hindu temple on the edge of the monsoon-swollen reservoir, then descended and disappeared into the water—the reservoir swallowed it whole.

The reservoir looked too bland to be capable of killing: it was a silent, waveless flood, a flood in slow motion. It receded during the long dry season, but thanks to the accreting dam, during each successful monsoon its level reached an unprecedented height. It had already swallowed whole villages; on behalf of the inundated inhabitants, the Andolan had waged valiant, doomed struggles in the lowlands that were now underwater. Mist-covered hills above the reservoir had taken on their monsoon shading, a vibrating, nearly fluorescent green nearest the water, so bright that it obscured variations in the land and foliage. In the distant, higher hills, where the moist air dulled it, green gave way to gray and then to the lusterless white of the overcast sky. Abruptly, the hills were bisected by the reservoir's edge, a line so exact it seemed drawn with a razor, unaccustomed to the terrain it so thoroughly invaded, oblivious to undulation, ungraced by sandy shore or wetland, ahistorical. Beneath the line, the hills' green reflection gradually darkened until it was overtaken by the brown gruel of the reservoir's water, a souvenir from the obliterated sediment-rich river. Through the reservoir's surface protruded monstrosities: half-submerged trees and bushes, most of their visible portions already gray and shriveled. A smaller number were still green, still resisting the inundation—activists of a sort.

Several of the human activists had already gathered in front of the temple, waiting for the boat. The stern-faced Shripad Dharmadhikary, who'd graduated from India's best engineering college and then threw over a promising career for Medha and the Andolan, suggested that Bob and I go into the temple, but showed no interest in entering himself. Instead, as soon as the boat arrived, I noticed our seven bags and three cases of bottled water making their way to the boat before us, in the arms of activists. I realized that Bob and I, too, were being borne along, from Joe through Champalal, then Shripad, all the way to Medha. I peeked inside the temple from the entrance and got a glimpse of what looked like eternity: a sleeping sadhu, a shrine behind him, and stone bathed in a murky golden sunlight. Eternity interrupted: the temple, called Hapeshwar, would be submerged if the dam was built to its full height. The authorities had considered this fact, and constructed a new temple on higher ground, but nobody visited it—and besides, the activists were saying that it would eventually be inundated, too. (The drollness of India is infinite.) The oldest of the Narmada Valley's thirty-two major temples, the archaeologically significant Shulpaneshwar, was the first to be submerged; now people worshipped there by taking water from a spot on the reservoir's surface directly above it. "So what?" a local politician dismissively told journalist Sameera Khan, when she asked him how he felt about the temple's inundation. "It still exists."

I encountered Medha Patkar for the first time in March 2000, at the World Bank–funded World Water Forum in The Hague. Neither Medha nor the event dazzled me. The event assembled water ministers from 115 nations for the ostensible purpose of ratifying a policy statement advocating World Bank–devised solutions to the world's burgeoning water scarcity problems. A lack of unanimity on the merit of this prospect became evident at the opening ceremony, which was disrupted by four antidam activists. Two disrobed on stage, one rappelled from the balcony to the auditorium floor, and one climbed an interior wall, all while shouting denunciations of water privatization and a dam under construction in Spain. After that, the conference followed the Bank's prodam, proprivatization script all the way to its sixth and final day. Ismail Serageldin, the suavely attired World Bank vice president who oversaw drafting of the proposed proprivatization agenda, proclaimed that the conference would mark "the birth of a movement." He gave every evidence of taking for granted the water ministers' endorsement of the agenda right up until the moment they voted it down, on the conference's last day. With this jaw-dropping development, the conference gained entry on a long list of World Bank water fiascoes.

With more than four thousand people attending, the conference was part debating society, part trade fair, and part showcase for disenfranchised politicians. Among those wandering the halls were Shimon Peres, Robert McNamara, and Mikhail Gorbachev, who held a press conference to extol environmentalism with his familiar mustachioed Soviet-era translator at his side. Expert panelists at about eighty overlapping workshops argued over dams, water scarcity, privatization, and full-cost water pricing. The emphasis on private enterprise resonated in the bizarrely festive World Water Fair adjacent to the conference hall, where corporate giants such as Vivendi and Nestlé's Perrier Vittel SA installed pavilions alongside those of United Nations agencies, while women in mermaid costumes danced and rock musicians sang "Water Makes the World Go Around."

Within this dystopian eddy, Medha and Arundhati Roy, the writer and pro-Andolan activist, circulated like drifting leaves; they made solemn, defiant appearances at colloquies on such grave subjects as globalization and privatization, and concluded with a press conference of their own. The two women were the odd couple of Indian antidam activism. Roy dazzled: the Booker Prize–winning author of *The God of Small Things* was making a celebrity appearance, a star turn. At thirty-nine, she was seven years Medha's junior, and beautiful. She wore stylishly casual Western clothes and spoke with a verbal command that Medha's English lacked. Her accent carried just a hint of her native Kerala, enough to make her seem tropical, flamboyant, unknowable. She lured her listeners with verbal intimacy, while holding them at bay with her politics and indignation. At the press conference, she explained her radicalization in lyrical fashion: "As my book sold and sold and sold, I watched the air grow blacker, the fences grow higher, the people pushed into dark places like lice, and still my bank account grew." The conference, she said, "is a meeting of powerful people who've come to stroke each other's backs"; a better alternative for participants would be to visit the Narmada Valley "if you want to see a really magical war being fought." Roy charmed her audience even as she cut it dead. She was witty, impetuous, fierce, and extravagantly self-involved, as evidenced by her announcement that she would attend no more conferences like this one. What surprised me was that that declaration mattered to so many people. From the floor, an Oregon State professor named Aaron Wolf, a specialist in international water politics, sweetly asked her to reconsider, as if he were a disappointed fan. She declined.

Medha seemed as relentlessly Indian as Roy was cosmopolitan; it was as if she were wearing plaid in a polka-dot universe. She was formal, ideological, impersonal, and unceasingly serious: she was the frumpy, graying scold in the faded sari. Many of her comments took the form of denunciations—of dams, the World Bank, globalization—all pieces of what I later understood as the Andolan's Gandhian/Chomskian/feminist/Mother Teresan worldview. She did not flirt. She made no one smile. Her speeches were variations on *The Little Engine That Could*: she spoke softly at first, as if headed uphill, then huffed and puffed, seeming to draw energy from the act of oration, until the pistons churned and her hoarse voice grew loud and the accusations flew. She said privatization didn't acknowledge the contribution of local people, the people who tended the forest and got displaced by dams, because they weren't investors. She said privatization rewarded cash crops, which meant that poor people, who depended on subsistence agriculture, went hungry. Nearly yelling, she said, "States now are puppets of the World Bank and the International Monetary Fund!" Then, the mountain climbed, the point made, she stopped.

It wasn't until a year later, when I spent a month in Cape Town, the headquarters of the World Commission on Dams, that I began to suspect I'd gotten things backward. The commission's Secretariat members were mostly water experts, few of whom believed that the dam debate was as one-sided as Medha did, and they weren't noted particularly for humility or guilelessness—yet they spoke about Medha with awe, a hint of reverence, as if simultaneously acknowledging her stature and vulnerability. Everyone called her by her first name. Even the engineers and capitalists across the ideological abyss, who are as enthusiastically prodam as Medha is anti, treated her gently. It was understandable, they said, that Medha would fight for displaced people, but their plight did not invalidate dams. Most people spoke of her toughness, her determination, her willingness to suffer. I was told that at the World Commission on Dams's first meeting in Washington, D.C., when the twelve commissioners stood tensely in the same room for the first time, Medha quickly dispensed with small talk. In front of Göran Lindahl, a fellow commissioner from Sweden, she spread out recent photographs of police beating protesters at the Maheshwar Dam, another conflict-laden Narmada Valley project; Lindahl was president and CEO of ABB Ltd., one of the world's largest industrial concerns, which owned a stake in the dam. (According to a third commissioner, Lindahl's response was a dismayed "Enhhhhhh.") I learned that Medha declined to eat her meals during a flight to a commission meeting in Cairo, and gave the food to Cairo street children after disembarking. I was told that Medha attended one commission session immediately after a week-long hunger strike, and threw up everything she ate. (A staff member eventually found her some plain yogurt, which she kept down.) And I was told that during the heated last weeks of negotiations over the commission's final report, Medha was the only commissioner who won, as the price of her signature, the placement of an individual "comment" at the report's end, in which she bemoaned the commission's refusal to consider dams' relation to the "unjust and destructive dominant development model"—globalization. Of the commissioners, Medha was the best known, the most outspoken, the most politically sharply defined, yet, against all the odds, she was the most admired and, in a certain way, cared for. The impression I gleaned from the staff members in Cape Town was that while Arundhati Roy wrote passionately about the immorality of the Narmada dam project, Medha embodied morality.

I knew that much, and still I was not prepared for our first conversation. It was at the commission's last function, when its "stakeholders"— engineers, investors, governments, antidam campaigners—convened for two days at a wine resort in Stellenbosch, outside Cape Town, to decide how wholeheartedly to disseminate the commission's final report. Even in absentia, Medha was under attack in India; the previous day, the Indian Supreme Court had announced its decision to try Roy and her on trumped-up contempt charges, and the threat of extended jail terms suddenly loomed over them. A good chance existed that if the two women stopped denouncing the Court, it would lose interest in the case, but Medha couldn't resist. From South Africa, she issued her customary statement: angry, principled, hopeful, and prolix. She declared that she hoped the contempt proceeding would give her "an opportunity to present, not just before a court of law but before civil society what we really feel about the judiciary and how we view justice . . . We hope that the judges, who once exhibited the broadness of their shoulders, will again express their openness of heart."

I had by now decided that I wanted to visit Medha in the Narmada Valley, and, Arundhati Roy's invitation to the contrary, I was not at all sure she'd like that news. I was, after all, an American, with unspecified, and therefore suspect, political views, and I was proposing to do extended reporting in her territory. I also wanted what she valued most: her time. During a break in the Cape Town deliberations, I asked Medha if we could chat for a few minutes, and a little while later she obliged. As she walked toward me at the rear of the auditorium, the woman whose voice boomed through every hall she spoke in seemed to shrink, until, by the time she reached me, I loomed over her by nearly a foot. We sat down on folding chairs, and I told her what I wanted to do. She responded so softly I could barely hear her. "You are welcome," she unhesitatingly said, and added that the Andolan was putting on a "program" in April. I mentioned the contempt case. She dismissed it with a laugh.

Medha is often compared to Gandhi, whose repertoire of nonviolent civil disobedience tactics she has embraced and refined. Both Gandhi and Medha used their own suffering to advance their cause: while Gandhi spent more time in jail than Medha has, Medha has gone on longer and more frequent fasts. Her longest fasts—twenty-six days, twenty-two days, seventeen days—have pushed her body to its mortal limits, with the result of permanent physical damage; she suffers from a urinary tract disorder and spondylosis, a degenerative spinal disease, and discreetly wears an elastic lumbar brace beneath her sari. Like Gandhi, Medha walks a knife-edge between violence and nonviolence, deploring violence as a political tactic yet repeatedly offering herself as its victim. It's a mistake to think of political nonviolence as the opposite of violence: "passive resistance" is anything but. Rather, nonviolence walks so close to the edge that it feels the power of bloodletting and feeds off it—it's Medha's hot-blooded restraint, her benignly channeled fury, that makes her compelling. No violence, no material comfort, and, at most, a two-meal-a-day regimen of rice and vegetables—the Andolan discourages even hating. The Andolan's menu of political "actions" is itself an expression of Gandhian suppleness: marches, sit-ins, dam occupations, office seizures, *gheraos* (the surrounding of officials, sometimes for days, until they agree to specified demands), *rasta rokhos* (road and traffic blockades), *gaon bandhis* (closing villages to prodam officials), *satyagrahas* (literally, "insistence on truth," expressed in such political actions as attempts to drown rather than move from rising reservoir waters), and *jal samarpans* (declarations to drown on a certain date unless officials accept specified demands). These actions all but invited police to respond brutally, and the police reciprocated. The Andolan's list of police-inflicted casualties includes a fifteen-year-old boy shot dead, two girls raped, most of the female activists assaulted, virtually all the activists repeatedly beaten.

Among activists around the world, suffering—even fasting until death—is not uncommon. In 2001, for example, hundreds of protesting Turkish prisoners and their supporters hovered near death for months by consuming nothing but salted or sugared water, until at least forty of them died; the Reverend Al Sharpton conducted an extended fast in Puerto Rico. Medha's distinction is not the extremity of her suffering, but its breadth: she piles one physical burden on top of another, as if stopping to recover from any one of them might imperil her momentum. On the nineteenth day of Medha's longest fast, police tried to end it by forcibly injecting her with nutrients; her response was to deny herself not just food but also water for the next twenty-four hours. When she finally ended the fast, after twenty-six days, she drank some juice, then stood on a car and gave a speech to a cheering crowd. Despite her various ailments, she habitually neglects her health. She moves around India relentlessly, never staying in one place more than a day or two, grabbing sleep in three- or four-hour intervals while Ajit hurtles her down some rutted highway, on her way to a strategy meeting followed by a protest followed by yet another late-night meeting. Even the Narmada's tribal people, who are used to walking for hours from village to village, are awed by Medha's energy: most people from the cities can't keep up with them, they say, but they can't keep up with Medha. Over fifteen years, she has traveled up and down the length of the projected reservoir, 133 miles, many times. She sleeps on the ground as many tribals do, and bathes in their fields, and eats what they eat. In appreciation, they call her *Medhadidi*—Elder Sister Medha. For Medha, who is lavishly gifted in elder sister skills, it's an apt honorific.

Whereas Gandhi's political and spiritual pursuits were intricately intertwined, Medha professes no such connection—she calls herself an atheist. Yet in her usage the word turns squirrelly, until its meaning is blurred—for the truth is that she's as close to being all spirit as anyone I've ever met. "I believe in and live on spiritual strength, with all humility," she once told me. "I don't pray and perform *pooja*"—rituals. "Spiritualism is much beyond gods and religions as rituals. It's human. I believe in everything human."

Prominence was never Medha's goal. If it had been, she surely would have found an easier way to attain it than fighting for dislocated tribal people. With its population of a billion-plus and boundless poverty, India is abundant with appalling cases of discrimination, abuse, neglect. A staggering third of the Indian population—about 350 million people, the U.S. population plus 20 percent—consists of ecological refugees. They've been pushed out of countryside homes by dams, mines, deforestation, and the other seemingly inexorable consequences of development, and now live on the urban margins as day laborers, hawkers, beggars. Indian dams alone have displaced between 21 and 55 million people—take your pick of estimate. In response, hundreds of tiny rights groups manned by earnest social activists much like Medha have arisen throughout India. They fight disheartening and usually futile battles against government-sanctioned injustice; for this, their typical rewards are police beatings and poverty wages. Killings aren't uncommon: during my month-long visit to India, for example, police in a village in the state of Bihar shot and killed six flood victims and severely injured at least twenty others after they protested their village's omission from a fair share of desperately needed food relief. The Andolan is only one of dozens of groups that have been formed to fight dam construction in India, yet Medha has enjoyed such uncommon success that her fate has been to stand for all of them. In 1993, the Andolan's resistance led to the World Bank's withdrawal of support for the Sardar Sarovar dam—the first time in its forty-seven-year history that the Bank abandoned an ongoing development project—and Medha became famous. A year later, the Indian Supreme Court accepted the Andolan's petition to assess the dam's adherence to resettlement and environmental regulations, and Medha seemed on the verge of an even more improbable victory—stopping the dam. At the very least, she'd slowed it down: until the Court reached a verdict six years later, no more construction was allowed. In 1991, she won the Right Livelihood Award, the so-called Alternative Nobel Prize; a year later, she received the Goldman Environmental Prize, the world's largest prize ($60,000) for grassroots environmentalists. Hers was the rarest of rights movements, the movement that tasted victory. When she overcame the World Bank's opposition to become a commissioner on the World Commission on Dams in 1998, the global antidam movement was effectively legitimized.

In October 2000, she seemed to have brought off yet another improbable victory, as the commission's final report validated many of her assertions about the destructive impact of dams. Her gratification, however, was short-lived. On October 18, 2000, the Supreme Court ruled that the Andolan's case had no merit, and ended the long hiatus in dam construction. Coverage of the Andolan in Indian newspapers dropped; journalists said the Andolan was in decline. Now the only question left was whether the Andolan could stop the dam short of its planned 455-foot height.

It was partly in the service of that diminished cause that Medha was again threatening to drown, but her targets now were more numerous: she fought not only against Sardar Sarovar, but against all ill-considered dams, and for that matter, all rapacious development, which to her was just another phrase for globalization. Now she was across the bloated Narmada, awaiting our boat.

The boat—a humble, mud-splattered motorboat big enough for forty or fifty people, covered by a blue plastic tarpaulin—was plainly utilitarian, but its provenance is literary. It represents the bulk of £20,000, about $30,000, that Arundhati Roy was awarded along with the Booker Prize, all of which she handed to the perennially impoverished Andolan. In the newly swollen Narmada, the purchase was a necessity, for the tribal inhabitants' modest hand-crafted boats were abruptly obsolete. Indeed, now that the roads were washed out, the motorboat was the only transport of any kind that could reach Domkhedi.

We stopped in Jalsindi (jal-SIND-ee), the site of another Andolan outpost where a few more activists boarded, then crossed the reservoir to Domkhedi. The boat now carried thirty-four people, and as we approached Domkhedi, many of them greeted activists onshore with raised fists and chanted Hindi slogans. The Domkhedi people responded with slogans of their own, until the two sides' chants merged. "We are—one!" they yelled. *"Zindabad!"*—freedom. Medha was onshore, smiling as she chanted. From one of the huts, a powder-blue Andolan flag waved: it showed four people—stick figures, essentially—holding hands.

We had reached the highlands, and felt as if we'd gone so far inside India that we'd somehow fallen through, as if we'd found India's obverse. Here were no diesel fumes, no billboards, no trash, not even roads or solid buildings. The heat was oppressive, the humidity was off the charts, but the sweat on our foreheads felt clean. The Andolan outpost—two large huts and a covered outdoor meeting place—was squeezed onto a patch of slope that was otherwise entirely covered with chest-high corn: the hills looked like a soft, green undulating brush. In our folly, Bob and I had brought a tent, but now we saw the way every square inch of land supported the common agricultural enterprise, and realized there was no room for something so self-centered. This was tribal land, occupied for generations by indigenous non-Hindus, descendants of people who once lived in the plains, then may have been driven into the hills by Aryan invaders. Within the rankings of the traditional Indian caste system, tribals are beneath even untouchables, beneath category, if you will, and therefore only nominally human. The Indian Constitution of 1949 magnanimously singled them out for preferential treatment, in a kind of permanently codified affirmative action plan, but for the last half century the government has still managed to overlook them. Almost all of India's seventy million tribals are illiterate, and their life expectancy is forty-seven years. They typically receive no government services, and when a state government feels like building a dam, it is often tribals who are forced to move, without thought given to where. Of the tens of millions of people displaced by India's large dams, about half are tribals, even though they comprise less than a tenth of the country's population. The result is that many tribal communities have been shattered. Medha Patkar, an educated half Brahmin from Bombay, has joined a short list of outsiders who have since colonial days fought to make the tribals visible: she has become a prism through which you can see the tribals' suffering.

In Domkhedi, Medha wore an amber cotton sari, and her hair was tied behind her back into a long, graying braid. She displayed no jewelry, no makeup, and no airs, and moved with a litheness that tempered her unmistakable gravity. Now I realized that Medha's oratorical bellowing had required great effort, for here she often whispered, her cracking voice alone suggesting exhaustion. She welcomed each arriving activist with an easy familiarity, as if she were the host of a somber countryside festival. This was a role she was accustomed to assuming. An Indian journalist told me that during Arundhati Roy's Rally for the Valley in 1999, when about 250 middle-class Indians descended on Domkhedi, Medha stayed up all night to greet the arriving hikers as they trickled into the hamlet. Now she showed us to our "room"—the smaller of the two huts, inside which coexisted Medha's office, several shelves full of copies of newspaper clippings and Andolan antidam and antiglobalization publications, banners with Andolan slogans ("Free the Narmada, Stop the Dam," "Let Them Live—Stop!"), and a raft of rusted lockers containing the activists' earthly possessions, which might or might not include a change of clothes. The lockers sat behind a bamboo partition, which offered all the privacy any activist ever got. At night the hut's hard mud floor became the bed on which six or eight people slept. Medha was among them, and now so were we.

Medha showed us spots close to the door where we could lay out our sleeping bags, and brought us a teakettle filled with water so that we could wash up. When raindrops penetrated the plastic tarp overhead, she adjusted the ceiling's bamboo framework. "We meet in a very different place," she said, shaking her head at the thought of Stellenbosch, the wine resort where the final session of the World Commission on Dams had taken place. I wondered for a moment whether she was bemoaning the poverty of Domkhedi, then realized she meant the reverse: the resort's opulence amid South Africa's destitution was all the more disturbing given the proximity of the grim townships a fifteen-minute drive up the road, while Domkhedi was something like home. The resort's guest rooms consisted of ersatz huts packed with the accoutrements of modernity, from televisions and king-sized beds to bathrooms equipped with oversized towels and ferocious state-of-the-art showerheads. In Domkhedi, the bathrooms were the cornfields, and people brushed their teeth with twigs; in Domkhedi, the huts were real.

Both our hut and the tentlike assembly place next to it were calefactory sandwiches, two slices of heat-trapping blue plastic tarp on the roof and mud floor with a slab of wilting humanity in between. In the daytime, the tarp lent the tent's occupants a garish blue cast—in that way (and that way alone), the participants looked like the stars of a gaudy Bollywood movie. The overhead tarps kept water out, but the monsoon was stalled, and rain was sparse. And the floor tarp was too thin to work as a cushion, which Bob and I found out that night, when we sat on it during an evening session of the youth camp.

The tent was lit by one of the two light fixtures in the hamlet— a naked bulb hanging from a cord. An Andolan-installed mini–hydroelectric plant up the hill provided the power. The plant, in fact, is a demonstration of "appropriate technology"—the application of grace and scale to machinery in pursuit of ecological balance—and that means, instead of monstrous impositions on the land such as Sardar Sarovar, a bevy of modest installations such as the five-foot-high check dam above the village. Water behind that dam turns a tiny 300-watt turbine, enough power for seven lightbulbs, which Medha posits is about all the energy any hamlet needs. Of course, this notion runs contrary to prevailing fashion in India, which aspires to American-style energy consumption. The Andolan's energy plan, calling for sustainable alternatives to large dams including such devices as rooftop water harvesters and modest check dams like Domkhedi's, could really work—if more Indians were willing to settle for a few lightbulbs per community. Instead, Indians are growing more electrically voracious, not less, and Medha has convinced few of them to go without.

Bedimmed by a halo of bugs, that single lightbulb cast a golden light that illuminated the craggy faces of a couple of dozen Domkhedi villagers who sat barefoot on the tarp. Their sandals, politely left outside, rimmed the tarp like the border around a child's painting. The activists talked, the villagers listened; then everyone joined in songs and chants. The talk was conducted either in Pawri, the local tribal dialect, or Hindi, in which case it was translated to Pawri. Medha spoke Pawri, in addition to another tribal language, Bhilali, plus Hindi, English, and Marathi, her native language, the language of Bombay. Without a translator, I couldn't be sure what she was saying, but from the occasional English phrases, I gathered that the subject was dams and politics. Any hint of harangue was gone now: she was the tribals' instructor, explaining the world beyond the valley with which they had so little experience. The voice that in The Hague had sounded sluggish and mechanical now took on elements of song. She smiled as she spoke. She moved constantly but unhurriedly—all her movements were curved. She'd sit cross-legged or with legs tucked beneath her, while her arms provided a kinetic counterpoint to her words. At the end of a sentence, one arm might be outstretched, with open hand and barely touching thumb and forefinger, as if to express a thought with exquisite precision. The teenaged Medha had been considered an adept dancer, but now, I learned, Medha danced only once a year, on Holi, the "festival of colors" celebrating spring's imminent arrival. On Holi, the tribal residents of Domkhedi stayed up all night to dance, and Medha danced with them.

Medha switched to English to ask Bob and me to introduce ourselves to the youth camp participants. I said I was there to write a book about three people intimately involved with dams, one of whom was Medha. Though Medha translated this into Pawri, she was not pleased: she told me later that I'd embarrassed her, that I should have said the book was about the Andolan, not her. She seemed to have nothing invested in a cult of personality; rather, she was afraid that attention focused on her promoted jealousy. It was true that the Andolan zealously courted Indian journalists, bombarding them with phone calls and press releases, and Medha knew dozens of their phone numbers by heart, but most of the Indian journalists I spoke to acknowledged that the publicity she sought wasn't personal: she was a conduit for the cause. Her response to the enticements of money and fame was to design an organization that in its egalitarian fervor lacked titles and that in its wish for unassailable purity banned donations from foreigners. It's a reflection of both the United States' looming presence in the Indian region and the paranoid strain within Indian politics that controversial people and groups across the political spectrum are routinely accused of being funded by the CIA. The Andolan devised its ban on foreign contributions to refute these claims, but the accusations fly regardless, and the Andolan is condemned to perennial poverty.

That night Medha talked to the villagers about dams. "We are not fighting against isolated projects," she said in Pawri. "We are fighting for the principles of opposing large dams." She briefly described the World Commission on Dams, and said the commissioners' unanimous support for its final report showed that people of different points of view could arrive at consensus. Then she led the group in chants.

"We are all—one!"

"The Narmada dam is treachery—now there's an opportunity to give it a blow!"

"To whom does the forest and the land belong?—It belongs to us!"

"The struggle is incomplete! One more battle is necessary!"

"No, no, no, we don't want Sardar Sarovar!"

The night was too hot for sleeping bags—it was almost too hot for sleep. Medha herself complained of the humidity, and said it must mean that a heavy rain was coming. That, I realized, was what she fervently wished for, and not even chiefly because it would make the tribals' corn grow tall. Rain meant the reservoir would fill until it inundated the Andolan huts. Then, once more, she could court death, she could step up to the abyss and look down. It was a grand game of chicken, which she played for high ideals: *Stop building the dam or I will drown.*

Now, a few dozen strides down the gentle slope from our hut, water at the reservoir's edge lapped up against former cornfield. Though the village was only thirty miles upstream from Sardar Sarovar, that distance was enough to ensure the region's remoteness: it would have taken a Domkhedi man a few days to walk those thirty miles, and walking was the only way to get there.

Bob and I crept into our sleep sheets, far enough in to keep the mosquitoes off our legs, not so far in as to make us unnecessarily hot. I placed the sleep sheet on top of my zipped sleeping bag, and put the bag on top of my foam pad. Then I lay on top of that tottering pillow of First World gear, and looked around. The hut felt as cozy as any room adorned chiefly with political banners could possibly be. ("We want development, not displacement," said the banner behind Medha's head.) A few feet away from us, a tribal woman and her small boy lay asleep on the plastic mat that covered the hardened mud floor. Medha was still at work. She'd arranged herself behind a writing table that was less than a foot high, tucking her feet beneath her in a way that reminded me of Gandhi at his spinning wheel. The lightbulb from Domkhedi's second and last electric fixture hung from a cord above her head, casting a glary light across the hut. Beyond Medha, in the corner farthest from us, a few other people lay on mats—I could barely make out their faces. Medha was trying to read a stack of documents Joe had sent from Baroda, but interruptions were frequent. Other activists stopped outside the door to remove their sandals, then entered and conversed with her, often at conversational volume. Even late at night in the hut, I realized, noise was the norm. It was after eleven when my exhaustion overcame all the sounds, and I fell asleep. Awakened in the middle of the night, I checked to see if Medha had ever gone to bed, and caught sight of her lying on a mat across the room. Nevertheless, when I awoke the next morning, at a quarter to six, she was already up, wearing a fresh sari and sweeping the hut's floor. It was clear that Bob and I should get up, too: at the exact moment that we lifted our sleeping gear from the ground, Medha swept the ground a few inches from our toes.

Led by Vikram, a young activist, Bob and I marched to a hillside stream above the village for morning ablutions. Vikram was the most accessible of the activists at Domkhedi. He was young—celebrating his twenty-third birthday that day, in fact—and he'd been in the Andolan only three weeks, so his responsibilities were modest. He also spoke excellent English, which put him in a distinct and valued minority among the activists. He described himself as a middle-class Bombay kid who got an economics degree and then had a severe change of heart. His parents didn't object to the Andolan's politics, he said, but when he joined, they were furious nevertheless. They warned him he'd regret choosing a life of poverty instead of a lucrative career.

We walked on soft mud paths between corn rows, and strode through a tribal hamlet thick with the stinging oversweet smell of burnt cow dung. In ten minutes, we reached a small waterfall and a rock pool beneath it. This was the men's washing place; the women's was farther uphill.

By the time we got back to the hut, the youth camp in the hut next to us was already in session, while Medha was stationed at her miniature table, continuing to sift through yesterday's delivery of documents. She was wearing plastic reading glasses, which otherwise dangled chest-high from a cord draped around her neck. Once she looked up from a document to announce, "One killed in the G-8." She was referring to a nonviolent antiglobalization protester who'd been killed by Genoa police during a summit of economic powers. Medha emitted a single *tsk*.

At a quarter to seven, Medha joined the youth camp session. Overnight its ranks had swollen with new arrivals, including many villagers who were not young. Now about fifty-five people—forty-five men at one end, ten women at the other—sat on the tarp. The dark-skinned tribal women wore oranges and reds; some of the older men wore gold turbans. The session began with singing. The women's voices were sassy and sonorous; at times they emitted plaintive, harmonious, shivery wails that seemed to express a communal suffering. After that, the group turned to face the river. A ten-minute silent meditation ensued.

The youth camp sessions were by turns poignant, saddening, and dull, an exotic outcropping at the intersection of leftist political organizing and village culture. The five or so young activists from outside the valley all looked resolute and grave and aspiring to be Medha-like; the villagers looked stunned, as much by the hardness of their lives as the dam's arrival—the dam was merely the crowning blow. It seemed to me that the two groups were trying to cheer each other up—the activists with their determination, the villagers just by showing up. Some of the villagers had walked four hours to get to Domkhedi, and some lived outside the projected submergence area. Even though July was a vital farming month, when their presence was needed in the fields, they took the time to attend the session.

Given that the tribals were illiterate, the political agenda chosen by the activists was ambitious and eclectic, hinting at both the activists' sophistication and naïveté. The session's topics were posted on a banner outside:

HISTORY OF TRIBALS
HISTORY OF ANDOLAN
IMPACT OF URBANIZATION ON TRIBALS
WORLD COMMISSION ON DAMS
DOCUMENTS ON LAND RIGHTS AND SURVEYS

The agenda comprised a neatly conceived tale. "History of Tribals" placed the villagers within the context of dams, for all over the world, the collision of dams and tribal people has produced a rich, grim history. Of the forty to eighty million people that the World Commission on Dams estimates have been displaced by dams, a substantial minority have been indigenous people, who, in the WCD's words, "have suffered disproportionately from the negative impacts of large dams, while often being excluded from sharing in the benefits." By splintering tribal communities and forcing their demoralized members to take residence in nearby cities, dams have reduced human diversity as clearly as they have depleted biodiversity. Within that framework, the Narmada saga fits comfortably, for 60 percent of the people facing dam-induced dislocation in the valley are tribal. The Andolan history connected the movement to the struggle to oppose the dams. In case villagers at the camp were reluctant to fight for a cause they perceived as lost, the urbanization discussion illustrated the consequences of giving up, as displaced villagers were forced to the urban margins of society.

It was the discussion on the World Commission on Dams that surprised me, for it suggested a belief that the deliberations of twelve world-class dam specialists, meeting every three months for two years, could by the authority revealed in the wisdom of their conclusions have impact on a place as remote, as relentlessly local, as Domkhedi. Here were activists risking their lives, abandoning their careers for the movement, and yet what they wanted was nothing bloody, not revenge or revolution, but merely that their country abide by the recommendations of an international commission. The villagers were meant to feel their connection not just to other tribals and the Andolan but also to an international antidam movement, and that in turn prepared them for the agenda's final topic, about how to ensure their rights in the face of the dam. The Andolan placed a maternal arm around the villagers, then tried to mobilize them.

The Narmada Valley development project includes thirty large dams, of which six were completed. Of those, Bargi Dam is the biggest, and Bablubhai, a villager displaced by it, was present to recount his experience. In 1971, government surveyors visited his village, he said, and explained that a "tiny" dam would be built. Nine years later, Bargi's construction began, and it was not tiny. Resettlement of the ten thousand villagers it displaced was administered cynically, if at all. Some villagers weren't told that their houses and land would be inundated; others were promised five arable acres and a government job, and got neither. While construction proceeded, the authorities closed the village school and took away a water pump. The officials said, "What good will water pumps be underwater?" Bablubhai's village relocated three times, each time driven by the expanding reservoir farther up the same hill. Even then, despite the inundation of their subsistence farms, the villagers were not allowed to fish from the reservoir, as a local contractor owned fishing rights.

I heard many stories like Bablubhai's in the next few weeks. Their features included a chaotic or nonexistent resettlement plan; indifferent, usually corrupt, government officials; fragmented tribal communities; and residents sent scattering into nearby cities, where they joined the already vast ranks of the urban unemployed.

When Bablubhai finished, Pravin Shivshankar, the youth camp organizer, tried to weave the threads of the narrative into a global morality tale. This led, of all places, back to Genoa, site of the economic summit. "Of course, the Third World is the victim of whatever the G-8 decides," Pravin said.

The villagers looked on impassively. I wondered whether Genoa, on the other side of the world, was beyond their understanding. Pravin explained that 150,000 people protested in Genoa, and that police broke into a school and beat up some of them as they slept. "This was done to prevent them from acting," he said.

The villagers still looked blank.

Medha had been quiet until now. "Who understood?" she asked, and a hint of impatience showed.

Fewer than half of the villagers raised their hands.

More explanation of Genoa ensued. "This was a meeting of the G-8 countries to decide how they can make more profit," a young woman activist said.

Medha said the G-8 countries controlled the World Bank, and the struggle in the Narmada Valley had been against the World Bank. "The World Bank is a loan shark," she said.

Then Medha asked everyone to stand and face the river—she wanted to commemorate the slain Genoa protester. Though most of these people had never heard of the protester or the G-8 or Genoa until that day, they stared at the river in silence.

The session ended with more chants.

"Water not Pepsi!"

"We are all—one!"

Bob and I had already decided to pass up the Andolan meals, out of our fear of illness. Instead, we retreated to the hut, sat on the mats, and tried to eat our own food discreetly. While Medha sat across from us, having skipped the meal to make another pass at her documents, we self-consciously consumed American-manufactured energy bars and multinational-bottled water—it was as if, in the heart of the heart of the antiglobalization movement, we were globalization moles. Eating became a surreptitious act. Out of our unease about being noticed, we discovered the noisiness of First World packaging. Removing an energy bar from its wrapper was a fundamentally loud act, issuing a crackling static that intensified in proportion to our increasingly frantic efforts to find weaknesses in the energy bar defense. In the end I pried the wrapper open with my teeth. To the energy bar's vibraphone, the plastic water bottle played bass. We'd gulp a swig only to hear the bottle boomingly start to implode; startled, we'd remove our lips from the bottle's mouth, and with an even louder thunderclap it would resume its former shape. Medha gave no sign of noticing.

Over the next couple of days I conducted three interviews with Medha. She was well-known for holding interviews en route—inside a jeep late at night, say, or while standing in the crowded second-class compartment of a train. The results were correspondingly hit-or-miss. Western journalists in particular had a hard time understanding her because her brand of English was so different from theirs, and the noise and disruptions of the journey didn't help. A small cottage industry in Medha interview transcripts seemed to exist: by the time I got back from India, I had a stack of them, culled from the Web, e-mail, and Indian newspapers. Yet Medha's comments in most of them were frustratingly unclear, as if the collectors' motive was finding one interview that clarified things. Maggie Black, a British journalist, kindly e-mailed me what I decided was the archetypal Medha transcript, of an interview she did in February 2001. What struck me wasn't the content but the preface, which warned readers that the conversation took place during a four-hundred-mile train ride, "late in the evening after a very tiring day. The noise on the train was deafening and parts of the tape were very difficult to hear or inaudible." With tactful understatement, Black added, "Meaning in some places is elusive and needs interpretation."

My interviews with Medha had the benefit of being held while both of us were stationary, yet they remained only intermittently audible, as if part of her wanted to defeat my tape-recording gear in all its manufactured First World glory. Instead of railcar click-clacks, my tapes contained chants and political perorations from the youth camp session next door. When the youth campers chanted, she raised her voice, but not quite enough; when they stopped, she resumed her customary whisper. Meaning was elusive on a regular basis.

Another part of the problem was language. Many of Medha's sentences were jumbles of phrases, frequently gerundial, often lacking subjects and predicates. Her vocabulary was broad and her reading comprehension was no doubt first-rate, but her spoken words got lost in syntactical fog. Sometimes she spoke like a lawyer out of *Bleak House*, wielding words like "howsoever" and "hence" and "thereafter." In a way, these archaisms revealed her life story, which was that she grew up in Bombay and fled, suspending links to family, home culture, and native language.

I began by asking her about her life before she reached the Narmada Valley. In response, she meandered through her childhood, her academic career, her near-Ph.D. in social work, her work in Bombay slums with untouchables (properly known as *dalits*), and her work with tribals in the hills. "In my family, my father was a freedom fighter and also a labor union leader," she said. "So in the family there was to be meetings throughout, you know, people come here. All the discussions to be on the social, political, economic issues." As a teenager, she said, she spent every vacation at Socialist youth camps—no doubt the model for the youth camp around us. Her version of her life was stripped down and sober, as if shorn of nearly everything but politics from a young age, all nonpolitical memories expunged in the service of the movement.

At one point in her soliloquy, she mentioned a schoolteacher-turned-activist in a district of Gujarat called The Dangs, where a dam was forcing tribals from their land. "I got associated with the teacher's family," she said. "And he died of cancer. That night I was sitting in the hospital doing the night duty for him because he was sick in hospital, and he died when I was there alone with him." Her nervous laugh was barely audible.

"He was a very old man, beyond sixty."

I asked her what effect the teacher's death had on her, but the question went right by her. "Not only that," she said in response, "but I was very involved in the work there which needed support."

When I asked her to describe how it felt to undergo a hunger strike, she answered obliquely. "Hunger strike is sometimes taken by the Gandhians as a means of cleansing oneself, of purifying oneself. Here in our movement, we did not do it from that point of view . . . It expresses both determination and commitment, and also it is a means of protest. It's not necessarily to be seen as torturing oneself, challenging oneself . . . The decision of fasting has to be taken by the person who does the fast, because it is a very, very personal decision. But . . . a group of people are involved in deciding whether to fast, how to fast, when to fast, how long to fast, etc. It is taken as a last resort in the political affairs of the movement."

I asked whether she'd ever been tempted to stop a fast simply because it was so difficult.

The question seemed to confuse her. "No. Never. Personal feeling, you mean?"

"Yes."

"No. Actually, it's a good torture."

Later, I asked whether being a woman had helped or hindered her work.

"I think it has helped me," she said. "In the first phase, you deal mainly with men in the patriarchal society. But once that happens, then there is much more access and much more faith. Basically, it all depends on how you present yourself. You must show knowledge of the issues, which is not normally considered feminine. Then you can reach out to the kitchen and talk to the women and children—at the same time you can win the people's confidence that you are able to deal with issues which are beyond women's domain. You can't appear to be very delicate, or appeal to men's other motives"—sexuality, presumably. "You must show that you can fight, that you can rough it out, that you can travel into days and nights—you have to be ready for it. You can't just pretend to do that. You have to actually do it."

She mentioned that she'd divorced her husband in 1983 after being forced to choose between him and her vocation, and that she was now "married to the work." When I asked whether she had any regrets about her decision, she professed none: she now had "a larger family, and no dearth of love and support." She was personal and impersonal at the same time; she was there and she wasn't there. It was only when she spoke of the World Bank that Medha's composure slightly lessened, and I thought I heard disdain. "Eight years of struggle with the World Bank," she said. "I know what it eats and what it drinks."

Over several hours of interviews, I had a chance to look closely at

Medha. Her small face kept reminding me of the phases of the moon: her eyes were heavy-lidded and her nose and mouth were wide, while her chin was modest, as if so many full-sized features couldn't quite fit inside the first-quarter profile of her countenance. But the full-moon Medha was present, too, and it was radiant. The changes in her expressions unfolded calmly, gradually, almost languidly, and yet nothing about them looked feigned. It was as if at any moment you could see in her face not just one emotion, but several. Her weight—which she later described as "somewhere around" 120 pounds, last measured during a fast—was all in her arms and legs, as if her body were all tool, all instrument. Her arms were solid, and her almond hands were chubby. Her waist was slim, and her chest and shoulders slight; whatever she denied herself registered there, in her core.

Medha's power lay not in her words, but in her actions, her example. She earned the affection of tens of thousands of tribal people by spending years walking from one tribal village to another, talking with the inhabitants, explaining the grievous impact of the dam. She learned the names of every member of every family, and could recall them all when she met them, five or ten or fifteen years later—she made them all feel they were important to her. Now I discovered that her graciousness extended to foreign writers, even ones with bottled water and unspecified politics. In the course of the interviews, she repeatedly offered me food: "You eat so little, and you are such a huge man!" When she discovered that I suffered from a bad back, she insisted that I sit in her spot, behind her tiny table, so that I could lean against the bamboo wall—never mind that she herself wore a lumbar brace. I sat uneasily in that spot, as a pretender sits on a throne.

On our second night in Domkhedi, the camp session went late into the night, and Bob and I tried to sleep despite it. This wasn't entirely unpleasant, for the villagers were chanting and singing. I was starting to doze off when I heard someone announce in English that it was Vikram's birthday. For this, Vikram was given the honor of singing a song. On that green hillside, in an isolated bastion of antiglobalization sentiment, Vikram sang "Blowin' in the Wind," an act that itself bespoke globalization's success. Vikram's exquisitely raw and delicate rendition, combined with the protest banners, the cornfield, and the starless sky, re-vivified those crusty lyrics, until I shivered at them all over again.

The next morning, Pravin asked me to speak to the villagers about "struggles you've participated in." I knew I couldn't say no, though I was hard-pressed to recall the sort of struggle Pravin would appreciate. I sat cross-legged in the middle of the blue mat, facing a semicircle of tribal men with bloodshot eyes and missing teeth. I talked about what I'd learned about dams' social and environmental destructiveness, and how, partly as a result, dam construction in the United States had ebbed and all but ceased entirely. Pravin translated. Then he asked for questions but elicited none. Finally, an old man with a gray beard and a pen in his white shirt pocket addressed Bob and me as Americans. "The way you have stopped building big dams and are decommissioning dams—we want our dams to stop in the same way. There are poor tribals who have been here for centuries, and they want to maintain their livelihood."

A villager asked about the condition of tribal people in America. I warned that it wasn't a happy story, but I told it anyway, in broad strokes, from 150 years ago to today: the slaughter of Native Americans, broken treaties, reservations, alcoholism. I felt the immense difficulty of communicating the difference between contemporary indigenous Americans (nonagricultural, racially distinct, living in a cash economy) and the contemporary indigenous Indians around me (agricultural, racially blurred, living in a subsistence economy). The difficulty became apparent when, after another silence, a man asked, "Where do the American tribal people get the money to buy alcohol?" as if they were unimaginably lucky.

At the end of the session, Pravin asked Bob and me to provide some revolutionary slogans. After a long, flustered pause, I was relieved to hear Bob produce "Power to the People."

"Power to the People" was duly chanted.

The water level still wasn't rising; the Andolan Web site warnings of imminent inundation seemed increasingly like wishful thinking. With the monsoon season already half over, a good chance existed that the reservoir wouldn't reach the Andolan huts this year. Medha kept saying that during the previous year the heavy rains didn't fall until late in the monsoon, as if she were consoling herself—she seemed disappointed that she might not get a chance to drown. At the same time, however, she was making alternate plans.

Bob and I decided to retreat to the urban comforts of Baroda for a day or two, then meet Medha in Badwani, forty miles upstream, where she was holding a gathering of Andolan supporters. Our last act in Domkhedi was to dispose of our trash. We were in the midst of carrying cartons of our empty water bottles back to the boat when Medha caught sight of us, and told us to leave them behind—villagers could use them to carry local water during their treks. I put the last of our energy bar wrappers in a tattered cardboard waste box inside the hut, then noticed that virtually everything in it—wrappers, dental floss, packaging for assorted pills and bandages—was ours.

We departed the valley by degrees, leaving behind first Medha and the villagers, then the river, then clean air. Our traveling companions were Joe, who'd arrived in Domkhedi the day after us, and Vikram, who was wearing a T-shirt that said "Quantum Mechanic." We recrossed the reservoir and met the Andolan jeep, but it took us only as far as Kadipani, the town on the rim of the valley where the terrain reverted to soiled plain. Joe negotiated a trip for the four of us on a commercial jeep whose seat padding had been worn to a boardlike hardness. The trip went slowly. The driver stopped twice for tea, once for gas, and once to make an offering at a roadside temple.

The ample value of the drive was in the conversation. After only three weeks in Domkhedi, Vikram was going back to Baroda for repairs. At the age of fourteen, he had fallen out of a window on the fourth floor of his school. Doctors told him he was lucky to live. He broke both legs, two vertebrae, and a wrist, and spent more than a year recovering. Not surprisingly, the hard life in Domkhedi stressed his back, and he was looking up a doctor in Baroda whom Medha recommended. Vikram didn't consider his situation unusual; the Andolan, he said, was "an army of the lame and the sick."

For his part, Joe was happy to hold forth on the Andolan, and his vantage point was unique, in that he'd traveled with Medha through most of the previous seven years. From him, I learned that while Medha never forgot names and phone numbers, she was terrible at remembering roads; that she could be tough on fellow activists, from whom she expected a sacrifice nearly equal to her own; that during the 1999 Rally for the Valley, she carried on without sleep for five days; and that Patkar was her married name, the last remnant of her unsuccessful marriage.

Joe told how thugs broke into the Andolan's Baroda office late one evening in 1994 and proceeded to ransack the files—provoking police to arrest several activists. Joe seemed bemused at the thought of the episode, which he believed offered a lesson in the superiority of Andolan tactics. The government knew how to handle militants—ruthlessly—but the Andolan's nonviolence baffled it, he said. If the government responded to Andolan acts with violence or blatant injustice, it looked bad; if it tolerated them, it seemed to validate the Andolan.

Terrain is precious in India; in slow traffic, vehicles bunch together as if fearing citations for wasted space, and failure to occupy an open space, even if only a few inches, is considered a sign of timorousness. By the time we reached Baroda, our driver had lost what little nerve he possessed. With a turn across thick streams of traffic required, he was unable to persuade himself that he could insinuate the jeep through the flows, and refused to budge. Joe kept spotting openings, encouraging him to go now . . . or now . . . or *now*, until at last Joe was shouting and the driver was visibly cowering. At last, he darted into the traffic like a cornered house cat, and, surprising himself, emerged unscathed on the other side.

The Andolan office was just down the street.

2 SARDAR SAROVAR

The Ganges is the Subcontinent's most famous river, but the Narmada is more sacred—it may be the most revered river in the world. Start with the name, whose derivation Gita Mehta elegantly described in her 1993 novel, *A River Sutra*: "It is said that Shiva, Creator and Destroyer of Worlds, was in an ascetic trance so strenuous that rivulets of perspiration began flowing from his body down the hills. The stream took on the form of a woman—the most dangerous of her kind: a beautiful virgin innocently tempting even ascetics to pursue her, inflaming their lust by appearing at one moment as a lightly dancing girl, at another as a romantic dreamer, at yet another as a seductress loose-limbed with the lassitude of desire. Her inventive variations so amused Shiva that he named her Narmada, the Delightful One, blessing her with the words 'You shall be forever holy, forever inexhaustible.' Then he gave her in marriage to the ocean, Lord of Rivers, most lustrous of all her suitors."

Mehta felt compelled to make the Narmada the setting of her spiritual tale even though she'd never visited the valley, for no other Indian river inspires as much imagining. (Mehta said she followed "Oscar Wilde's principle that it was easier to imagine Japan from a bench in Hyde Park than from the middle of Tokyo.") The Puranas, a holy Hindu text, says that a person can gain salvation by bathing in the Ganges, but the same goal can be achieved merely by catching sight of the Narmada. Lingams, the gray phallic stones worshipped as Shiva, are found only in the Narmada's beds. Until reservoirs inundated many of their temples and trails, each year hundreds of loincloth-clad mendicants known as sadhus and a roughly equal number of their female, sari-clad counterparts set off on treks of precisely three years, three months, and three days to circumambulate the river, from its mouth at the Gulf of Khambhat to its source in the Deccan Plateau and back, more than eight hundred miles each way. To valley tribal children, these spiritual seekers looked intoxicated and terrifying. As Vijay Paranjpye explains in his book *High Dams on the Narmada: A Wholistic Analysis of the River Valley Projects*, their mission was to trace "the progress of the material and spiritual history of mankind, beginning with the remains of pre- and proto-civilizations with their simple, animistic faiths, the worship of nature and all the elements, to the gradual development towards more settled and complex cultures, and finally into the realms of recorded history."

It is tempting to think of Medha Patkar as a kind of pilgrim, or even as Narmada herself, despite her professed lack of interest in spiritual issues. It is not just that she chose to work in the Narmada instead of numerous other regions of India where tribals suffer, or that she has walked the valley for much longer than three years, three months, and three days. What caused me to shiver when I first read it was that the Narmada is considered so holy that even during British rule, the crime of attempted suicide was overlooked if the offense was committed in the river.

Into this hotbed of devotion, Indian officials decided to insert the largest water-development project in the world. If the project is completed, presumably around midcentury or later, the Narmada and its 41 tributaries will be home to 30 large dams, 135 medium dams, and 3,000 small ones. The world's largest canal, which we crossed on our trip to Domkhedi, will traverse 19 major rivers and 244 rail lines and roads on its way toward delivering water to a 45,000-mile network of canal branches. The biggest dam in the project, Sardar Sarovar, alone will dislocate a few hundred thousand people, which will make it first in displacement among all the world's dams but China's. The Three Gorges Dam in China is better known in the West, chiefly because of its even more mammoth size, yet Sardar Sarovar, not Three Gorges, became the world's most contentious dam project, the most complete embodiment of the international battle over dams. That's because Sardar Sarovar touches on every dimension of controversy raised by a half century of large-dam building. The displacement of indigenous people, environmental impact, resettlement, water scarcity, energy production, spirituality, political and corporate corruption, human rights—the Narmada project encompasses them all. Unlike the Subcontinent's other major rivers—the Ganges, the Indus, and the Brahmaputra—which all originate in the Himalayas, the Narmada is all Indian: the valley occupies a seismic fault that practically cuts India in half. Its history contains the quintessential Indian experience, the meeting of settled indigenous people and nomadic Aryan invaders, a two-thousand-year-long dance of conquest and absorption that climaxed a few centuries ago when the tribals fled abuse in the plains for the hardscrabble security of the mountains. The Narmada development project embodies a tug-of-war among three Indian states, the central government, and the Indian Supreme Court; between the visions of Nehru and Gandhi; between the World Bank and just about everybody. The valley is where the evidence of dams' destructiveness that has accumulated over decades finally collided with the optimism of engineers.

The need for water storage in India is obvious. The country's water deficit—the difference between its demand for water and available supplies—is the largest of any country in the world, three or four times greater than China's, which ranks second. In the mid-1990s, the deficit was estimated at more than one hundred billion cubic meters per year, a fifth more than the average annual flow of the Nile River. India already diverts so much water from its biggest river, the Ganges, that water often fails to flow to its mouth during the dry season. Throughout India, groundwater depletion is causing a drop in aquifers of three to ten feet per year; at this rate, some aquifers will soon dry up, and the water level in others will drop so low as to make pumping prohibitively expensive. Either way, as much as a quarter of India's agricultural production will be threatened. At the same time, India's population is booming, and will probably overtake China's by midcentury. All this has caused Indians to turn reflexively to dams to store water. Though India's forty-three hundred completed large dams place it third among countries in the world, behind China and the United States, it is now the world's most prolific dam builder, with somewhere between seven hundred and a thousand large dams under construction. Yet, according to a study commissioned and published by the World Commission on Dams, virtually all of India's dams have been planned and built without any consideration of less costly and damaging alternatives, and most have paid only cursory attention to the social and environmental disarray they cause. "Farmers with large landholdings" got most of the irrigation benefits, and the urban rich and rich farmers got a "disproportionate" amount of the electricity. As a result, "dams have not only helped to maintain the current inequities in the Indian society but, in some ways, have exacerbated them."

Even before Indian independence, dam engineers coveted the Narmada, India's fifth-largest river, but were constrained by a political reality: the river winds through three states, all of which laid claim to the envisioned dams' benefits. Moreover, the most promising dam site was in the coastal state of Gujarat, which would therefore enjoy a preponderance of its benefits, chiefly drinking water and irrigation, while its adverse consequences, such as the displacement of a few hundred thousand people, would register primarily upstream, throughout much of Madhya Pradesh and the hilly northern tip of Maharashtra.

Of the districts of the three states affected by the planned dam, Gujarat's was by far the richest, and the state's inhabitants looked on the dam as both a colossal instrument of development and the "lifeline" to its arid northern districts, Kutch and Saurashtra. In return for this bounty, Gujarat agreed to provide the two other states a modest sum of hydropower as well as resettlement sites for people displaced by the dam—never mind that most tribals didn't want to move to Gujarat. As the dispute escalated, so did the proposed size of Sardar Sarovar, from a height of 162 feet when the Indian Planning Commission approved it in 1960 all the way up to a towering 530 feet, then back down to its current and still lofty projected height of 455 feet.

When Jawaharlal Nehru, India's first prime minister, laid a foundation stone at the dam site in 1961, he set in motion a process that forced the dislocation of two thousand families in the village of Kevadia to make way for the engineers' offices—the dam's first "oustees." (In fact, according to Thayer Scudder, the World Bank's chief resettlement consultant through the '80s, the money paid for the engineers' quarters was more than the project's entire allocation for resettlement.) For their pains, these people were offered no homes, no land, and in many cases, no compensation. This was indicative of prevailing attitudes both in India toward tribals and among international developers toward displaced people: they were simply the unfortunate victims of a project that would benefit many times their number—end of story. The final indignity meted out to the Kevadia oustees was that construction didn't even begin until twenty-six years later, in 1987. In the interim, the Indian government convened the Narmada Water Dispute Tribunal to settle the states' claims for Narmada water, but the dispute was so politically charged that it took a decade to reach a decision—one that, among other features, assumed a much greater annual water flow than many scientists believe the river generates.

The tribunal's decisions were, in fact, not scientific decisions at all, but political ones. It settled on 28 million acre-feet as the Narmada's annual flow because that was the figure the three states agreed on, even though the fragment of hard hydrological evidence that existed suggested that the true flow was closer to 23 million acre-feet. Nor did the tribunal consider economic optimization when it specified that the Sardar Sarovar's height should be 455 feet—that figure, too, was the result of political compromise. In fact, the last nineteen feet of the dam's height were added not for water storage, the project's main goal, but for power generation in Madhya Pradesh, as payback for the state's acceptance of the inundation of its rich agricultural land. For those nineteen vertical feet, Madhya Pradesh will pay dearly in a horizontal direction, as the reservoir enlarges by more than a third, to about a hundred square miles—four times the size of Manhattan. Worse, those nineteen feet will increase submerged cultivable land by half and the reservoir-affected population by three-quarters, yet the power generated by nineteen vertical feet will increase the dam's electrical output by only a tenth. Given the vast impact on villagers of such a small increment, it was no wonder that Medha was courting her death to limit Sardar Sarovar's height, even though she'd already lost the fundamental battle over its existence.

The result of the tribunal's negotiated settlement was a plan for a massive dam whose structural superlatives puffed the chests of patriotic Indians, while comprising a political bounty in the form of $6.5 billion worth of patronage and bribes. Alas, on a functional level, Sardar Sarovar's parts didn't quite fit. Both the dam and the canal were bigger than necessary to meet their stated water-delivery goals, while the reservoir was far too small. The tribunal allotted Gujarat 9.5 million acre-feet of the 28 million acre-feet it optimistically deemed the river capable of delivering, yet the dam and the canal were so big that they could divert more than 8 million acre-feet—nearly all of Gujarat's annual allocation—in the course of a three-and-a-half-month monsoon. The World Bank was aware of the problem: in 1982, Professor Nathan Buras prepared a study for the Bank that said the canal would flow at less than half its capacity two-thirds of the year and would achieve its designed flow rate on about one day in twenty.

The smallness of the reservoir presents a particularly vexing problem. The Narmada Valley receives 90 percent of its water during three or four months of sporadic but intense monsoon rain. Storing this water until it can be doled out in regular allotments throughout the dry season requires more capacity than the Sardar Sarovar reservoir possesses. Project designers opted for an expensive solution: the simultaneous construction of three large upstream dams—Indira Sagar, 185 miles upstream; Omkareshwar, 160 miles upstream; and Maheshwar, 135 miles upstream—that together will store water to augment Sardar Sarovar's supply. The four dams are a package, "a technically and economically interdependent project," in the tribunal's words: take away the three upstream dams, and the rationale for Sardar Sarovar develops severe cracks. The tribunal said that 85 percent of the water diverted to Gujarat would be provided by the release of water from the three dams, but it soon became clear that that wouldn't happen, since two of the upstream dams faced even greater political and financial obstacles than Sardar Sarovar. The biggest problem was that the three dams were to be built in Madhya Pradesh, a poor state that couldn't afford them. Without the upstream dams, the Sardar Sarovar reservoir's water level fluctuates dramatically, and more water must be released down the spillway, which the dam engineers consider a waste. (To be sure, until recently this view of undiverted river water as wasted has been the Indian conventional wisdom, accepted even by Mahatma Gandhi. Of course, to downstream flora and fauna and the people who depend on them, the water is anything but wasted. In its wisdom, the tribunal allotted precisely zero acre-feet for the downstream river, thereby depriving about ten thousand downstream fishing families of their livelihoods.) With less water available for the canal, the "lifeline of Kutch and Saurashtra" is even less likely to flow to those far-flung districts, though the sugarcane growers and the water park at the head of the canal should still get their share. A World Bank memorandum in 1992 concluded that without the three upstream dams, the area irrigated by Sardar Sarovar would drop 30 percent, and generated power would decrease by 25 percent.

None of this deterred the World Bank. The Bank was the world's largest international loan-making institution, and India was its biggest client; one derived status from providing loans, the other from receiving them. The Bank had never been a stickler for such tangential concerns as resettlement and the environment; the idea was to get the dam or tunnel or power plant or irrigation system or chemical factory up and running as quickly as possible, to start revving the engines of development. The Bank had done exactly that in Ukai, the river valley just south of the Narmada and parallel to it, where the Bank funded the installation of Gujarat's largest irrigation system. Ukai displaced seventy thousand people, mostly tribals, and offered them no resettlement. Until the irrigation system was constructed, people lived on subsistence crops, millet and barley and corn, but with irrigation, the land's capacity was altered: suddenly it could support cash crops such as sugarcane (never mind that sugarcane is a notorious water guzzler) and wheat. Alas, the Ukai Valley's poor farmers couldn't afford to buy the pesticides and fertilizers that the cash crops required, while the wealthy farmers could. It was Robin Hood in reverse. The wealthy farmers got wealthier, and the value of their land appreciated; most poor farmers eventually had no choice but to sell their farms to the wealthy ones, and many ended up as day laborers and beggars. The engines of development churned all right, but to the benefit of the rich, at the expense of the poor.

The World Bank agreed to lend India $450 million to build Sardar Sarovar in March 1985, the same year Medha Patkar moved to Gujarat. The head-on collision between the two took five years. Medha was then a social science researcher, known for her enormous energy and deep melancholy, who was looking for a cause, preferably in defense of downtrodden tribal people. In Gujarat, she found what she was looking for. At first she worked to secure a just resettlement for the tribals, until she concluded that the problem wasn't only the lack of resettlement, but the dam itself and the concept of development it represented. She used her walks to organize an improbable coalition of the soon-to-be-displaced, including wealthy Hindu farmers in the eastern plain and poor tribal farmers in the hilly west. When she wasn't plying the valley, she visited every major Indian city. By the time the Andolan took shape in 1988, it was a national organization, including environmental and human rights groups, scientists, academics, and displaced people.

India, meanwhile, was on its way to becoming the world's most prolific dam builder. All that stored water generated up to 25 percent of India's food supply, according to dam advocates, but even Rajiv Gandhi, India's prime minister from 1984 to 1989, was skeptical. "We can safely say that almost no benefit has come to the people from these projects," he said in 1986, referring to irrigation projects, including dams. "For sixteen years we have poured money out. The people have got nothing back, no irrigation, no water, no increase in production, no help in their daily life." Yet a year later, Gandhi approved Sardar Sarovar, apparently out of fear that rejecting the dam would further weaken support for his tottering party. He overrode the objection of his own Ministry of Environment and Forests, which gave the dam only a "conditional" clearance because dam officials failed to conduct required cost-benefit studies. Construction on the dam began in May 1987.

By 1989, the Andolan was gathering momentum. During one protest, ten thousand people stormed the dam site; at another, sixty thousand people representing 250 organizations gathered at the remote Madhya Pradesh village of Harsud to denounce "destructive development." The Indian government responded by invoking the Official Secrets Act in the dam region, effectively banning gatherings of five or more people—that alone led to hundreds of arrests. The Andolan's international allies lobbied the Bank's twenty-four executive directors to abandon Sardar Sarovar, and Medha testified at a U.S. Congress oversight subcommittee hearing on the project. New York City Democrat James Scheuer, the subcommittee chair, was so impressed by her that he extended her ten minutes of allotted speaking time to forty-five. He told her, "I've seen many lawyers who charge five hundred or a thousand dollars an hour, but you surpass all of them." Scheuer wrote letters to environmental legislators around the world asking them to support the anti–Narmada dam movement. In many parts of the valley, Bank and dam officials weren't welcome, and were forced to leave if they tried to conduct engineering or resettlement surveys. Thayer Scudder, the Bank's resettlement consultant, had already written reports for the Bank in 1983 and 1985 that deplored the Narmada resettlement effort, and still calls the project he found when he arrived in India in 1983 "the worst resettlement I've ever seen anywhere in the world." With opposition to the dam growing in 1989, he was sent to India one more time and wrote an even more scathing report. Scudder advised the Bank to cease loan disbursements until Indian officials took action on nine key points related to resettlement. The Bank's India Department, whose raison d'être was loaning money to India, dealt with the report the way bureaucracies customarily treat information they abhor. According to Catherine Caufield, whose valuable book *Masters of Illusion: The World Bank and the Poverty of Nations* devotes a chapter to the Narmada, the India Department blatantly misrepresented Scudder's report as an endorsement of the project, alleging that Scudder concluded that all major resettlement problems had been resolved. Lacking evidentiary proof of this charge, Scudder acknowledges only that the Bank "ignored" his report.

The collision occurred at the end of 1990, a year in which the central government's own Commissioner of Scheduled Castes and Tribes called for a stop to Sardar Sarovar, and the Ministry of Environment and Forests revoked its 1987 conditional approval of the dam on the grounds that the studies, surveys, and action plans it had required remained undelivered a year after the deadline. Nevertheless, construction continued.

After months of planning, Medha responded on Christmas Day by launching the Andolan's "Long March": in Rajghat, deep inside Madhya Pradesh, three thousand people strode under a banner that said, "No one will move; the dam will not be built," and then trekked down the valley all the way to Gujarat. It was typical of the broadly leftist Andolan to apply a Maoist label to a Gandhian tactic: the idea was to march 150 miles to the dam site, until the authorities agreed to a comprehensive review of the project. The Andolan's plan was touching in its earnestness. It was carrying out a nonviolent protest that risked arrest, beatings, and jail on behalf of something as ethereal and conceivably toothless as an independent review. On the one hand, the activists brimmed with confidence in their case: they thought, *If we can just show the evidence, we'll win.* Yet the act hinted at their desperation, for they had no detailed strategy for victory beyond rousing the Narmada Valley to unified opposition. The idea of a project review presented a potential opening, so they seized it. In a way, they were fighting for time. Unless the Andolan could quickly stop construction, the dam would become a fait accompli. What the Andolan needed, short of outright victory, were construction delays, which increased the dam investors' costs. Pile on enough delays, they hoped, and the investors would pull out.

Gujarati politics was no more than an incidental concern of the Andolan, and that was a serious oversight. For one thing, because of a preponderance of merchant-caste banias in Gujarat, the state "is the part of India where capitalism finds its most natural acceptance," as political writer Harish Khare puts it. The state also enjoyed a reputation for contrariness, as when, in 1975, it briefly resisted Indira Gandhi's dictatorlike emergency declaration and suspension of civil liberties. Moreover, its new chief minister, Chimanbhai Patel, was a skilled and happily corrupt spokesman for the state's rich farmers and big industrialists. His grip on power was tenuous, owing to the weakness of his ruling coalition, and he saw in the Narmada issue a chance for consolidation. His strategy was simple. Medha and the other leading dam opponents were all from outside Gujarat, so he portrayed them as anti-Gujarat: he equated opposition to Sardar Sarovar with state sabotage. Many Gujaratis supported dams as vociferously as the Andolan opposed them. A cement company went so far as to appropriate Sardar Sarovar as a logo, presenting the dam wrapped in the arms of a muscle-bound (but oddly pin-headed) giant, whose image looms in monstrous billboards over main roads in the state.

The Andolan procession never managed to penetrate the state border, let alone reach the dam site. After a week of walking, the protesters reached Ferkua, a town on the Madhya Pradesh–Gujarat border, and found their way blocked by a comparable number of participants in a prodam counterrally organized by the Gujarati government. The confrontation was something rare: an intranational border incident. As usual, the Andolan came tactically prepared. It sent twenty-five volunteers, all with arms crossed in front of them and wrists bound together to make plain their nonviolent intent, across the line of demarcation. Gujarati police stopped them from crossing the border, so the protesters sat on the road. They didn't just sit: they settled in. Other volunteers brought them food, tea, water. For two days, more and more self-bound volunteers joined them. Back and forth across the border, the two sides exchanged heated rhetoric: the prodam forces yelled slogans over loudspeakers, and the antidam forces sang songs about the Narmada and the tribal struggle. By the third day, the police had had enough. They loaded the seated protesters into buses and drove them far from Ferkua before releasing them. More bound protesters quickly took their places.

It was high drama, but it wasn't enough. To break the impasse, the Andolan raised the stakes. On the Madhya Pradesh side, Medha and six other activists lay down on mattresses next to the road and announced their intention to conduct an indefinite hunger strike: they would fast until the authorities agreed to a project review. It could not have surprised anyone that Gujarati officials refused to yield. The dam with its promised bounty was more popular in Gujarat than the state's most famous native son (and subject of countless Gujarati monuments), the Mahatma himself. The fast stretched beyond three weeks, until Medha's kidneys were failing, and her doctor was saying that she must stop the fast immediately. In the end, the activists decided that the survival of Medha and her fasting colleagues was more important than whatever outcry might follow their deaths. Medha reluctantly ended her fast.

Soon, however, the seeming defeat turned into victory, for the drama made her a national figure, adored in some quarters and loathed in others, and galvanized support for the Andolan throughout the country. In the face of the burgeoning opposition to the dam, the Japanese government, which had pledged $200 million in financial and technical support for Sardar Sarovar, withdrew from the project (according to Caufield, only after having become involved long enough to ensure that the dam used Japanese turbines). The controversy reverberated within the Bank. Indeed, one way of looking at the dispute is that it was the price the Bank paid for scores of bad Indian projects, for its historical disregard of the environment and dislocated people. Caufield cites numerous World Bank officials who, while split on whether Sardar Sarovar was a worthy project, all agreed that it was far from being the Bank's *worst* project in India—what distinguished it was its size and organized opposition. An unnamed Bank official told Caufield, "It was really becoming an embarrassment to the Bank, and there were tremendous tensions between the executive directors and the senior management." On March 14, less than two months after the Ferkua fast ended, World Bank president Barber Conable announced that the Bank would commission an independent review of the project. It was the first time in the Bank's forty-five-year history that it had asked outsiders to evaluate one of its projects.

The reviewers weren't supposed to perform an actual critique of the project. As Caufield tells the story, it was understood within the Bank that the review would be cursory, a kind of glorified consultant's report—"a couple of weeks in India and a typewritten document, stapled at the corner," as a Bank official put it. But if that was what the Bank wanted, Conable chose the wrong man to lead the review team. Like Conable, a New Yorker, Bradford Morse was a former Republican congressman from a northeastern state—Massachusetts—who'd left the House to work in international development. He was that now-extinct breed, a liberal Republican, whose most noteworthy act as a congressman was expressing opposition to the Vietnam War. He resigned his House seat to become a U.N. undersecretary general, then served a decade as administrator of the United Nations Development Program—at the very least, his résumé suggested that he possessed a broad understanding of development dynamics. As his deputy, Morse chose Thomas Berger, a Canadian jurist with a reputation for evenhandedness and experience in reviewing controversial projects. After consultation with Berger and Bank critics, Morse confounded Bank officials who expected a perfunctory report: he insisted that the Bank give him access to all relevant documents, freedom to publish an uncensored report, and a $1 million budget. Cornered, the Bank agreed to all three demands. "We think it unlikely," Morse and Berger later wrote, "that any other international aid organization has ever established a review with a mandate as sweeping as ours in connection with a project, no matter how controversial."

Instead of two weeks, the reviewers—Morse, Berger, anthropologist Hugh Brody, and hydrologist Donald Gamble—spent nine months. In Washington, they reviewed Bank documents and spoke to Bank officials and critics from the nonprofit groups that tracked the Bank's activities. In India, they talked to ministers in all four governments, leaders of volunteer groups including the Andolan, and people facing dislocation. They visited not just the reservoir region but also the downstream river, the area to be watered by the canal, even the distant arid regions in northern Gujarat that the canal would supposedly transform. They scrutinized villages facing inundation and the sites where they were told to relocate. In the valley, they were welcomed where Bank officials had been shunned. Morse suffered from emphysema, which made hiking difficult, and on two occasions he reluctantly agreed to sit on a charpoy while tribals carried him from riverbank to village. It's a stunning, subversive image: on the face of it, the white man was still in the sedan chair, playing his customary autocratic role, yet in pursuing a set of hidden truths he had become the tribals' servant.

The result of this effort was a 363-page report that remains the best of the many treatises written about Sardar Sarovar. With plain, unequivocal language and devastating detail, the report showed how the Bank and the four Indian governments—the central government and the three state governments—ignored resettlement and environmental policies that they were obligated by treaty or regulation to uphold. For example, in 1958, India become one of the first countries to ratify International Labor Organization Convention 107, which laid down provisions for the treatment of indigenous people dislocated by development projects. Such people, the convention said, "shall be provided with lands of quality at least equal to that of the lands previously occupied by them, suitable to provide for their present needs and future development." Similarly, a 1980 World Bank operational manual statement on involuntary resettlement said that "the Bank's general policy is to help the borrower to ensure that, after a reasonable transition period, the displaced people regain at least their previous standard of living and that so far as possible, they be economically and socially integrated into the host communities . . . Measures to be taken in this regard should be clarified before, and agreed upon during, loan negotiations." In February 1982, another Bank operational manual statement went further, by specifically addressing tribal issues. "The Bank will not assist development projects that knowingly involve encroachment on traditional territories being used or occupied by tribal people, unless adequate safeguards are provided . . . The Bank will assist projects only when satisfied that the Borrower or relevant government agency supports and can implement measures that will effectively safeguard the integrity and well-being of the tribal people." Both ILO 107 and the Bank policies were responses to the growing acceptance of the concept of human rights and the understanding that tribal people suffered disproportionately from the effects of development projects.

Unfortunately, the policies existed only on paper: neither the Bank nor the Indian governments paid them much more than lip service. In part, this reflected the common belief among Hindus that tribals were inferior. Thayer Scudder said that I. M. Shah, one of Sardar Sarovar's four chief engineers, told him that the solution to the tribal problem was sterilization. A more mainstream but no less prejudiced view was expressed by Vidyut Joshi of the Gandhi Labor Institute in Gujarat. The review quotes Joshi as saying that tribal displacement was part of the changes that other peoples have welcomed "in the name of progress, development or modernization. This being so, why should anyone oppose when tribal culture changes? A culture based on lower level of technology and quality of life is bound to give way to a culture with superior technology and higher quality of life. This is what we call 'development.' "

The 1979 tribunal award, the fundamental Sardar Sarovar document, shows a careless disregard for tribal practices, if not outright prejudice. The tribunal's purpose was to adjudicate an interstate water dispute, but in doing so it casually set up the terms of Sardar Sarovar resettlement. Without making an effort to understand the potential impact of dislocation on tribals, without even an estimate of the number of people who would be affected, the tribunal stipulated the rules by which displaced people would receive land: this, too, had been the subject of a negotiation among the three states, never mind that it violated ILO 107.

The tribunal ruled that landowning oustees should receive the same amount of land they lost to inundation, five acres at a minimum. It said that the "major sons"—sons at least eighteen years old—of the landowning families should also receive land. And it provided for house lots for landless oustees. These provisions left out most tribal people.

For starters, they left out people dislocated by the canal and irrigation system, which occupied twice as much land as the reservoir, and they made provision only for Madhya Pradesh and Maharashtra resettlers, not those who lived in the fourteen villages in Gujarat that would be inundated. The 165 landowning families and 120 landless families in Kevadia who nearly two decades earlier had been forced from their land to make room for dam infrastructure and engineers' quarters were not acknowledged. Neither were the people in five villages who'd been driven away since 1961 because of continued construction, nor the thousands of people downstream who depended on fishing for hilsa, a popular food fish. The lower Narmada's hilsa fishery was the largest on India's west coast, but the dam would drastically shrink it or wipe it out.

Equally unjustly, the exclusion of landless oustees disregarded tribal agricultural practices, which thoroughly blurred the distinction between landowner and landless. For centuries, the tribals used slash-and-burn agriculture, working plots for a few years until fertility declined, then moving on to other fallow forestland. Though this was an effective form of crop rotation, the British believed it harmed the forest and disrupted orderly administration, so they banned it. Instead, they insisted on permanent fields that constituted "revenue lands," whose owners were subject to taxation. The idea of taxation did not thrill the tribals, and many took advantage of their remoteness to elude registration; entire villages lay outside the revenue system even after independence. The tribunal treated people without registration as encroachers, even if their families had worked the land for generations: they'd become encroachers on their own land. And the land records were wildly unreliable. The names of sons who inherited land were rarely recorded, and land farmed by several households was often registered as belonging to a single person, who in many cases was long deceased. By the tribunal's ruling, the independent review estimated, between 60 and 80 percent of tribal oustees did not qualify for resettlement land. The review concluded: "The result of classifying encroachers as landless oustees means that people who are in fact cultivating land they regard as their own will become landless laborers. This is not rehabilitation. It does not leave them at least as well off as before."

Even the seemingly generous provision of land to sons at least eighteen years old was highly problematic. Madhya Pradesh, for example, decided that it would include sons who were eighteen by 1987, when dam construction began. However, the long delays in dam construction meant that in most cases resettlement didn't occur for more than a decade. Thus, a son who was thirty-one in 2001, when some resettlement was carried out, was excluded because he was seventeen back in 1987.

The reason resettlement was even mentioned in the tribunal award was that it was part of the deal. As another payoff to Madhya Pradesh and Maharashtra for the loss of Narmada Valley land, Gujarat agreed to house most of the oustees in its arid regions that presumably would be watered by the canal. Madhya Pradesh, the state with by far the most oustees, grandly assumed that most dislocated people would move to Gujarat and planned to provide only 10 percent of the land needed for its oustees. Unfortunately for dam officials, most oustees preferred to stay in their native states, as close to their original homes as possible. Among the valley's tribal groups, communal life was still valued, and in many, marriage partners usually represented neighboring villages: the proximity of a potential mate's village strongly influenced his or her desirability. To scatter each cluster's residents to dozens of different small plots in an alien region, as resettlement authorities intended to do, was a sure way to destroy tribal communities. Indeed, Scudder cites evidence that the inhabitants of Gujarat's nineteen villages were dispatched to 175 relocation sites.

People experienced in resettlement know that the suffering dams cause starts not when resettlement occurs, but long before, at the very moment the projects are announced. Upstream from Sardar Sarovar, tribal families began fighting among themselves for the elusive registrations, and all weighed the alternatives of resisting the dam against the likelihood of failing to stop it, the benefits of accepting what meager land was offered against the risk of getting nothing at all. The mere fact of the planned dam set tribal against tribal.

Even in 1992, seven years after the World Bank agreed to the loan and construction, Madhya Pradesh still had done nothing to prepare for resettlement, and Maharashtra was struggling to produce a plan. Stories of villagers who were deceived became commonplace. Villagers would be shown an arable plot of land in Gujarat, which they'd agree to settle on, but when they'd return to the resettlement site after dismantling their houses in their home states, they'd be taken to a different plot, with poor soil or no water. Some of them split time between their resettlement site and their original homestead, living fully in neither one; others just went home. The report said Sardar Sarovar "offends recognized norms of human rights—human rights that India and the Bank have been in the forefront to secure."

The dam's environmental record was even worse. Dam officials did not provide required environmental impact information in 1983 when they applied for clearance from India's Ministry of Environment and Forests, which caused the ministry to withhold support. The Bank overlooked the absence of an environmental impact assessment when it agreed to the $450 million loan in 1985. In 1987, under political pressure, the environment ministry agreed to a clearance on one Kafkaesque condition: the builders had to conduct environmental studies during construction, when it was already too late for the findings to affect construction plans. The builders failed to carry out even these studies. "The history of environmental aspects of Sardar Sarovar is a history of noncompliance," the report said.

The report piled on one dam failing after another. Upstream sedimentation would be far higher than officials had estimated, which meant that flooding would be more extensive, and the life of the dam would be reduced by as much as half, to less than a century. Waterlogging and salinization were likely to be serious problems in the areas watered by the canal. Though cited as the fundamental purpose of the dam, water delivery to Kutch and Saurashtra still was "only in the earliest stages of development." And by transforming a moving river into a lake, the dam would create malaria "death traps."

It wasn't that the Bank didn't know these things. The review noted the irony of finding confirmation of many of its conclusions in studies conducted by Bank technicians and consultants. "All this information was in the Bank's files," Morse is said to have told Lewis Preston, Conable's successor as Bank president. "You could have saved a lot of money just by looking in your own files."

"It seems clear that engineering and economic imperative have driven the Projects to the exclusion of human and environmental concerns," the report said. "Social and environmental trade-offs have been made that seem insupportable today." The report's terms of reference instructed the reviewers to provide helpful recommendations to amend whatever flaws it found, but in view of the project's many defects, the reviewers considered that option "irresponsible." Instead, they advised the Bank "to step back from the Projects and consider them afresh"—in other words, to consider ending support for the dam.

Caufield cites a "Bank insider" who said, "There wasn't a soul in the department that had any idea what an impact the Morse Report was going to have . . . It was like watching a ship hit an iceberg in slow motion. The Bank thought it could ride roughshod right over it. They were unconcerned up until the day it arrived." Within the Bank, the report created a sensation, which it chose to acknowledge ten weeks later, in a fashion. In a document called "Narmada: Next Steps," the Bank assured its directors that since the review was released, the Indian government had adopted "a comprehensive set of actions" and had demonstrated "its seriousness of purpose . . . by several important discussions." A suspension of funding was therefore unnecessary, the document said.

When Morse and Berger found out about "Narmada: Next Steps," they erupted. From a hospital bed, where he was suffering through a bout of emphysema, Morse told several Bank executive directors over the phone that "Narmada: Next Steps" had misrepresented the report. In a letter to Preston and the executive directors, he wrote, "The Bank may decide that overriding political and economic considerations are so compelling that its Operational Directives are irrelevant when decisions have to be made about the Sardar Sarovar Project. But it should not seek to reshape our report to support such decisions." The anger spread to nonprofit environmental and human rights groups around the world: 250 of them bought a full-page ad in the *Financial Times* to publish an open letter to Preston. The headline said, "The World Bank Must Withdraw from Sardar Sarovar Immediately."

The Bank was once more cornered: a review headed by a man chosen by the Bank itself had painstakingly detailed the sham behind the dam. Under pressure from hundreds of nonprofit groups and some of its executive directors, the Bank had no choice but to withdraw from Sardar Sarovar. Even so, it acted halfheartedly. Instead of suspending the loan immediately, the Bank gave India five months to comply with terms of the loan. Then, in an orchestrated face-saving move on March 31, 1993, the day before the deadline, India asked the Bank to skip the last $170 million of the $450 million loan—thus, the decision to end Bank support for the dam appeared to be taken by the Indian government, not the Bank. India lost another $440 million in loans under negotiation, but soon after canceling Sardar Sarovar, the Bank announced $2.3 billion in new Indian loans for other projects.

The Andolan had reached the crest of its endless roller-coaster ride. The tiny group had forced the World Bank out of the Narmada—the first time the Bank had ever left a project unfinished. Its victory had heartened antidam movements around the world. It had shown governments everywhere that indigenous people, resettlement, and the environment could no longer be disregarded. But it still hadn't stopped the dam. The Indian government vowed to continue Sardar Sarovar without the World Bank's help, and kept on building it. The Andolan first threatened to commit a mass submergence a few months later, but did not carry through because monsoon waters did not rise high enough. On the local and regional front, the Andolan conducted more rallies, fasts, satyagrahas, and other actions, and beatings and arrests of Andolan activists reached their peak.

On the national front, the Andolan petitioned the Indian Supreme Court in May 1994 to halt dam construction until a comprehensive review evaluated its worthiness. It was the same strategy that the Andolan had used against the World Bank, but now it was turned on India itself. In recent years the Indian Supreme Court had made a reputation for itself as the one national institution with a spark of responsiveness to social issues. The Court's "judicial activists" issued directives to government agencies on issues as diverse as municipal garbage clearance, dengue fever prevention, automobile emission standards, even the saving of the Taj Mahal from pollution damage. All this suggested to Medha and her colleagues that they had a chance at legal victory, and the Court gave them reason to be optimistic. In January 1995, the Court ordered a halt to construction while it considered the case, and then took six years to decide.

There was one more front, of course—the international one. To Medha, the dam was a monumental example of globalization. The world's biggest international lending institution, purportedly designed to relieve world poverty, instead cast the planet's poorest citizens aside so that the wealthy could prosper. In India, this dynamic was well established. Indeed, two Indian scholars, Madhav Gadgil and Ramachandra Guha (who later became embroiled in controversy when he disparaged Arundhati Roy's antidam screeds as careless and hyperbolic, and advised her to stick to fiction), went so far as to sort the Indian population into three groups based on development's impact on them. In *Ecology and Equity: The Use and Abuse of Nature in Contemporary India*, published in 1995, they described how development transformed natural resources into material goods, with the help of such instruments as dams, mines, and factories. This enabled about a sixth of India's billion people—whom Gadgil and Guha call the "omnivores"—to participate in the consumer economy of cars, computers, televisions, and so on. Alas, the process omitted the rest of the population, who enjoyed none of the benefits of development: about half were "ecosystem people," who continued to rely on the local environment, while the remaining third were "ecological refugees," who'd been displaced from the land as a result of development and now lived marginal lives as day laborers and beggars. Dams such as Sardar Sarovar, they suggested, enriched the omnivores, while turning millions of ecosystem people into ecological refugees. Instead of banishing poverty, which the World Bank described as its mission, the process seemed to reinforce it.

The plight of Sardar Sarovar's oustees struck global chords, for the process is hardly limited to India—resettlement opens grievous wounds in virtually every project that displaces people. As a result, the Andolan was able to establish links with international environmental groups. Despite professing hatred for airlines and hotels, Medha traveled around the world in a quest for political (but not financial) support, and her international allies—most notably, the International Rivers Network in Berkeley, California—advanced her cause. Perhaps most significantly, in 1997, the IRN and a few other antidam groups secured her a seat on the World Commission on Dams, where she energetically promoted her views. Her presence on the commission worked in two directions, lending legitimacy both to it and to nonprofits like hers.

As the November 2000 release date of the WCD report approached, the Andolan dared to hope that the antidam cause was about to flourish. Then, on October 18, the Indian Supreme Court published its long-delayed judgment. In a two-to-one vote, the Court decided that the Andolan's case lacked merit and allowed construction on Sardar Sarovar to resume immediately. Justice J. Kirpal's eighty-seven-page judgment is marvelously dense, apparently in hopes of obscuring its illogic. Whereas previously the Court responded enthusiastically to cases brought by public interest nonprofits, now it turned icy. "Public Interest Litigation should not be allowed to degenerate to becoming Publicity Interest Litigation or Private Inquisitiveness Litigation," the judgment rather injudiciously said, as if vanity and greed were Medha's motivations, not the dam advocates'. The judgment even scolded the Andolan for filing its petition so many years after the project was launched, even though the evidence of the project's failures of implementation took a few years to accumulate.

More fundamentally, the judgment showed an obsequious embrace of the government position, uninfluenced by hard data. The majority's mind-set was evidenced by the judgment's second paragraph, which refers to the Narmada River's 4 percent rate of "utilization"—i.e., the percentage of its flow diverted for human purposes—as if that low number were by definition a cause for regret. To Kirpal, it clearly meant that 96 percent of the water was being wasted, for he next referred to the river's "huge potential." The judge even got his figure wrong: the Narmada's true utilization rate was double or triple Kirpal's 4 percent. Not that Kirpal was disposed to ponder statistics for long: in fact, with a sweep of his judicial arm, he dismissed the Morse report, a cornerstone of the Andolan's documentation, on the strangest of procedural grounds: "The Government of India . . . did not accept the report and commented adversely on it. In view of the above, we do not propose . . . to place any reliance on the report." In other words, because the Indian government, the very party under scrutiny in the case, disapproved of the report, the Court would not consider it. Having dismissed the Andolan's evidence, the Court then resorted to an Alice-in-Wonderland flourish, declaring that no instance of the dam's having done harm had been brought to its notice. At a press conference the next day, Medha cried, and then apologized for crying.

The judgment fell short of handing the government a complete victory in only one respect: it stipulated that after each five-meter increment in the dam's height, the builders would have to get approval from the dam authority's environmental and resettlement units. Yet even at the dam's present height, the builders hadn't lived up to their resettlement and environmental obligations, for which the Court provided no reprimand. Considering that, what grounds would the environmental and resettlement units have for holding up construction in the future?

Nevertheless, it was that slender reed that the Andolan clung to. If it could not stop the dam, it would resist each new increment of it on the grounds that the states had yet to implement a resettlement plan, and at the same time it would fight for the dislocated to get resettlement. The Andolan had run out of options: now it was working against itself.

Bob and I wanted a tour of the dam, which only the government could provide. The main office of the agency responsible for building the dam, Sardar Sarovar Narmada Nigam, commonly called the Nigam, is in Baroda. The structure that houses the Nigam is a typical Indian government building, except perhaps with one less layer of grime. Bob and I were directed to the second-floor office of the Nigam press officer, Mr. Desai. A tour of the dam would be no problem, he said. And though he warned us that the dam site was filled with signs banning the use of cameras, he said he would make an exception in Bob's case: Bob could simply disregard the signs. Mr. Desai never explained why Bob qualified for this act of largesse.

We drove to Sardar Sarovar the next day. We knew we were close to the dam when we saw big transmission towers—evidence of the dam's electricity output—and an overgrown helipad, apparently once used by officials but now a grazing area for cows. Soon afterward, we were stopped at a checkpoint and turned back to the Kevadia engineering compound, four miles from the dam. Built in the '60s, the compound was already on its way toward succumbing to decay. If I squinted, blurring the walls' soot and stains, I could make out an element of grandeur in the buildings' design: it remained visible chiefly in the generosity of the landscaping. This was a compound with suburban, even Western ambitions, with intriguingly un-Indian ideas of privacy and family size, with outdoor space allotted to each residential unit. I wondered how many villagers were relocated in the interests of this spaciousness.

Our presence in the compound provoked a flurry of consultations, which concluded with our being ushered into the office of a senior dam official. His pinched severity suggested that he wasn't happy to see us. He clearly aspired to a kind of majesty, with long fingernails on both pinkie fingers that seemed to put keratinous exclamation points on his disdain for us. He immediately got down to business. We could have a tour of the dam, he said, but of course Bob wouldn't be allowed to take photographs. We cited Mr. Desai's assurances to the contrary, but the official said he didn't know who Mr. Desai was. Of course, he said, Bob was free to use the Nigam's own glossy photographs of the dam. We could also visit the dam model next door, and Bob could take pictures of that. When I asked him for his business card, he refused to provide it. As we left, I took the liberty of scribbling his name and title down from the shingle over his door: D. D. Parmar, Superintending Engineer.

Whatever the Nigam hoped to gain by prohibiting first-class photographs while distributing its blurred glossies was beyond me. Clearly it was a reaction to Andolan pamphleteering, which often featured anti-dam slogans plastered over images of the dam, but this response could please only bureaucrats; it would have no impact on the content of anti-dam publications. Nevertheless, instead of the dam photo op, we got the dam *model* photo op. We were taken to a large, sweltering room with fluorescent lights and a checkerboard linoleum-like floor, which we could walk on only after removing our shoes. We were in a kind of temple, after all, where schoolchildren were taught to worship: several classes funneled through during our visit. The room contained four installations, the biggest of which, maybe fifteen feet square, showed the dam area in three dimensions. The installation resembled a giant sandbox whose filling had been replaced by a congealed, garishly painted plastic that failed to support the intended theme of heroic structure. You could see the dam and the powerhouses, the beginning of the canal, even some unconvincing puddles meant to represent bodies of water. It reminded me of the electric train sets I'd coveted as a kid, without the electricity and the trains.

Our guide to the model was Ashok Gajjar, another engineer, as congenial as Mr. Parmar had been sour. Mr. Gajjar was happy to offer me his card, which in addition to containing the usual information such as name and job title (Executive Engineer) conveyed the news of his blood group, which is O. Mr. Gajjar was small and round and bald, and he wore white Reebok sneakers. He was an enthusiastic supporter of the dam. In fact, what distinguished him was the guilelessness of his belief in the dam, which he justified in both scientific and philosophical terms; indeed, his most interesting arguments possessed considerable charm, though somewhat less rigor. When I asked him what effect climate change might have on the estimates of the Narmada's flow (my round-about way of addressing the notion that the tribunal's flow estimate was grossly inaccurate), he answered disarmingly but not reassuringly. "It is very, very difficult to predict the rainfall patterns throughout the earth," he said. "My answer is to leave this in the hands of God. Let God decide whatever is required."

Mr. Gajjar ran through the dam's problems one by one, dispelling each with a kind of technological confidence. Waterlogging and saliniza-tion in the canal area would be averted because "we engineers are aware of this problem," and the appropriate mitigative steps would be taken. Malaria would not occur because the reservoir would flow continuously, depriving mosquitoes of a home. (That would have been news to numer-ous valley inhabitants who, according to Joe Athialy, contracted not just malaria but gastrointestinal and skin diseases as a result of the reservoir's construction.) Mr. Gajjar didn't even mention resettlement.

In his summation, Mr. Gajjar resorted to metaphor. "You have to weigh advantages and disadvantages. In any scheme, you can't have 100 percent advantages. Even marriage is a risk. You may get a good wife or you may not. Does that mean you should not marry?"

We heard the dam long before we saw it. It sounded like an unceasing earthquake, part rumble, part roar—it was the loudest dam I ever heard. We drove around a hill, taking note of one in a series of "No Photogra-phy" signs, when suddenly we found ourselves looking down on the dam. It is gargantuan. In the model, it looked tame and inert, like a thick inverted V, but the real thing oozes power. Here is contemporary cul-ture's version of a pyramid, cold and geometrical, the antithesis of life's curvy essence. The perfect sheet of water that spilled over the dam's wide lip broke into thousands of angry white rivulets that fell twice the height of Niagara Falls before hitting the bottom. The water's force had already eaten holes into the dam's cement catch basin, and forced the construc-tion of "humps" atop the spillway to soften the falling water's impact. At the dam's base, the water spewed and hissed, and sent up vaporous white clouds that obscured the dam's bottom third. Engineers call this water "hungry" because it has shed its sediment in the reservoir, and thus will corrode the downstream riverbank in its search for a new load. The intensity of the green in the hills behind the dam deepened from farthest range to stoppered river. I tried to imagine the river without the dam.

Our guide, Mr. Shah, a "deputy engineer," was even shorter than Mr. Gajjar and even prouder of the dam. I asked him a question or two about its technical aspects, and he was delighted to expound. He got so wrapped up in explaining his colleagues' unique solution to the dam's foundation problems that he didn't notice when Bob slipped away, retrieved his camera from the car, and shot some images of the dam through the guardrail.

3 BADWANI

We rehired Ajit in Baroda and set out again in pursuit of Medha. The idea was to meet her the next day in Badwani, where she was holding a satyagraha—in this case, a public meeting. She and Joe arranged for our day-long journey to Badwani to include stops at two relocation sites, and Vikram was coming along to translate. That meant we'd have another chance to chat. Although foreign travel was beyond Vikram's means, I learned on this trip that he could talk knowledgeably on subjects as diverse as Duke Ellington, Bruce Springsteen, and the difference between American and Indian patent law. Even when he taunted us with anti-American barbs, the broad smile that accompanied them nullified the sting. "I read an interesting statistic," he said. "For the whole world to reach the consumption level of the United States would require three worlds!" Yet he volunteered that globalization wasn't entirely one-sided, and went on to describe the loophole in Indian legislation that Indian pharmaceutical companies exploited to reproduce expensively developed American drugs without paying for the right. Here was a fair-minded activist.

Resettlement site #1 consisted of a grid of huts on a swath of low-lying terrain. All the huts were identically rippled with corrugated-tin walls and roofs, supported on frames of gaunt tree trunks. Vikram invited me to select a hut. Ten minutes later, Bob and I were seated on a charpoy inside it, while nearly a dozen of its inhabitants sat or squatted on the cement floor around us. All the men sported mustaches, and some wore beards; the two old men wore white khadi, while the others wore Western pants and shirts. One man leaned forward on a knee, his arm draped around the shoulder of the kneeling man next to him, whose leg brushed against the hip of a third man, sitting on a thin blanket. A few men near the back of the room rolled cigarettes and lit up together. At the rear, a young woman wearing a flame-red sari, bangles, and nose ring sat with a two-year-old asleep in her lap. The walls were adorned with an electric clock and fluorescent light fixtures, none of which worked.

Clothes were draped over a hewed tree limb. The place was blast-furnace hot.

Over the next forty-five minutes or so, I conducted an interview, or tried to. I'd ask a question, Vikram would translate it, and the response would ricochet among the villagers, back to Vikram, sometimes back to the villagers again, before Vikram would deliver a terse summation of whatever he'd managed to glean. In this manner I learned their story.

In Gadhail, their original Gujarati village, they were farmers, they said—until the dam started scattering them, nearly all Narmada tribals farmed. In the valley, they could produce everything they needed except clothes, which they often declined to wear anyway; they rarely touched money. But all that ceased in 1991, when officials told them that Sardar Sarovar would inundate their lands, forcing them to move. At first, most villagers were excited: after all, the government promised them irrigated land. But the land turned out to be unfit for agriculture. Just as bad, the people of Gadhail were dispatched to thirty-seven separate resettlement sites, neatly splintering the community. Without other options, some became factory laborers and hated every minute. With the Andolan's help, they demanded arable land. The government took its time responding, then gave new land to half of them, thereby creating another fissure within their community. After five years of resettlement, the twenty-two remaining families decided they'd had enough. They dismantled their houses, loaded the usable tree limbs into trucks, and headed back to their old homes, which still weren't underwater. Gujarati officials were not pleased. The police blocked the villagers' way and confiscated their belongings. The villagers abandoned all their possessions on the spot and started walking home. Men and women, old and young—all walked for the next twenty-four hours, until they reached a tribal village where many relatives lived. For the next two weeks, the inhabitants shared food with them while they recouped their strength. Then they walked the rest of the way to Gadhail.

They stayed in Gadhail even after their leader, a man named Shekhjibhai, was mysteriously murdered and five of their huts were burned down. But in 2000, the government made them a new set of re-settlement promises—not just land, but schools and jobs—and, sensing the dam's inevitability, they moved to their current site. Once more the promises went unfulfilled: many families received less than the pledged five acres of land, and the schools and jobs hadn't materialized. A decade since their initial resettlement, their land in Gadhail still hadn't been submerged.

Site #2 was a reprise of #1, except that the inhabitants had been displaced by the canal, not the reservoir, and their huts were even hotter. In the one we entered, the walls were hardened mud, and the door was made of flattened tin cans. Once more we heard a story of unkept government promises that ended with the transformation of farmers into day laborers. Only old men were present—the rest were struggling to survive as migrant workers, moving from job to job every week or two. Because they were canal oustees, unacknowledged in both the 1979 tribunal award and the 2000 Supreme Court ruling, the Gujarati government saw no need to heed their pleas.

By early afternoon, we were back on the road, minus Vikram—he was returning to Baroda for more back treatment. Now the jeep was down to six passengers: Noble and Anil, visiting Andolan supporters from the southern state of Kerala, plus Champalal, Ajit, Bob, and me. It was a predominantly nonverbal ride, at least in the backseat where Bob, Noble, and I sat. For one thing, Champalal and Ajit didn't speak English, and while Noble and Anil were familiar with the language, Noble rarely spoke in any language, and I could only intermittently make out Anil's thick Malayalam accent. As Ajit resumed his quest for Mach 1, I stared purposefully out the side window, trying to spare myself the sight of the recurring near collisions straight ahead. Bob's strategy was to wedge himself into the space where the door and backseat met—he did this so efficiently that his elbow developed a blister, which turned into a parable of Westerners in India. He daubed the elbow with alcohol that night, but it got infected anyway.

At a tea stop in Chota Udaipur, a grim town just west of the Madhya Pradesh border, Champalal passed word that we ought not to mention our Andolan connection: people here were so enthralled with the dam's promise that we'd probably be beaten up if we did.

Once we crossed into Madhya Pradesh, the road's potholes multiplied, and Ajit oddly sped up. It wasn't until the traffic became nonexistent and Ajit had achieved a driving speed unprecedented in our acquaintance that I thought to ask what was happening. The one word I understood in Anil's explanation was "bandits": we were going fast to avoid being held up by the bandits who roamed this area.

We reached Badwani after dark, and spent the night at the wry City

Heart Hotel, which occupied the second through fifth stories of a building whose ground-floor tenant was a cycle and auto parts store. From my room, I looked across the street into the shell of a three-story building: it had floors and ceilings but no outer walls. Inside the rooms, I could see wood planks holding up one of the floors and stacks of bricks and a worker leaning against a wall, hand on hip. The scene looked post-tornado, except that it exuded normalcy. Between that empty hull and the jeweler's shop facing it, a shepherd was leading a herd of goats down the street. My room's distinguishing characteristic was its rat, which scampered from beneath the bed to a perplexing wood and pipe fixture attached to the opposite wall, inside which it disappeared. The room was volcanically hot, and the one movable window lacked a screen. Forced to choose between heat and mosquitoes, I chose heat. I can sleep through this, I told myself, and I might have, if the room's ceiling fan, its invaluable saving grace, hadn't succumbed to a power outage. The man who'd checked us in downstairs obligingly came by with a candle, which he affixed directly to the room's only table with the candle's wax. In the squalid, unlit, toiletless bathroom, I poured cold water over myself with a bucket, soaped, and rinsed. I took two sleeping pills, blew out the candle, and climbed onto the bed. If Sardar Sarovar had been intended to augment Badwani's intermittent electricity supply, I might have consoled myself with this thought, but the opposite was true: the reservoir was going to inundate the town. I lay in bed until the power mercifully came back on, and went to sleep.

The Andolan's Badwani office is at the heart of a dense compound that combines shops facing the street and residences turned away from it, toward an inner courtyard. As with any labyrinth, a guide was required. Champalal met us at the hotel and led us down a main street, which was just wide enough for a single car. A few flat stones, stacked totteringly one atop another, marked our first step up to the building, into a dark covered staircase bracketed by shops on both sides. We walked up a dozen steps, walked down a dozen more, performed a complete U-turn around a corner, and climbed yet another dozen steps, taking care to stoop beneath a low overhang. For a moment we were in the courtyard, which reeked of its communal latrine. We glimpsed families through open doors and noted that few people looked twice at the lumbering

Westerners in their smoky midst. The Andolan office was distinguishable by the number of pairs of shoes politely placed outside it: they filled up the landing and were creeping down the stairs, while their owners stalked the office barefoot.

No enclosure has ever held my attention as raptly as the office's main room. It was decrepit, artful, slovenly, unself-conscious, and inspirational all at once. It was the antidam movement's Sistine Chapel, the nucleus of the cell that triggered the organism that exuded the magnetic force that enlivened activists and dam refugees around the world, in a thousand other towns as remote as this one. Posters, awards, photographs, children's drawings, and printed political slogans, all decaying to varying degrees, mingled on the walls without contradiction. Spiderwebs stretched across the ceiling. Accumulations of finger marks formed thick brown smudges two-thirds of the way up every doorframe. When the wind blew contrarily, the room caught whiffs from the latrine below. The bottom half of one wall was a faded and pockmarked powder blue that yielded to a dingy white above it. One poster macabrely displayed the blood-drained face of a dead child found in the rubble of the 1984 Union Carbide explosion in nearby Bhopal; the poster said, "Resist Globalisation, Combat Communalism, Defend Democracy." A Gandhian saying was the subject of another poster: "Earth provides enough to satisfy every man's need, but not every man's greed." Still another poster showed photographs of victims of the famines of 1877 and 1942, and argued that by promoting cash crops for export at the expense of subsistence crops, free trade was now causing another famine. Some posters were curled at the corners. Above them was a framed, crookedly hung photograph of the exquisite eighteenth-century Maheshwar Fort, which overlooks the Narmada. The wall oozed with bulbous electric fixtures and wires strung in all directions.

The wall directly opposite comprised a perfect mural. Here the broad blue swath along the bottom half contained a darkening horizontal smear, signifying the number of backs that had leaned against it over the years and at the same time evoking the Narmada, as if the river had already risen to this line and left its unruly mark. At the junction with the ceiling, much of the white paint had fallen away, revealing the bare wood beams above. Just below them was a single horizontal fluorescent tube, unlit, and beneath that, a row of Medha's awards: the Right Livelihood Award, 1991; the Goldman Environmental Prize, 1992; the Calcutta Chamber of Commerce Award, 1993; and so on. One of the awards was an imitation scroll, covered in plastic and hung crookedly with yarn. Its lower left-hand corner hid part of a child's painting, attached to the wall with masking tape, which depicted a village submergence: only the top half of trees and huts was visible above a blue lake; birds fled from the trees. The painting was flanked by more sayings of the Mahatma. Below them, someone had painted an inscription directly on the wall. In neat red and green Hindi letters nearly a foot high, it said, "To struggle is to live."

Medha, the principal struggler, was kneeling on the floor, alternately reading her mail and talking on the Indian equivalent of a '60s-vintage Princess phone. She was playing her usual role of administrator/organizer/headmistress/advocate/host, while around her swirled a ceaseless whirlpool of volunteers, oustees, admirers, and bystanders. As always, she was in constant motion, even as she sat. She knelt, crossed her legs, then resumed kneeling; she leaned forward, then back, then to the side; her free hand gestured emphatically, cupped her cheek, rested on her hip. Beyond her, a rotating cast of visitors stood or sat by her, waiting to get in a word. Two young girls leaned against the wall beyond her. Medha explained that they were sisters, whose mother had sold them into a circus, where, not entirely unexpectedly, they were beaten. They ran away. Medha somehow found them. Now she was arranging to reunite the girls with their mother, who'd had a well-advised change of heart. The younger girl wore barrettes and a blouse bearing the motto "Work hard and you will feel relief at heart."

"The tribal minister will come on the seventh, and will confirm by the thirty-first," Medha looked up from the phone to announce. She asked us if we'd like some fruit, and when we declined (having consumed our energy bars in our rooms), she seized on the uninspiring specimen of banana in front of her to provide a lesson in globalization. "This is the best banana-growing area," she explained, "but the best bananas are shipped outside India, and the worst ones stay here." Clifton, a young activist from Bangalore who betrayed signs of hipsterism—he wore hexagonal wire-rimmed glasses and a Che T-shirt—was explaining that working for the Andolan was so hard that many activists dropped out. Behind him, Arild, a gangly Norwegian filmmaker who was in the midst of a six-month project filming and researching Medha, stooped by the doorway. Thirty years old and full of energy, Arild was expressing his amazement at Medha's stamina. On a trip with her, he found he couldn't keep up, and gave up trailing her after six days. "She'd sleep only four or five hours a night—in cars. Then she'd network during the day. Some of her meetings started at one a.m. She'd ask village men if they gave their wives chances to go out. She talked to all layers of society—you don't find that anywhere else."

Among the people Medha had talked to were Badwani's inhabitants, who themselves faced submergence of their homes. The resettlement plan offered plots to about a third of them, but even those lucky recipients didn't get replacement houses. "The traders, the laborers, the kiln workers—they get nothing," Medha told us.

Somehow, our chat with Medha returned to bottled water, a subject our presence seemed constantly to summon up. Medha said she'd attended a national conference where each participant received a neatly packaged bottle of designer water. When she examined hers, she discovered that it proclaimed to be "fresh from the headwaters" of a river in Kerala. Medha said, "We drink water fresh from headwaters already, so why do we need bottles?"

It was a bad day for a satyagraha. By the time we left the office, at just before noon, Medha had absorbed the news that this was a festival day, which meant that most people would stay home. Today's "insistence on truth" would take the form of a meeting to discuss dams—but it would have negligible impact without a crowd. Sukumar, a simultaneous full-time activist and successful law student with a long crew cut that made his head look like a brush, told me later that Medha scolded him for failing to get the word out. Her manner must have been gentle, for Sukumar added, "She will never get angry, even if you wake her up at midnight. That's a problem with her. Her health is deteriorating. She has leg pains. Her hair is nearly white, and she's not even forty-five." (In fact, Medha was forty-six.)

During our stop in the Badwani office, I pieced together Medha's itinerary of the previous twenty-four hours. She'd left Domkhedi; stopped at another satyagraha at Maan, the site of a partly constructed large dam; slept in a village hut for three or four hours; then was driven three hours to Badwani starting at six a.m. She gave no sign of fatigue.

"I wish there would be heavy rains," I heard her say.

It was ordinarily a fifteen-minute drive to the site of today's satyagraha, in the village of Kasarvad, but this was not an ordinary day. Even loading the jeep was a protracted process, involving extensive rearrangement, for eleven people climbed in, among them Medha, Bob and me, assorted activists and oustees, and the two circus escapees. A carton of our Aquafina water was lashed to the roof. We drove a block or two before encountering a slowly moving procession of dalits mourning the recent unsolved shooting death of Phoolan Devi, the renowned Bandit Queen turned demagogic Member of Parliament. In a way, Devi was India's O. J. Simpson, if Simpson had had the wit and ambition to become a successful populist politician after his trials. Devi was the anti-Medha, violent and vain and vacuous. Medha watched the procession without comment.

We stopped to pick up another rider, an old freedom fighter. He was dressed in Gandhian khadi, Nehru cap, and worn black loafers, and his cheeks had started to disappear into the spaces once occupied by his upper teeth. For my benefit, he issued a kind of official statement. "We fought for freedom, but India has not turned out as we hoped," he said in nearly incomprehensible but highly grammatical English. "The Andolan is giving us hope."

Then the jeep broke down.

The driver knew exactly what the problem was, because he'd been meaning to get it repaired: the clutch was broken. We debarked. The driver crawled underneath the jeep and began fiddling. It was absurdly hot, and no shade was in sight. On one side of the road, a cornfield disappeared into the horizon; on the other side, a field of cotton did the same. We were on the verge of walking to the satyagraha in the midday sun, a prospect I earnestly hoped to avoid, when the driver unexpectedly emerged from subjeep to announce that the vehicle was fixed, sort of. We got back in the car and stopped only once the rest of the way, to pick up a few more satyagrahis. There was no room inside the jeep, so they hung off its sides.

The satyagraha turned out to consist of a modest knot of people sitting under a tree. Medha knew all of them. They were arrayed on yet another blue plastic mat, on a gradual slope that spilled downhill into the Narmada.

"All of us!" Medha chanted in Pawri.

"One!" answered the Pawri chorus.

"All of us!" she said in Hindi.

"One!" came the Hindi reply.

A woman poured Medha a cup of tea. She arranged her legs beneath her on the mat. Medha was still hoping that other satyagrahis would arrive—hope is a vital currency in this kind of work—so she vamped for time. She chatted about her journey from Domkhedi, the circus sisters, even the balkiness of a certain journalist's back. I was given the honor of leaning against the tree.

It took me a while to get my bearings. We faced upstream, parallel to the river and a couple of hundred meters away from it. Across the gully in front of us was a vast blue tent, adorned with Andolan slogans, that was the satyagraha's intended site. The tent was so ovenlike and the crowd so paltry that the satyagrahis wisely moved to their present shady locale, while the tent loomed over the proceedings like a reproach, a reminder of the size of the miscalculation.

I wandered down to the Narmada. Here it resembled a natural river, as we were substantially upstream from Sardar Sarovar, beyond the reservoir's sinewy reach. But "resembled" remained the operative word, for the river here still wasn't natural: the construction in the 1980s of the Bargi Dam, 350 miles upstream, made certain of that. In fact, what directly threatened Domkhedi wasn't heavy rainfall, but the release of water from the Bargi reservoir that a downpour would likely induce. Even so, the Narmada here at least flowed like a river: it was broad and languid and unseasonably low, reflecting the dearth of monsoon rain. Closer to the river, the gully narrowed and revealed bare curry-colored slopes, too steep even for the lush vegetation that otherwise covered the hill, except where a herd of goats and cows was avidly consuming it. The brown of the river yielded to identical low cliffs on the opposite bank, which, bathed in the slanted sunlight, looked golden. Above them loomed majestic clouds, dark enough to evoke rain but not supply it.

I walked to the edge of a low cliff above the river, rounded a corner, and found myself looking directly at a temple, modestly proportioned, exquisitely spired, dangling over the river like a tarnished earring. Ordinarily, at this time of year, its stone steps descend into the river, but right now the Narmada flowed ten or fifteen yards beneath the lowest rung. No matter: children in the temple smiled and waved. Beneath them, at the river's edge, squatting women produced the timeless *thwack* of cloth against stone, and fishermen tossed nets into the river with swings as graceful as Joe DiMaggio's. It took effort to pull my eye from these slow-motion discus throwers to a point much farther downstream and yet above us, where towers loomed that signified the impending death of all of the above. Because the reservoir would eventually inundate the existing Badwani bridge, a higher bridge was being constructed. Now its two dozen still-unlinked stanchions marched across the river like scouts of an invading army.

I got back to the shade tree in time to hear Medha tell the group that "nature has decided not to fill the reservoir" this year. The disappointment in her voice was evident, as if she hated losing another opportunity to drown. The group temporarily dispersed.

Now Medha led us in the direction opposite the tent, into a garden filled with flowering trees, sparrows, and chipmunks. It was as if we'd somehow crossed dimensions into a parcel of idealized India, a Gandhi-Disney coproduction. This was the realm of Baba Amte, an eighty-seven-year-old lepers' advocate, garbage-scavenger organizer, endangered-forest preserver, tribal defender, and, for more than two decades, antidam activist. A revered Gandhian, Baba was the winner of even more prizes than Medha, including the Gandhi Peace Prize, conferred by the president of India himself.

In ways both good and bad, Baba Amte (OM-tay) represented what Medha was becoming. While leading a life of extraordinary selflessness, he had worn out his spine, and had spent the last fourteen years chiefly on his back. Sitting was beyond him: he was "withering away petal by petal," as he put it in a poem. Amte was born into a wealthy Brahmin family whose expectations he continually confounded, starting as a child when he ate with the servants. He later offered himself as a tester of leprosy medications, and he established Anandwan, a center in Maharashtra where lepers received medical care, rehabilitation, and skills that foster self-reliance. He consulted experts in agriculture, administration, engineering, and the social sciences, then promoted small-scale schemes to conserve energy and develop renewable energy sources—rainwater harvesting was a prominent example. The theory was that he'd decompressed his spine while traversing so many of India's back roads—some of the same roads Medha was traveling now, with similar results. Amte's ailments had not prevented him, eleven years earlier, from establishing his home in a run-down cottage overlooking the Narmada and declaring that he would not budge from it even if submerged. He named the cottage Inner Strength and went about transforming it into a shaded outpost of appropriate living. To Medha, he served as mentor and adviser.

Threatening submergence near the end of one's life is the act of a man who doesn't want to waste a minute of it, and Amte drew energy from that spark. He was lying on a wire-frame bed at one end of the veranda behind his house. We walked up a couple of steps and caught glimpses of another unintended mural: the wall in front of us displayed a functioning clock, electrical snarls, a crooked calendar courtesy of the East India Paper Co., and, best of all, assorted photographs—such as shots of Amte with the Dalai Lama and of Amte's son embracing a leopard. Once on the veranda, we turned left to face Amte. He was lying on his back, perpendicular to us, one arm leisurely behind his head, one calf poised over the other leg's bent knee. He wore a sleeveless khadi tunic and loose shorts, as well as a thick lumbar brace—braces might as well have been official Andolan gear. Amte's wife, Sadhana, also an accomplished social servant, sat nearby. Farther down the veranda, a couple of activists were reading the paper. Medha got on the phone.

I was given a chair no more than an inch or two from the bed, a distance that suited Amte. I leaned over him when I couldn't understand him, and he touched my arm when he had something to say. Whatever else Amte had lost—and at this point I wasn't sure of his mental faculties—he still had a sense of humor. When Bob asked for permission to take pictures, Amte said this brought to mind a comment he'd heard about photography: "Life in the close-up is a tragedy, but from the long shot, it's a comedy!" He laughed gaily.

After a while, I realized that Amte often repeated himself—in fact, he spoke in loops. He'd come back to the same five or six points over and over, as if they were his daily speaking points. They didn't flow one from the other, but individually they were coherent.

Amte reported what he said to police in 1994, as they arrested him to prevent his drowning, while the water level climbed up the bamboo trees outside: "I'm not jumping in the river. I'm lying in my bed. I'm not breaking any law. Why do you disturb my privacy?"

He spoke of himself in the third person, the way American politicians and athletes do, but the subject was unmistakably Indian: drowning for the cause. "If Baba Amte is drowned like this, mercilessly, the thing I could not achieve in my lifetime—stopping the dam—will happen. The victor is he who even in defeat never surrenders. That's the moral sanction for the satyagraha."

With concerted effort, Amte sat up and put on a more complicated brace. Then, to my amazement, he stood and walked. For a moment I wondered whether I was witnessing a miraculous event, until Amte explained that his condition prevented him only from sitting, not standing or walking. "I can walk three or four kilometers like a soldier," he said. As if to prove it, we strolled through his garden, around the banyan tree that sent vertical roots from its limbs all the way into the soil, toward the cliff overlooking the river. At that moment the river looked flat and brown and undistinguished, like a football field turned to mud. Amte compared it to the dry season river, when it was "blue like the sky."

When we got back to the veranda, we learned that Medha had gone into the house to take a bath.

That afternoon, Medha held a private meeting at the house with Baba and a few other activists. Then she reconvened the satyagraha in the shade of Baba's garden. The crowd had grown to thirty-five people. Medha did not look like a person short of sleep or satisfaction, or like a person whose spine was crumbling: she looked as if she still had all her petals. Her voice had a lilt in it, and she smiled with her eyes, and her arms moved with a dancer's grace, as if no better conceivable afternoon existed than one on which she could tell a few dozen people about mini-hydroelectric projects and the process of making soap to advance small-scale village industry. As daylight disappeared, a swarm of unseen insects began their loud chafing, and Medha raised her raw voice over them.

Medha was leaving for Domkhedi that night; we would return to Baroda the next morning. That evening, as the satyagraha broke up, we thanked her for her hospitality. She replied that she had done nothing, she was just doing her work. Then she looked around for a moment, leaned in, and said, "Shouldn't you say good-bye to Baba?"

On the trip back, we stopped at a few villages in the broad Narmada Valley plain of Nimad, where the landowners were Hindu and, by Indian standards, affluent. Like the tribal people downstream, they faced impending inundation. One said, "We believe in Medha more than Mother Teresa and Mahatma Gandhi. If there were no Didi"—no Elder Sister Medha—"there would be no antidam movement. If there were no movement, we wouldn't be here"—that is, they would have been forced to leave years earlier.

At another village, the letters "NBA"—for Narmada Bachao Andolan—were painted on the front walls of many houses. Eventually, a group formed outside the village temple. A dozen men formed a half circle in front of me; behind them gathered another dozen women. That arrangement made me curious about what impact Medha's stress on women's rights had had, so I asked. The men did the talking. They said they agreed with the idea of empowering women, though no evidence around them supported the claim. I directed my question to the women, but they responded with silence, broken by one nervous giggle. A woman whispered something to a friend, so I asked her what she had said. "The women who talk are in the fields today," she said.

We got lost on the way out of the village, and had to stop several times for directions. Earlier in the day, a resettlement official and an accompanying policeman had arrived at the village to conduct a survey, but the residents considered the survey a couple of decades late, and wouldn't allow it to proceed. (Defiance of surveys was nothing new: Badwani villagers once dug up dam-related stone markers from the submergence area, transported them more than 150 miles to Bhopal, Madhya Pradesh's capital, and deposited them outside the state legislature.) Now, in our confusion, we found ourselves overtaking the two men, who were marching out of the village empty-handed. As the policeman walked, he swung his lathi. They knew who we were, and looked at us unsmilingly. We chose not to ask them for directions.

4 BARODA, NEW DELHI, AND BOMBAY

Part of what makes Medha unusual is not the extent of her suffering—hundreds of millions of Indians suffer profoundly on a daily basis—but its voluntary nature. My interviews with her cast little light on her reasons, so when I had time at night, I called up her friends all over India, and I dropped in on them in New Delhi and Bombay. Fortunately, they were willing to talk. After a while, I gave up on scribbling their comments into a notebook and started typing their words directly into my laptop computer. They didn't mind.

One night in Baroda, I called Girish Patel, the human rights lawyer who for a time was Medha's close adviser. Patel was only sixty-five miles away, in Ahmedabad, Gujarat's largest city, but the weak phone link suggested he was much farther away. Patel was sixty-eight but sounded older: his voice was raspy, and he seemed constantly out of breath. He was a man of considerable eminence. He'd been a graduate student at Harvard, a law professor, and a practicing advocate in the Indian Supreme Court and Gujarat High Court. He was a pioneer of public interest litigation in Gujarat who'd spent his career fighting for the rights of the state's poorest people. Given his own stature, I was surprised that he spoke of Medha with a respect that verged on reverence.

Patel first explained how he met Medha. On behalf of sugarcane workers, he filed a petition in the Gujarat High Court in 1986. "The petition was mainly concerned with the overall impact of a major dam on the Tapi River, the second largest river in Gujarat. The dam was constructed around 1962, and the government at that time promised all the benefits they are promising now for the Narmada dam. But what happened was that it displaced about seventy thousand people, almost all of them tribals. There was a rehabilitation package, but it was never seriously implemented. A few people got land, some got compensation, and others were neglected.

"The second impact was that the crop pattern changed, from coarse grains and wheat to rice and sugarcane. In the beginning, sugarcane wasn't acceptable to the farmers because it's slightly more expensive, it consumes more water, it yields just one crop per year, and it's slightly risky. But the big and middle-range farmers established modern sugar-cane factories—when I filed the petition, there were nine sugarcane factories in south Gujarat. Every year in about October, thousands of workers were brought to the factories, where they worked until April or May. A Dutch scholar named Jan Breman wrote two articles that appeared in the *Economic and Political Weekly*, an internationally known magazine published in India. He concluded that these workers had lost their land to the dam, they were not even being paid the minimum wage, and they were living in an almost bonded condition for eight months a year in the workers' camps. We filed the petition on the basis of those articles. It caused the High Court to appoint an inquiry committee, and one of the members was Medha Patkar.

"Medha and I weren't very clear about whether a major dam should be opposed or not, but we started working on the problems of displaced people, and slowly we came to the conclusion that if you really want to compensate people displaced by dams, if you really want to do full justice to them, it's an almost impossible task unless the society and the state are fully committed to resettlement and rehabilitation. If you treat displacement as merely a consequence of the project, not a part of it, you are likely to do great injustice to the people . . .

"Sardar Sarovar has become almost a psychological problem for Gujarat—it's just like Vietnam for America at one time or Kashmir for India. The middle class in Gujarat always looked on the Narmada as the solution to Gujarat's water problems, and that's why the Narmada issue has become very emotional for these people. Even if internally they know that there are many problems with the project, they are not ready to give up, and no political party has the courage to say openly that we ought to. Most of the judges believe in development projects. I've filed a number of petitions for tribals affected by Narmada dams, and the judges ask me, 'Mr. Patel, why do you oppose such a good thing? Is it because you are antidevelopment?' "

Patel spoke slowly but unhaltingly, as if consciously facilitating my efforts to type his words. It was time to ask him what he thought of Medha.

"I have never seen a person so fully committed to one issue. I have rarely found her talking about anything other than Narmada, except on a very few occasions when she sang tribal songs. She wrote almost all of the important movement songs, and she coined all the important movement slogans. She has supervised every part of the movement personally. Even for a small meeting in some village, she would plan what should be discussed and who should speak. Every day she would hold five or ten meetings. It might be nine o'clock, ten o'clock, eleven o'clock, and she'd fall asleep almost in the middle of a discussion because she was so tired, and as soon as she got up, she'd start talking about last night's topic—no break, nothing. I've known Medha since 1986, and in that time we never stayed in a big hotel or ate a big meal in a restaurant. When we're in Baroda, we eat and sleep in the office there. Medha takes what she gets.

"She's also a disciplinarian. Except during the people's festivals, she does not allow loose talk or merriment. She is particular about the clothes activists wear. They're working with villagers, so she wants very simple dress. Her brother is an architect in Bombay, and I distinctly remember that he came to see Medha wearing an attractive T-shirt and pants—she asked him to change his dress and then come back. She leads this very simple life, but she is fully conversant with all the technical aspects of the dam. I used to tell her that she was working for tribals with the most sophisticated communications system—the Internet, telephones, faxing. When the Morse report was released in Washington, our members there faxed us the main recommendations, which we received at eleven o'clock at night. Medha immediately faxed the recommendations to the press. As a result, the press got them two days before officials in Gujarat did. The officials couldn't understand how this happened, and accused us of being in collusion with the Morse report."

Patel mentioned that he was Medha's chief secretary when she went on her twenty-two-day fast at Ferkua, on the Gujarat–Madhya Pradesh border. "Even on the last day of the fast, when she was terribly weak, she wanted to talk to every tribal who came to pay respects. If a new person came, she wanted to convince him of the antidam position. We would ask her not to speak, but she would speak. If she couldn't speak, she would write on a slate . . . I am older than she is, but she has changed our life. I have never lived in the open sky, but she taught us at Ferkua, where it was extremely cold, so cold, people couldn't sleep. She has totally changed our perspective on life."

When Medha's kidneys began failing as a result of the fast, the time for a decision arrived. "It was I who had to decide whether to give up or not," Patel said. "I was under such great pressure from all sides. 'How can you waste such an important life? Ask her to give up!' We told her that she should not die at this time. Even to this day, she blames me for this decision. She says, 'Why did you not allow me to die?' "

The conversation ended soon afterward, but I couldn't get Medha's recurring question out of my head. It was as if she would allow herself only one alternative to defeating the dam, and that was to die opposing it.

A week later I called up Patel again. The timing was good, for the next morning he was leaving to visit Medha in Domkhedi. It wouldn't be an easy trip, he told me, as he suffered from arthritis and a bad knee. I asked him to explain the Andolan's decision-making process, a subject that so far remained opaque to me. He happily assumed dictation mode, and described a complicated hierarchy of valley tribal leaders and "core groups" that periodically held sober, two-day-long meetings with fixed agendas. Medha was the "dominating personality," but several activists acted independently of her.

What about jal samarpans and satyagrahas? I asked. How did Medha go about deciding to drown herself?

"That decision remains a very mysterious process," Patel said. "There is always some controversy about that, because most of the members do not support the idea of her sacrifice. She believes that this is a personal decision. We've always carried on a discussion about whether you can call this a personal decision or a collective decision. We say that Medha is part of the movement, and whatever decision she takes must be a decision of the movement. She believes that a time comes when somebody has to take a personal decision. She says, 'Leave this question to me.'

"On all other questions—agitations and negotiations—there is no difficulty. The controversy arises because the others generally ask whether her death would be a sufficient shock to the nation's conscience, whether she'd be able to achieve in death what she cannot achieve alive. Her thinking is that no, someday one has to die with the movement. I believe that one day she will die with this movement.

"Medha does not simply want to die—she wants her death to be the successful culmination of the movement. Every time she's made this decision, she has planned out a whole program of awakening throughout the country—mass mobilization, writing letters, demonstrations, protests involving political parties—that would follow her death. She's not afraid of death, but she doesn't want to die futilely.

"My personal opinion is that she has become part of the larger national movement, and the issues she has raised are no longer confined to the Narmada Valley. They've become national and even international issues—about the control of natural resources and who makes decisions for what purpose. She has challenged the very concept of development, which has become much more relevant in modern India. We still have some choice—after five more years it will be too late. Therefore we tell Medha that she is needed at the all-India level.

"She has discussed that question with me. She wants to be part of the national movement, and she started the National Association of People's Movements. But many hard-core members also complain that she devotes too much time to national problems and not enough to Narmada problems. Her feeling is that others should come forward to take her place, so she can devote more time to the national problems, but it is very difficult to replace her. Most of the others find it difficult to keep pace with her. We tell Medha that she cannot be so demanding of others, because she has reached a level that is practically impossible for others to reach. You have to be satisfied with a normal commitment.

"She feels torn, but she can't give up the Narmada issue because it's a very important part of her life, and the movement is in critical condition. Many people think it has reached its end, while she considers it the time of the last-ditch battle. Certain ground realities face us—for example, the Supreme Court judgment. The only door open to us now is a political movement, but there are certain imponderables there also, because people are bound to be tired. For example, the Kevadia villages have fought from 1987 onwards, and almost every core person there must have faced twenty-five or fifty criminal cases for demonstrating. If the government wants to harass participatory movements, the police can harass them like anything, and there's a limit to what a movement can endure. All programs have been exhausted—Gandhi's satyagrahas and other forms of protest have become part of the Establishment. People are in search of a new creative approach. We can't collect the number of people we used to collect for a demonstration. This is a natural limitation of a long, drawn-out battle."

In another city, I talked with a man who'd worked closely with Medha. We met at his home. He had a bad cold, and periodically burst out of the room in fits of coughing. For nearly an hour, between attacks, we talked about the Andolan events he'd been involved in. When I finally turned the subject to Medha, he paused. He wouldn't let me connect his name to what he was about to say—he wasn't even sure I ought to use the information, he said. Furthermore, it might take me a long time to understand his comments, for in the beginning Medha might simply dazzle me.

Medha has many remarkable qualities, he said, but she has a number of weaknesses that became more apparent the longer he worked with her. "It has become a ritual to have a satyagraha," he said. "The same thing has happened so many times. The water rises, and the Andolan's subordinate supporters send out a frantic message. It has become farcical, and it isn't having any effect. It is three months of the year wasted in an absolutely useless activity. It's a symptom of having run out of ideas and imagination. You mechanically do the same things again and again.

"I keep telling her that the pace she is leading is good neither for her health nor her mind. She must have time to rest and relax, physically as well as mentally. Otherwise she has no chance to reflect, to think about how to take this movement further. Throughout his life, Gandhi had significant periods of rest and relaxation, where he would think and chalk out his next move. She doesn't do that. She is just living life minute to minute, second to second. She seems to feel a compulsion to perform all the time. She doesn't need to. She's reached a stature where nobody would think anything amiss if she were to take a month off.

"Of course, she has some very, very remarkable qualities, which make her an outstanding person, but not everything is what it appears to be. Tribal people are losing enthusiasm, after doing the same thing year after year. It's as if you have to take everyone out on a drill every day, morning to evening, in order to keep their enthusiasm going. But it won't keep their enthusiasm going—they get tired.

"The Andolan is not in very good health, unfortunately, even if that might not appear so to an outsider. Medha is somewhat insecure as a leader. There are fewer activists of the caliber of those who were involved in it earlier. At one point there were some ten first-rate activists, each of whom was capable of becoming a leader in his own right. Each of them left, or was marginalized, or was confined to Maheshwar. If a leader cannot develop a second rung of leaders around her and cannot trust that second rung to take charge, it's a very serious flaw. Gandhi had the ability to do that. Indira Gandhi did not."

I visited Harish Khare, a distinguished senior journalist in the New Delhi bureau of *The Hindu*, South India's most prominent newspaper. He'd covered the Sardar Sarovar dispute from 1985 to 1992. He said that when Medha had launched the "Long March" back in 1990, she didn't anticipate her impending celebrity. "Chimanbhai Patel, Gujarat's chief minister, had to find a villain. Baba Amte was too respected a man to be turned into a villain. He was physically challenged and a true Gandhian, so it was difficult to demonize him. Medha became an easier target for this contrived, state-sponsored process of demonization. Once that happened, she attracted support from outside Gujarat and a very small segment of the population within it. Medha has not looked back since.

"I'm quite convinced that until then, she was not extremely conscious that she was onto something very big. She was just like any other honest, sincere NGO activist, trying to make herself heard against the insensitive state. But when the state itself politicized her, she became much more than she wanted to become. From that time onward, there was a qualitative change in the nature of her support and her opposition. The big weakness in the whole movement is that Medha doesn't enjoy any support in Gujarat. Her support comes from areas outside it, mostly in Madhya Pradesh."

I asked him what impact her drowning would have.

"Everyone would write agonized editorials, but—this may sound insensitive on my part—I'm afraid it won't have the kind of moral impact she thinks it will, because from 1991 onward, India has become an extremely materialistic, consumer-oriented society. We're a Coke-driven civilization. We have joined the American slice of market economy. The most creative people in this country think of selling McDonald's to unsuspecting Indians. Medha's movement arose at a time when politics here still involved mass movements. Our politics was influenced by Gandhian values, with a moral purpose, for a collective purpose. Today there is no more moral purpose for a collective purpose. Today whatever is good for corporate India is good for the rest of the country.

"I personally would be very, very unhappy if anything were to happen to Medha. We have seen her grow from a very innocent young person to a dedicated, hardworking, passionate advocate of the people's cause, but I'm afraid that if she were to die, some extremely sensitive MP would force Parliament to hold a special hearing or something like that, and that's all. Maybe Ms. Roy would be moved to write one more passionate, long essay."

I called up Arundhati Dhuru, a former activist who held the unofficial Andolan record for most times jailed—fifty-four—resulting from more than sixty arrests. What a vivid array of legal adventures she collected. One district labeled her a "habitual criminal" and evicted her. She was often beaten up. Police once assaulted her and tore off her clothes, and probably would have raped her if a senior officer hadn't intervened. Another time, to keep her from shouting slogans (which she insisted on doing), a policeman jammed his lathi into her mouth until she vomited blood. She was never convicted of anything, even when she and three other activists were charged and tried for murder. That case was an exception: ordinarily, the charges against her didn't reach trial. Once arrested, she was usually offered freedom on bail, which she invariably rejected as a matter of Andolan policy. And for good reason: if she paid the bail, it was automatically forfeited. Besides, the activists liked following the legal process as far as it would take them, in hopes of making their case in the biggest possible arena. So Dhuru went to jail, usually for ten or twenty days, until the authorities dropped charges.

As she spoke, I could hear the dare in her laugh. She exuded a kind of tough gaiety as she described her jailings, as if the follies of the legal/correctional system barely needed mention. Beatings, assault, murder charges—ha! These ordeals were minor irritants, insignificant contributions to the cause. She was particularly upbeat on the subject of jails, or at least central jails. "We didn't mind going to them because you can sleep and you can get proper food. The people in them get longer punishments. They hold two hundred or three hundred convicts, and the convicts clean them. Most of the time they're okay. It's when you're kept in police custody, in lockups, that it's really bad. You're in a very small room where twenty or thirty people are kept. There are no toilets, or they're clogged. There's no door and no light. You can't have a bath."

She met Medha back in 1985, when Medha gave a talk at the Tata

Institute of Social Sciences in Bombay. "She looked like a very, very typical professional social worker, with a leather bag and a cotton sari. Her clothes were starched and prim and proper." Even so, Medha's talk on the tribals impressed Dhuru enough to visit her in Gujarat in 1985. "I was a typical Bombay girl—I'd never seen a river before." She tried to walk with Medha and discovered that she couldn't keep up. She felt so incapable that after eight days she decided she was done with Medha and returned to Bombay. "But after three months I was back." She became a full-time activist in 1987.

"When we started, it never occurred to us that we were building an historic movement," Dhuru said. In the beginning, she and Medha slept at bus stops or temples because they had no money and knew no villagers. Many villagers would ignore them just because they were women—male villagers would talk to Shripad, but not to them. Once they displayed their commitment by persevering, they garnered invitations to eat in villagers' homes, but even inside them, they weren't allowed to see the faces of their hostesses.

They held their first meetings in temples, until they realized that they'd never reach dalits that way: dalits couldn't enter the temples because they were considered spiritually unclean. Instead of challenging caste practice directly, they held their meetings in central squares, where everyone could attend, or else in the dalit areas themselves. That caused some higher-caste people to drop out, but they eventually came back because of the urgency of the cause. Now poor tribals and wealthy Nimad farmers shared food and houses, she said; the movement shattered every caste barrier but marriage.

Medha changed, too. She gave up fine clothing, jewelry, and animal products, and began using only locally made goods. Back in 1985, she had a room of her own in Ahmedabad, but she gave that up, too, along with the last vestige of privacy. "I don't think there's anything of her own anymore," Dhuru said. "Her whole concept of ownership has undergone a change."

Medha's perfectionism often surfaced. "Medha doesn't shout, but she gets angry," Dhuru said. "She gets angry more and more nowadays. Even small things anger her. She's finicky about documents—she can't stand it if somebody doesn't fold a document properly. Obviously, you take that in stride."

After nearly a decade of the Andolan life, Dhuru retreated into marriage and domesticity in the faraway city of Lucknow, but even with two small children, her withdrawal from the Andolan sounded less than absolute. She didn't have her first child until Andolan activists had discussed the issue for a year. Even now, she said, if Medha asked her to participate in a jal samarpan, she'd leave her husband and children to court drowning. "My sister still thinks I'm not fit to stay in a family," she said, laughing again, as if she appreciated her sister's reasoning. "There has been nothing so close to our heart as the Narmada."

It wasn't until I'd circulated in Bombay for a while that I realized I had to go backward into Medha's life to see her whole. Bombay was her hometown, where she'd once dressed fashionably, worn earrings, even driven a jeep; it was the megalopolis she fled for the country. Somehow, for Medha, Bombay had been a dead end. Now I tried to understand why.

I hired a taxi and went for an hour's drive from my hotel in Churchgate to Chembur, the district where Medha grew up and where her brother and mother, Mahesh and Indu Khanolkar, still lived. The drive was solid city all the way. I saw building after building exuding a hint of yellow or green or purple, whatever their original color happened to be, now streaked with unidentifiable viscous liquids and all but obliterated by a layer of tropical gray/brown/black crud. I saw flowers growing lushly out of balcony pots and tin cans, and ground-floor shops filled with silk and cotton and mobile phones. I saw a billboard advertisement for a Go-Cart park, an Internet café, a man wearing a striped polo shirt and dhoti. I saw a one-room welding works open to the street and a tiny temple wedged between stores where sari-clad women sat and chatted. I saw a Domino's Pizza. I saw a sign that said, "No Urination/Defecation in Public." I saw television antennas on nearly every roof. I saw a man, either unconscious or dead, being hauled across the street by four others, each of whom clutched a separate limb. It was raining. They carried him facedown, head dangling, until they reached the wet sidewalk, and let him down gently, which suggested to me that he was alive.

Medha's brother's solution to the squalor of Bombay was to rise above it—that, too, is a kind of flight. His architectural firm occupies the sixth and highest floor of its building. Its windows face the north and west, away from Chembur's tallest buildings and overlooking nothing but low-rises—the effect "makes you feel on top of the world," as Mahesh put it.

The office itself is spiffy but cramped, as if the accoutrements of a Madison Avenue design studio have been squeezed into half the requisite space. Mahesh, Medha's only sibling and five years her junior, greeted me somberly and ushered me into his miniature conference room, where his mother soon joined us. They looked notably un-Medha-like: he wore stylish Western clothes; she wore earrings, a gold necklace, and gold bracelets. I worried that they'd consider my request to talk about Medha odd, an invasion of privacy, but they clearly had something they wanted to say. It seemed less a set of ideas than an inchoate mixture of emotions: intense pride in Medha's achievements, worry about her health, regret over the cultural-political-economic divide that now separated them from her.

On the one hand, Medha was a star, and always had been, from the time she was a small child. In addition to her many oratorical triumphs, she'd won prizes for her acting, and she excelled in art and poetry, her mother said; she was first in her class from the fifth grade on. In one month, Indu said, in what surely was a flight of maternal exaggeration, Medha read 250 books. She might have become a doctor or an architect, her childhood ambitions, but her own predilections seemed to get in the way. "Whenever she saw beggars, she asked me to give them money," said Indu. "Helping poor people is her nature."

Her nature got plenty of reinforcement. Her father, Vasant Anant Khanolkar, a labor leader and freedom fighter, often held political meetings at his house. Mahesh said Medha got her values from her father, and then he rattled them off: simplicity, straightforwardness, ethics, absence of materialism. His political perspective was Socialist. (As opposed to Communist: the Maharashtran socialist movement of the 1950s and '60s combined Marxist class analysis and Gandhian nonviolence.) The "turning point"—that's Medha's mother's phrase—occurred when Medha was sixteen and joined Rashtriya Seva Dal, a Socialist Party youth organization. "She was exposed to grassroots-level social work, which the RSD did in villages and downtrodden areas," Indu said. From that time on, even if the family desired her to study medicine or things like that, she was on a track to do grassroots social work."

Even then, Medha was relentlessly serious. Her brother called her "a headmistress kind of sister" who scolded him when he showed any sign of neglecting his studies or getting into mischief. I asked for an example. "If I brought too many of my friends home and created a lot of noise,"

Mahesh said, "or if I wasn't ready on time to go to school. She was very much the older sister."

Apparently, she still is. Mahesh mentioned that she'd importuned him to lead a simpler life, but he couldn't, he said. "I am an architect, so my entire world is different from hers. Our motto is, 'The user has to be satisfied.' Imagine Medha's reaction if I say I'm designing a five-star hotel. So it's better for us not to discuss that. I am true to my profession, but it may not be on the lines Medha would desire." He didn't seem to begrudge her this, for he added, "Whatever I am doing in my profession is no match for the nobility of her cause."

This didn't sound like cant. Medha clearly astonishes, even humbles Mahesh. "I have never come across someone eating, drinking, sleeping, and living a cause as Medha does," he said. "From the time she discovered the Narmada, from 1985 until today, I've never seen her take a personal break—not a single day in more than a decade. People on normal schedules have weekends and holidays, but not her. Even if she's feeling ill, she'll be continuously writing. This is something superhuman."

Mahesh's pride in his sister clearly overrode his uneasiness over the disparity in their lifestyles. When I alluded to Medha's skill in remembering names and faces, he corrected me: "It's not a matter of 'skill.' It's a matter of her extra concern. When you have concern for somebody, you remember his name. She remembers these people because she has a bond with them. She has a bond with millions. All this comes naturally."

On the other hand, the state of Medha's health worries her mother and brother unceasingly. Indu visited her during the Ferkua fast, and tried to persuade her to stop long before she did. "My thoughts about Medha are accompanied only by worry," she said. "There is pride definitely, and there are good wishes and appreciation for her cause, but it's connected to worries, because all this activity will take a toll, won't it? . . . We feel she must take a rest."

Mahesh said, "My mother never stops worrying about her. Medha does get irritated by her, and there's no use trying to dominate Medha."

"I am not dominating her," answered her mother, with a flash of the willfulness that found anchorage in Medha. As if she were trying to convince herself, Indu added, "Somebody should work for the nation."

At the end of our discussion, Mahesh summoned his driver and instructed him to take Indu and me to her home, so I could have a look. Indu is a tiny woman, who disappeared into the plush front seat of Mahesh's car; I was given the honored spot in back. Her home is only a few blocks from Mahesh's office—she is as close to her son as she is far away from her daughter. Mahesh even designed her current residence, which she moved into after her husband died. It's deluxe. It occupies the entire sixth and highest floor of the tallest building in the neighborhood, and therefore offers a 360-degree view of Chembur. The walls of the elevator landing are covered in marble; the servants' quarters are down the staircase a few steps. Its spirit owes more to Donald Trump than to Medha Patkar.

The place is immaculate. We sat at a glass dining room table, on which Indu placed a plastic mat, then a tumbler of orange soda atop that. A tall vase nearby erupted with giant sunflowers, which Mahesh had bought for her. On the wall was a photograph of Vasant standing in front of a microphone. His nose was wider than his mustache. When he died of a heart attack two years ago, Medha felt stricken and immediately came home.

Medha's visits, however, are short and unsatisfying, Indu said. Medha comes home about once a month, but, aside from sleeping, she is rarely there longer than fifteen or twenty minutes at a stretch—and then she is usually talking on the phone. Sometimes Medha brings villagers with her, and they all sleep on the shiny floor. "We have no time to speak, ever," Indu said.

Before I left, Indu showed me the flat's two bedrooms, the garden terrace, and the television room, dominated by a giant Sony. Whenever Medha comes, Indu said, she doesn't sleep in a bedroom, but there— and she pointed to the floor beneath the Sony.

I called up Pradeep Patkar, Medha's former brother-in-law, a practicing psychiatrist near Bombay. Medha, his brother Pravin, and he all attended the same elementary school; even as a child, Medha was so accomplished that Pradeep's relatives constantly used her school performance as the standard for judging his. "I was always good," he said, "but she was always better."

Pradeep bore Medha no ill will for her divorce from his brother. On the contrary, he supported her because the marriage "almost imprisoned" her. It lasted four or five years, he said, and it foundered on the conflict between Medha's work and Pravin's requirements. "I thought she had lots of potential and great compassion for people's plights, but it wasn't possible for her to continue because she had lots of household responsibilities," Pradeep said. The divorce was painful, he said, for the families as well as the principals, but his brother was now happily remarried, and Medha—well, he gushed over Medha. "I am extremely pleased to talk about her," he said, indulging himself. He was thrilled that she had kept the family surname, for her fame "has given my surname a global understanding . . . I don't know if we have given any happiness to her, but she has given us great pride in being associated with her for a period of time."

On the day before I left Bombay, I had one last conversation about Medha. By the end of it, I knew I'd reached the last leg of my Narmada circumambulation. Vijaya Chauhan was Medha's close friend and fellow social worker, who'd known Medha during both her major phases, before and after she discovered the valley. Chauhan first knew Medha from a distance, as the well-known orator whose father was a Socialist labor leader, and then Medha walked into her office, looking for a summer job. She got the job, working with slum children. The two women became friends.

Chauhan and I met in the lobby of my hotel, then walked down the street to a café. Like Medha, Chauhan eschewed both makeup and pretense. She had a kindly face and round body, enclosed in a plain tent dress. She seemed to possess the full Patkarian quotient of tenderness, but without Medha's ceaseless movement and single-mindedness. She talked earnestly and directly. I mostly typed.

Medha and Chauhan met in the mid-1970s, when both were in their early twenties. They quickly realized that they understood each other, and they had intense conversations about poetry, literature, films, politics, and social trends. Medha was married then, so Chauhan, herself divorced several years earlier, often visited Medha's home. The disparity between Medha's home life and her workplace was obvious, Chauhan said.

On the one hand, Medha's work stimulated her. For a time Chauhan and Medha were director and deputy director respectively of an NGO urban development program. "The work was so challenging that we used to spend hours together talking about it," Chauhan said. "We were trying to develop a community out of a slum, and our entry point was providing educational support to children. Through them we could identify poor families and communicate ideas about development, health, cooperative groups. We were working with almost two thousand families. Many children who would otherwise have dropped out of school continued because of our program. We could create aspirations for better lives. We could tell the people that it was their right, that they could call on the government to provide the services they needed. At times we had confrontations with officials, but most of the time we were mobilizing people, insisting that they meet officials and make sure they were heard. This process of empowering people was really exciting to us."

At home, however, Medha felt frustration. It showed up even in the poems that Medha wrote and read to Chauhan. "The most common topic was quarrels between husband and wife," Chauhan said. "Her husband and she had different points of view, and it was a matter of great disturbance to her. Her husband was a chauvinistic person, who preferred that his wife be devoted to him totally and get involved in things according to his convenience. She being more talented, people called on her and asked for her services, and he would blame her and say, 'It's you who invites all these things.' It all appeared in her poetry, in a very subtle way.

"Medha was getting exhausted because work demanded a lot of time, and that was adding to the family tension. She was getting totally fed up. That was when she thought of quitting the job and getting out of Bombay. She wasn't really happy in the city, because she didn't feel that the politics could be dramatically changed there. And she knew that if she stayed in Bombay, the tension between her and her husband would continue. I myself left my husband, but for her, the process was much more painful and complicated because of her nature and her husband's nature. They had a long relationship before marriage, and he was extremely possessive of her. On one side was Medha's feeling of losing the person next to her, of feeling defeated. On the other side was the conflict in their values and ideas and ideology."

Medha's mother was another irritant. "Medha's mother is a perfectionist. When Medha participated in oratory competitions, her mother would make her stand in front of the mirror and get her ready, deciding on proper gestures and poses. Her mother would make her do this again and again and again. Today Medha is a very perfectionist sort of person. Even the tiffin box she brought to the office was so neat: she would never get the rotis mixed with the vegetables. She gets really disturbed when things don't happen the way she would like them to."

Chauhan paused to answer her cell phone, then continued.

"Medha resigned her job in 1984, and we started to travel together. This was when she was at her lowest level, very depressed. What sort of work should she get? Where should she settle down? She would cry quite often. I remember evenings spent in big parks where we would sit and chat. She was very nonplussed. She couldn't see what she could do." Chauhan's grave expression underlined her words.

The idea of a young, devastated Medha astonished me for a moment, and then it answered a question: Medha's suffering preceded her valley career. She did not suffer because antidam activism required suffering; she suffered first, then found a meaningful expression for it in the valley. Suffering became her fuel and her power and her validation, the proof of her commitment to the cause and the source of her magnetism. Take away her suffering, and she might wonder what was left of her. Take away her suffering, and she'd choose death instead. In a way, she already had.

Chauhan said Medha joined a social research nonprofit called SETU, headquartered in Ahmedabad, but quickly became disenchanted. She realized that while SETU paid lip service to the rights of the deprived, it avoided upsetting the political establishment at all costs. Accordingly, when SETU received an offer to study the impact of Sardar Sarovar on children displaced by it, Medha was eager to take it up, while other SETU officials were indifferent.

"She started working in the hills on the border of Gujarat and Maharashtra. She'd go away for two or three weeks, collecting information. She started to realize the scale of the disaster that was about to happen. She learned the tribal languages and dances, and how tribals had been deprived of everything involving so-called development. Then, in one of our meetings, she said, 'There is no return now. If SETU doesn't allow me to continue, I will do the work independently.'" That happened soon enough. Medha resigned and in 1986 formed the Narmada Dharangrasta Samiti, the Narmada Dam-Affected People's Organization, a precursor of the Andolan.

As time passed, Medha's life apart from politics ebbed and threatened to disappear. "Other things gradually became meaningless to her. She once loved movies, but if I took her to a movie, she wasn't able to sit for two hours at a stretch. She'd say, 'How the hell can I sit here when there are so many things to accomplish?' "

Chauhan said she twice persuaded Medha to take a short vacation. Four years earlier, they went to Ajanta and Ellora, the caves east of Bombay that contain spectacular ancient religious paintings and sculpture, but Medha never relaxed. "If we sat outside in the evening, looking at the sky, she couldn't do that for a long time. There were always letters to be answered, articles to be written, books to be read, people to be reached. She had to do something for the movement."

Chauhan tried again two years later, when she took Medha to the birthplace of Vinobha Bhave, an honored freedom fighter and Gandhian scholar-activist. The two friends stayed in a room set aside for trustees of a center there, but instead of relaxing, Medha helped the kitchen cook.

"Though she is there," Chauhan said, "she is not there."

5 INDORE

I intended to see Medha one last time, at a protest in the city of Indore, but that depended on logistics. The Andolan had to decide on a date for the protest, and then I had to get to Indore. Bob and I were in Baroda, two hundred miles to Indore's southeast. By American standards, two hundred miles isn't much, but in India it can be a chasm. We could take a jeep to Indore, the memorable five-hour Baroda-Badwani trip times two, or we could take a train, nine hours overnight, with no possibility of sleeper berths, and not all that much prospect of seats. Only one other alternative existed: a flight from Baroda to Bombay, then one from Bombay to Indore, carving a neat V in the sky. The price—$165 per person, compared to $4 for the train—was exorbitant by the average Indian's standards, but we could pay it, and we did. We flew to Indore, and hoped no one would ask how we got there.

Indore is Madhya Pradesh's commercial capital, another in the endless series of Indian cities overtaken by modernity, with clotted air and swarming traffic. The protest was in Indore because it was the headquarters of the Narmada Control Authority, an agency with representatives from all four involved governments whose many administrative responsibilities include implementing resettlement. In its ruling a year earlier, the Indian Supreme Court had reaffirmed the NCA's role, which prompted the Andolan to conduct a one-day seizure of the NCA office a couple of months later.

Vikram once explained to me how the Andolan carries out an office seizure. It sounds like a heist in reverse. The object isn't to flee with the goods but to lock oneself up with them, to make use of their value for a little while. One by one, he said, activists enter the designated building and circulate inconspicuously, until the appointed moment: then they dash to the doors and block them. Officials trapped inside are usually so terrified that they ask only for deliverance from the building, which the activists are happy to provide. Then they bolt the doors from the inside. Negotiations—what Vikram called "the question-and-answer period"—

ensue. Police and officials demand the activists' immediate departure, and the activists demand specified steps toward a just resettlement. This is the only way, Vikram said, that activists can get the officials' attention. A year earlier, Andolan activists seized the NCA building and held it until four the next morning. That exploit resulted in suspensions of five policemen accused of manhandling female activists and elicited a number of promises related to resettlement, all of them unkept.

Information on the protest was scant. When, two days before the event, Joe finally confirmed the date in e-mail, he added, "It is not told to the press or anybody, to catch the officials unaware. So please keep this to yourself."

On the night before the protest, I asked our hotel deskman to find the address of the Narmada Control Authority. He wrote it into my notebook, in both English and Hindi script: "Nagar Nigam, Near Daly College." At nine the next morning, Bob and I flagged down an autorickshaw driver and showed him my notebook entry. It is a rule of Indian drivers-for-hire, I have learned, that if a driver does not recognize an address, he must never admit it, since doing so risks losing the fare. Our driver stared blankly at the entry, then nodded unconvincingly. Sure enough, we caromed around the city like pinballs. We drove from point A to point B to point C and back to B: we couldn't believe that the driver took us twice to the same mistaken place.

At last we found ourselves driven inside the very gate of the Narmada Control Authority building, as our chastened driver would settle for nothing less than a compensatory grand entrance. We were an hour late, yet the only hint of confrontation was a police jeep discreetly parked down the street. Was the protest canceled? Strangely, our arrival drew effusive greetings at the door from a couple of authority officials. Discretion still required that I not mention the protest, so I said I'd come for an interview. The officials said, "Yes, yes," disbelievingly, and conducted us to the backseat of an official sedan. Shades over the back windows shielded the occupants from the glances of nonofficials, and sheathed masts behind the headlights appeared to serve as tiny flagpoles. We'd mistakenly gone to the authority engineers' office; now we were taken to the authority resettlement office a couple of blocks away.

When we stepped out of our conveyance and onto the premises of the resettlement office, we found that Joe's secret was no secret at all. Half a dozen cameramen paced around the building. Inside, a couple of dozen police and half as many local journalists roamed through the rooms and up and down the staircase. Bob and I were led to the second-floor office of N. D. Tiwari, the office chief. It was hard to say which was wider, Mr. Tiwari's mustache or the smile he pasted permanently below it. I was asking Mr. Tiwari my first question when he stopped me to confer with a policeman. Then he announced, "We have been advised to lock the building." His smile looked conspicuously thin.

The resettlement office occupied a two-story residential house, surrounded by other residential houses with actual residents inside. Some were looking out their windows or over their walls to see what the commotion was about. The street was clogged with police jeeps, and more and more police kept arriving. Some carried lathis and wore padding on their chests, like umpires. Others wore dented helmets with wire-mesh face guards. The five or six female police sported purple berets.

It was almost festive. The police decided to lock the building only after emptying it, so office underlings placed wicker chairs in the front patio and invited the journalists to sit in them. We exchanged cards. The photographers huddled around Bob's Leica and took turns looking through its lens, but they ignored his more unusual large-format Mamiya. To my amazement, I detected no tension between police and journalists; on the contrary, Town Inspector B. S. Parihal chatted amiably with several reporters—they'd been through this before together. A few feet away, Mr. Tiwari and his deputy, Afroz Ahmed, stationed themselves in chairs next to the front door and sat with tightly folded arms and legs.

Whatever sort of protest was coming clearly would be no seizure: this was more theatrical, more choreographic. When Clifton, the young activist I'd last seen in Badwani wearing a Che T-shirt, and Devram, his older colleague, asked for admittance to the compound, the police blithely complied. Clifton and Devram headed straight for the journalists and explained the purpose of the protest: they were activist advance men, Andolan spin doctors. They sat down a few yards from Mr. Tiwari and Mr. Ahmed, both of whom looked purposefully in another direction, and began enumerating a long list of injustices in the implementation of resettlement. The Andolan was wearing its improve-implementation hat, not its stop-the-dam hat. The journalists took notes sparingly.

At ten to one, we could hear chanting from down the street. Both the police and the journalists ran for positions. Five policewomen took up places immediately in front of the gate. Women customarily led Andolan processions, so if there was to be a confrontation, this way it would be woman-to-woman—there'd be no repetition of last year's male-on-female brutality. But confrontation didn't seem the point this time around. The chanters marched up the street shaking their fists and carrying banners ("No Submergence—Struggle for Life," "Unity, Self-Reliance, Simplicity," the all-purpose "We Are—One!"). One man waved a stick on which was affixed a tiny Andolan flag. They marched nearly up to the front gate but obligingly stopped a few yards from it. Among the women leading the way, some tribals wore nearly incandescent saris in variations of red, from orange to cerise to puce. The crowd was just big enough to hide Medha momentarily: I found her near the rear of the procession, walking alone. She was wearing a faded gold sari. She looked tired, distracted, as if rushing despite herself to catch a bus. Within a minute or two she made her way to the front of the crowd, as beads of perspiration formed on her forehead. The photographers stood on walls to snap pictures while Medha hoarsely chanted and her followers chanted back, still shaking their fists in the air. The police formed two rows in front of the office gate, but the marchers never threatened them, and they never threatened back. Through all this, Mr. Tiwari and Mr. Ahmed did not budge from their seats at the back of the patio. A truck inched its way through the marchers, forcing them to brush up against one another to make room. I assumed at first it was a police truck, for the police were restricting vehicle entry at both ends of the street, but when I looked closer, I saw that it was the Andolan jeep, which the police had kindly allowed in. Now Medha stood on its running board and chanted into a microphone linked to a loudspeaker on the jeep's roof. When a misguided follower offered Medha a swig of bottled water, of all things, she waved it away, annoyed.

"Let us stay!" Medha chanted. "Let us live! Let humanity flow!"

After a few minutes, Medha handed the microphone to a male colleague and crossed the street to a vacant lot to hold a press conference. There, the activists had erected a canvas roof, which kept out direct light at the cost of amplifying the body heat collecting beneath it. Journalists crowded around Medha until sweat ran in rivulets down her cheeks and neck. With shouting no longer necessary, her voice softened to her customary raspy whisper. She called for a review of all the large dams in India, and charged government officials with abetting the felling and smuggling of trees from the valley in the name of rehabilitation. She answered every question calmly, even when a journalist asked whether the Andolan was in decline. Alluding to calls by the Madhya Pradesh deputy chief minister to outlaw the Andolan, she said, "If the Andolan is declining, then why is there a hue and cry to ban it?" Of course, calls to ban the Andolan could as easily have arisen from its weakness as its strength.

By the time Medha got back to the front gate, the protesters had spread another expanse of blue plastic tarp on the street in front of it, and sat down. Those who couldn't fit on the tarp—probably a majority of the three hundred marchers—squatted or stood behind it. The street was paved in humanity: people covered half a block's length of road, and across the street they sat on walls temporarily festooned with Andolan banners. By now the residents all seemed to have disappeared behind latched shutters, apparently having concluded that ruffians occupied their neighborhood. Yet when a family showed up in its four-door sedan and saw the assemblage blocking the path to its driveway, the protesters uncomplainingly squeezed together to make room for the car. For all the desperation of their cause, they were an unrelentingly courteous group. The protest even attracted a water truck down the block, where dehydrated protesters could replenish themselves. Some of the water made its way to Medha in a vessel she approved of (a tin cup), and she took a swig between chants. "Let us stay! Let us live! Let humanity flow!"

It was at this moment that I spotted Vikram in the crowd, and made my way to him. He looked exhausted. He was again wearing his "Quantum Mechanic" T-shirt, but now it was filthy, and his flip-flops and feet were caked with mud. He said his journey had started at Domkhedi the previous day, and included a five-hour boat ride up the Narmada, making frequent stops to take on more protesters. (An Andolan advance party earlier had gone through the villages to spread word.) From the riverbank, the protesters walked for an hour in darkness, until they reached a spot on the road where two large trucks met them. Presumably *very* large trucks: 150 people boarded one of them, and 150 people boarded the other. For the next ten hours they stood, while the driver contended with hypothetically paved roads. One truck broke an axle, which delayed it an hour or two. The protesters reached an ashram just outside Indore at midmorning. Then they marched and chanted for a few more miles, right up to the authority's front gate. The trip sounded as strenuous as any Olympic event, a kind of triathlon of protest. The protesters who surrounded me, who'd left their fields in the middle of growing season, had barely eaten or slept for more than a day.

Medha made the same journey, except that she traveled in a compact jeep packed with a half dozen people, not a truck packed with 150. Now she sat down on the blue tarp, at the center of the concentric circles formed by her followers: together, they formed a columbine, of which she was the gold pistil and they the multicolored petals. The officials now dared to step outside the gate, took off their shoes, and sat in two rows nearly beside Medha but facing perpendicularly from her. In their Western clothes, the two bureaucrats looked foreign and vulnerable, and sitting cross-legged exposed the strangeness of their stockinged feet. You could call this meeting an Andolan victory if its fruits weren't so modest: instead of forcing NCA officials to talk to them by storming the building, the protesters were peacefully received on neutral ground.

The Andolan tactic was to press that tiny advantage. If commitment was the protesters' strongest asset, they'd use it as a tool: they'd outlast the bureaucrats, they'd outtalk them. Medha spoke; other activists spoke; actual and projected oustees spoke. They talked while the sun bore down on them, even after a few tribal women crawled beneath the jeep in search of shade and other protesters covered their heads with scarves or saris. They talked into midafternoon, when it briefly rained and Vikram fell asleep on the front seat of the jeep, and they talked after the microphone broke down an hour later. They talked into dusk, by which time all the reporters had left to write their stories and the police sat on their chest protectors. They talked when only a few lightbulbs from the surrounding houses tempered the darkness, until they themselves were silvery silhouettes.

Of the seven hours that the protest lasted, Medha spoke for maybe three and a half. Was this evidence of her extraordinary energy or her inability to develop other leaders among the tribals themselves? Both, I suspect. From time to time, the other activists took short breaks from the meeting, to stretch, take a short stroll, or simply clear their heads—but not Medha. During those seven hours, she was the conductor of the Andolan orchestra: she sat, she stood, she constantly changed microphone hands, she brandished documents and chuckled and cajoled. The longer she spoke, the louder her voice became, the stronger she seemed to grow, as if total immersion was the only effective antidote to suffering. Clifton spoke, too, denouncing the bureaucrats with such venom that his eyes seemed to bulge, but next to Medha, he looked callow, histrionic.

When the villagers spoke, they told personal stories. One man said he was shown a relocation site whose agricultural land was entirely water-logged, so he refused it. Nevertheless, when he got back home, he found that his family was considered rehabilitated.

"In my village," another man said, "the government survey doesn't show the houses below mine as submerged, but mine is. Other houses aren't shown at all."

A man from Maharashtra said the first time his home was submerged, he moved to higher ground. In time, that place, too, was inundated, so he moved to even higher ground. "Then I discovered," he said, "that in government records my house wasn't even affected by the dam."

The same thing happened to an entire village, someone else said. It wasn't shown as an affected village, yet in 1994 it was completely submerged.

In response to all this, the officials adopted a rope-a-dope strategy, passively outlasting the protesters. An authority functionary passed out biscuits and water. "For most of the questions posed by the Andolan activists," said the Hindustan Times the next day, "the NCA officials were seen fumbling for answers." Mr. Tiwari and his deputies spoke as little as possible, and when they did speak, they emphasized the weakness of their office, its insufficient number of employees, the constraints imposed by the four governments and the Supreme Court. Mr. Ahmed said, "If you have any doubts, if you feel there are hamlets which haven't been counted as inundated, please let us know"—as if the Andolan had not been trying to do precisely that for years. (The next day, when I interviewed him, he dismissed Medha as "a nonconformist to the core.")

Near the end of the meeting, Mr. Tiwari made a small tactical concession: he and Mr. Ahmed would visit two villages to check the veracity of government land documents there against Andolan records. At a quarter to eight, Mr. Ahmed said, "We have heard everything you've said. We understand your genuine problems. We are understaffed—we are only four officials with responsibility for three states. We know it's difficult for farmers to come here." With a grace that startled me, the protesters applauded, out of politeness. Nearly seven hours after it began, the meeting ended.

Afterward, I talked to Medha. She finally looked weary. "We've achieved something because the officials will come," she said. "But when they come, they are always in a hurry. The secretary takes notes, but nothing happens. They will come, but whether they will bring policy changes . . ." Her voice trailed off. Medha may yet die for the cause, but not this day, not this year. It was just a very long day in the struggle.

The street emptied so quickly that I missed my ride to the hotel. A temporary amicability, the protesters' last residue, had settled over the neighborhood, and when Medha asked a policeman to take me to the hotel in his jeep, he cheerfully consented. As we drove off, Medha climbed in the front passenger's seat of the Andolan jeep and started the journey back to Domkhedi.

PART II: **SOUTHERN AFRICA**

ONE GOOD DAM

SOUTHERN AFRICA

AFRICA

Atlantic Ocean

Indian Ocean

area of detail

ANGOLA

ZAMBIA

Lusaka   Lusitu   Zambezi River

New Mazula   KARIBA DAM AND RESERVOIR

Okavango River

Zambezi River

Old Mazula

Harare

MOZAMBIQUE

Okavango Delta   Thamalakane River

Maun

Boteti River   Lake Xau

NAMIBIA

ZIMBABWE

BOTSWANA

KALAHARI DESERT

Gaborone

Pretoria

Johannesburg

Atlantic Ocean

Orange River

'MUELA DAM

LESOTHO   KATSE DAM   SWAZILAND

Maseru   MALOTI MTNS.

Indian Ocean

Orange/Senqu R.   MOHALE DAM

SOUTH AFRICA

Cape Town

0 Miles   500

0 Kilometers   500

© 2005 Jeffrey L. Ward

6 PASADENA

Professor Thayer Scudder, the world's leading dam resettlement expert, did not want to go with me to Zambia. More precisely, he did not want to accompany me to Mazulu, the village in Zambia where for nearly half a century he'd chronicled a social catastrophe set off by a dam. At its completion in 1959, the dam, an artless slab of arched cement named Kariba, rose the equivalent of a forty-eight-story building above its Zambezi River floor. It was then the biggest dam in Africa and formed the world's largest reservoir, which instantly became the world's largest man-made thing. Kariba is the first large dam financed by the World Bank, which lavished on it the largest loan in the first decade of its history. From the air, the reservoir resembles a giant caterpillar with tributaries for legs; it obliterates 175 miles of upstream river and an average of ten miles of terrain on either side. On that alluvial soil, enriched with sediment borne from Africa's core, the Tonga of the Gwembe Valley had survived for thousands, perhaps hundreds of thousands, of years. To make room for the reservoir, the dam's British colonial builders uprooted fifty-seven thousand Tonga, two-thirds their entire number, and forced them into resettlement sites on barren land.

In 1956, before the dam was finished, Scudder spent nearly a year in Mazulu, conducting an anthropological study of its six thousand citizens. In the four and a half decades that followed, he paid eighteen extended visits to them in their resettlement village, called New Mazulu. Elizabeth Colson, Scudder's mentor who led the study, and he took turns visiting the resettlement area until they'd conducted the longest-running study of a dam's social and environmental impact. In the beginning, Scudder had no inkling that he would document a disaster. He believed in dams, even as he constructed miniature Karibas in termite mounds to show Mazulu villagers how their land would be inundated: they reacted with "great merriment and disbelief" at the thought that a structure sixty miles away could affect them. It wasn't until the mid-1970s, after the villagers had seemed to recover from the shock of reset-

tlement only to founder, that Scudder realized his job consisted of chart-
ing the disintegration of Tonga culture. Now he calls involuntary resettlement "the worst thing you can do to people other than kill them."

In the spring of 2002, when I was proposing the trip, Scudder's
Mazulu informants reported that village residents were demoralized and
hungry, and Scudder lacked the stomach, or perhaps the heart, for a first-
hand look. He said we could go to Lesotho, the tiny African kingdom
where he was ending a thirteen-year stint as a consultant on a huge dam
project. And we could visit Botswana, where he'd played a crucial role
in stopping a dam that would have harmed the Okavango Delta, a
Connecticut-sized puddle of astonishing wildlife. But at Mazulu, he
said, "There is little to see." I'd have to go there without him.

Scudder's distress was as much intellectual as emotional. It arose not
just from watching villagers of long acquaintance suffer and die, though
that registered; it was, in a way, more personal, more fundamental, because it ate at the assumptions on which his career in dams is based.
Again and again, he admired a dam in plan, lent support to its construction as consultant, and then, after documenting its failures of implementation and its social and environmental damage, grew disillusioned. His
disillusionment has spread to his frequent employer, the World Bank, for
failing to enforce its own resettlement and environmental policies, but it
stops at the threshold of economic development itself. The conflict in his
mind pits the concept of development, in which he has always believed,
against all the evidence he has gathered of its failures. If dams, the cen-
terpieces in every development project that includes them, repeatedly
fail to produce environmental stability or even a modicum of economic
justice, then what is left of development? And if development fails, as it
so spectacularly has in Africa, then what will fill the gap between rich
and poor, what will prevent a billion or two of the world's poorest people
from unrelenting suffering and hopeless, squalid death? For Scudder, a
kind of Dag Hammarskjöld internationalist, a successful dam would confirm a worldview. At seventy-one, he was still looking for one good dam.

Dam building is a civic-minded industry, whose best practitioners
possess a genuine concern for public welfare and take pride in the water
and electricity they have bestowed on a previously thirsty and benighted
populace—it's a Rooseveltian vision, arising out of the New Deal, built
into Hoover Dam and the Tennessee Valley Authority, enthralled with its
seeming capacity to foster prosperity by subjugating nature. Though not

a builder, Scudder embodies this strain of thinking: the fondest belief of this man who eschews spirituality is the good dam. The price of dam builders' pride has been their reluctance to notice, not to mention take responsibility for, the damage that dams cause. Here, the difference between a dam builder and an anthropologist is probably crucial: Scudder noticed. As a result, he found himself in the very center of the dams debate, yet somehow isolated there. Of the three categories into which commissioner nominees for the World Commission on Dams were sorted—prodam, "mixed," and antidam—Scudder was drawn from the middle group, and no commissioner was more in the middle than he. If Medha Patkar delivered to the commission a vehemently argued vision of dams drawn from a single controversial project, Scudder brought a data bank, consisting of the knowledge of hundreds of dams that he'd stored in his head. Where Patkar is narrow, he is broad; where she is single-minded, he tempers with qualifications. In some ways, he seems split down the middle, between his two careers, as anthropologist and consultant; between the two kinds of anthropology, academic and "applied"; and, most urgently, between two emblematic adversaries in the international dams debate: the resurgently prodam World Bank, which Scudder advised through most of his career, and the ardently antidam nonprofit, the International Rivers Network, to which he pays dues and occasionally passes documents. He holds that 70 percent of the world's forty-five thousand large dams should never have been built— and so infuriates dam advocates. But he also maintains that for the other 30 percent, the dams he considers justified, the problem isn't dams but the way people plan and build them—for which he reaps antidam campaigners' scorn. Scudder denounces both sets of critics as "fundamentalists," possessing true-believer casts of mind that warp their judgments. He calls himself an "optimistic pessimist," which he defines roughly as believing that (1) tools exist to solve humans' most pressing resource issues, and (2) some form of apocalypse is likely to occur before humans get around to using them.

Instead of going to Mazulu, Scudder gave me a Mazulu slide show at his office on the campus of the California Institute of Technology. By then, we'd already spent ten days in Lesotho and Botswana together, and I'd gone to Mazulu with one of Scudder's former research assistants. I arrived at his office on a Tuesday morning at a few minutes after eleven. For thirty-six years, Scudder was the sole anthropologist at an institution dedicated to training natural scientists. Despite that distinction, or possibly because of it, Caltech's students in three separate years voted him their favorite teacher. He retired from teaching in 2001, when he was seventy, but only so that he could focus on writing three books. His absences, including the four months he typically traveled every year as a consultant, have been an issue between his wife, Molly, and him, and now, reluctantly, he has resolved to spend more time at home. Yet his definition of home includes the office, which he occupies seven days a week. Tucked into a corner on the second floor of the boxlike Baxter Hall (behind a gaudy auditorium Scudder referred to derisively as "the Taj Mahal"), the office is supremely cloistered, even by the standards of academe. "I probably eat lunch in my office twenty-eight days out of each month," Scudder told me with what seemed like monastic pride. "I could die in this office, and the only person who'd find out about it would be the custodian."

The office is satiated with exotica, both biological and anthropological. From low-hanging pots above his desk, ivy tendrils reach up to the ceiling and all the way back down, forming a cellulosic web. A seven-foot-high ladder leans against the wall, which is covered with floor-to-ceiling files containing field notes and project documents. The only space free of files displays half a dozen framed certificates, including the Bronislaw Malinowski Award and the Solon T. Kimball Award, the two most prestigious awards for applied anthropologists. (Scudder objects to his work being labeled "applied," but he happily accepts applied anthropologists' awards.) The opposite wall is nearly all window, overlooking a pleasant garden.

Both office and occupant look rumpled. Scudder detests formal clothing—he admits to owning five ties, one of which he wore at every session of the World Commission on Dams—and he customarily travels with three shirts, rotated in turn; now I noticed that he was wearing the same green plaid shirt he wore every third day of our Africa trip. The sleeves were rolled to the elbow, but I knew they'd start working their way down his forearms by midafternoon. He stands an inch or two under six feet, and he's grown stocky over the years, but he still walks briskly, and his body still possesses a hint of swagger, the residue of the mountain climber he once was.

Scudder's taste for wilderness is most thoroughly registered on his face: he has spent much of his career studying semiarid river basins, and now his visage has become one. It's a sun-parched river delta, as triangular as the Nile's, narrowing from his broad forehead to his soft, round chin. His small green eyes, modest nose, and fleshy cheeks form the prominent islands, and the channels across his forehead look deep enough to have been dredged. His fabulous inch-long eyebrows are themselves alluvial fans. (The Okavango Delta is another.) In up position, the triangular eyebrows intertwine with his usually tousled hair to form a filamentous mesh, which during our conversation occasionally seemed on the verge of becoming entangled in the ivy over the desk. More commonly, the brows seem to hover over his thick-framed tortoise-shell glasses like cantilevers.

The day before, Scudder had spent six hours culling the most relevant eight hundred of the two thousand slides in his Kariba collection. Within ten minutes of my arrival we were installed in a downstairs classroom, which he had reserved for the exhibition. It was a typical Scudder performance — infectiously energetic, gleefully self-involved, and, at six hours, very, very long. I'd just visited New Mazulu, and I'd seen how bad things had gotten. I was hoping Scudder would show me how good things once had been.

At the rate of a slide every few seconds, the images swept by kaleidoscopically. A thin, taut-skinned Scudder, sporting a mustache, holding up a specimen of the world's most ferocious fish, the tigerfish, caught in the undammed Zambezi with a gill net. His circa '56 Land Rover, mired in mud. Containers made of gourds. Assorted black research assistants, cooks, Men Friday, smiling, looking up at him. And bare-breasted women carrying water, a task in Africa as fundamentally female as childbearing.

As the slides rushed by and Scudder talked, the pieces of his biography meshed. He was twenty-five when he arrived in the Middle Zambezi River Valley. Before that, he'd "majored in mountain climbing" as a Harvard undergraduate, done graduate study at the Yale Divinity School, and returned to Harvard as a Ph.D. candidate in African social anthropology. To supplement Harvard's meager mid-1950s Africa offerings, he took courses across the Charles River, in Boston University's first-rate Africa Studies Department. The most famous teacher in it was the stern, implacable Colson, who'd been doing research in central Africa since

1946. Colson told Scudder that the British would soon build a mammoth dam across the Zambezi River, and the Rhodes-Livingstone Institute, the pulsing heart of Central African anthropological research, was conducting a benchmark study of the people who'd be resettled. Colson had already studied the Tonga people of the Northern Rhodesian Plateau, and now she would focus on their relatives, the Tonga who lived beneath the two-thousand-foot-high plateau, in the Gwembe Valley, straddling the Zambezi. She would do the social anthropology, she said, but the project also required a geographer. Did Scudder's adviser at Harvard have any recommendations? Scudder knew nothing about dams and resettlement, and he'd always assumed he'd do his anthropological work in mountains so that he could continue climbing, but after a month he told Colson he couldn't find any geographer candidates and offered himself. "She looked at me and said, 'Hmmph, I didn't think you were interested.' So that was that."

Six months after signing on, he was living in a Gwembe village where the maximum daily temperature averaged above a hundred degrees Fahrenheit over the full year; in October, the month of his arrival, temperatures often reached 115. One of Scudder's first slides shows his canvas tent and, side by side, a nearly identical tent occupied by Colson. Colson is thirteen years older than Scudder, and the difference felt biggest then. She was an eminent professor, and he lacked even a Ph.D. She was at ease in the village, while he was overseas for the first time in his life. She spoke ciTonga, the local language; he was learning his first words. After two weeks of fieldwork, he almost quit.

Instead, he did a simpler thing: he moved his tent. An ensuing slide shows the tent in its new location, beneath a spreading tamarind tree in the precise center of Old Mazulu—sixty Tonga lived on one side of the tent, and sixty Tonga lived on the other. From that magnificent vantage point, Scudder could see the Zambezi, Africa's fourth-longest river and the longest one in southern Africa. From its source in what is now northwest Zambia, the river curled through eastern Angola, reentered Zambia, meandered and widened to as much as two miles; then, 130 miles upstream from Old Mazulu, it tumbled over Victoria Falls, the world's widest waterfall, and fell eastward, past the village, past the dam gathering in the Kariba Gorge, to Mozambique and the Indian Ocean. At closer range, Scudder watched from his tent as villagers came and went; closest of all, he saw scorpions copulating on his ground cover. He awoke daily to the sight of dozens of small black human feet, glimpsed from beneath the tent's skirts, all attached to children fascinated by the white man in their midst.

In 1956, Northern Rhodesia was part of the British-controlled Central African Federation. Anthropology itself was an academic adjunct to the colonial enterprise, whose purpose was to enlighten colonial leaders about the emphatically unfamiliar people they presumed to rule. The doors of Scudder's Land Rover prominently bore the insignia of the Rhodes-Livingstone Institute, effectively a branch of the Northern Rhodesian government. Just by arriving in Mazulu with a Land Rover and establishing a staff of three, he became the richest man in the village. When Adam, his first Tonga research assistant, contracted a horrific hemorrhaging disease called *onyalai*, Scudder drove him three hours to the hospital, where he died. The smell of Adam's decomposing flesh lingered in the Land Rover for a week, and the village headman rebuked Scudder for not delivering Adam to a witch doctor to cure an obvious case of witchcraft.

Scudder is no artist, but his photographs powerfully convey what he saw. He watched initiation ceremonies in which pubertal girls with bones in their noses and cowrie shells in their hair were slathered with castor oil, and he attended three-day burials that began with hundreds of people wailing over wrapped bundles containing the deceased, and ended with ritual bathing and dancing on graves in sackcloth and ashes. He watched drum-cadenced dances to exorcise spirits from the possessed. Villagers deployed what he calls a "tremendous technology" for catching animals: ingenious snares for catching doves, partridges, and mongooses. They showed him how they stripped the bark off certain trees, boiled it until soft, cut it into small pieces, and tossed it into the river; then they reached into the river with baskets to catch the fish stunned by a toxin the bark released. In the beds of receding Zambezi tributaries, Tonga women stood in a row and caught fingerlings with woven baskets or felt for catfish with their toes.

The Gwembe Valley is a wide trough defined by two-thousand-foot-high escarpments that enclose it and narrow gorges at its ends—it's the perfect shape for a reservoir. The proof is that nearly half a century since its construction, some forty thousand large dams later, Kariba remains in a virtual tie as the world's largest reservoir. Kariba is a little more than half the height of Hoover Dam, and its crest is less than half the length of the Grand Coulee Dam in central Washington, yet its reservoir could hold four Lake Meads plus Grand Coulee's Lake Roosevelt. In the early 1950s, when the dam was under consideration, Kariba Gorge was fifty miles from the nearest road and two hundred miles from the closest city; at Kariba, as in countless other dam projects, the first job was building not a dam, but a road. But remoteness did not discourage dam engineers, who coveted the site because the gorge confined the Zambezi to a narrow bed flanked by seven-hundred-foot-high rock walls—an ideal setting for an arch dam. The white-settler businessmen and political leaders were equally enthusiastic for one overriding reason: high water-storage capacity means high power-generating capacity. On rivers like the Zambezi, whose flow in the dry season dwindles to a twentieth of its rainy season level, high water storage is necessary to ensure the flow of electricity during the dry season. To the British colonial leaders who ruled from the plateaus on both sides of the Zambezi, the valley was a disease-ridden and insufferably hot morass, whose inhabitants' misfortune was to live there. If they thought at all of the ninety thousand Tonga who lived in the valley, they figured that expulsion from it would improve them. Resettlement, a Southern Rhodesian provincial commissioner declared, would bring the Tonga "into close contact with civilization for their benefit and for that of the Colony." As Scudder and colleague David Brokensha commented, "The federation government viewed the lake as merely a by-product of the dam, relatively unimportant in itself, and the people requiring relocation as an expensive nuisance."

From the two plateaus, the dam's beneficiaries radiated outward, across the Mediterranean and the Atlantic. Colson writes: "The project was a monument to the international nature of the contemporary world, partly financed as it was by a loan from the World Bank and other international banking houses, built by an Italian firm from plans supplied by a French engineering company and with labor drawn from Italy, Tanzania, and Malawi. Its primary customers were to be the mines controlled by financial interests based largely in the United States, Britain, and South Africa." The dam would turn alien terrain into a producer of wealth, enriching everyone but the people directly affected by the dam, the Gwembe Tonga.

If plateau people disdained the valley, the Tonga embraced it: it was their refuge from menacing inhabitants on both sides. To them, the valley's isolation was one of its two outstanding assets, along with the Zambezi River itself. The Tonga were willing but ineffective warriors, so seclusion was a better bet. The valley was the Tonga briar patch, shielding them from black slave traders in the nineteenth century and their white colonial rulers in the twentieth. Until the dam came, most Tonga saw white government officials only once or twice a year, when they came around to collect taxes. The Tonga were so profoundly ensconced in the valley that their legends contain no stories of antecedents arriving from somewhere else—they'd *always* lived there. Scholarly estimates of the length of their habitation in the valley range from a few thousand to half a million years. The arrival in the valley in 1860 of David Livingstone, the missionary-explorer, as he made his way down the Zambezi, presaged the end of Tonga tranquillity a century later. Livingstone himself seemed to sense this, for afterward he said of the Tonga, "If such men must perish by the advance of civilization, it is a pity."* He reported that the valley "teemed with people," and that "along the river bank, every damp spot is covered with maize, pumpkins, watermelons, tobacco, and hemp."

Even so, Tonga life was harsh. When the slave traders raided the Tonga, they killed the men and took women and children captive. A smallpox epidemic occurred every twenty or thirty years. Other diseases—malaria, dysentery, bilharzia, yaws, leprosy—claimed victims annually; an influenza epidemic in 1918–19 was calamitous. Colson wrote in 1960, "The death toll is unknown, but Valley people say that no one then alive can forget the time when each day they began to mourn new deaths."

Even in rainy years, farming was hard. Sometimes the Zambezi flooded insufficiently, or at the wrong time of year. Thick hordes of crickets dined on the crops; hippopotamuses wandering from the river trampled them. Livestock was not a reliable option, for tsetse flies lethally infected cattle with animal sleeping sickness. In the first decades of the twentieth century, famines were frequent, but by the '50s the villagers usually experienced nothing worse than what Scudder dryly calls "sea-

---

*At the time of his visit, Livingstone was repulsed by the Tongas' nakedness. "They walk about without the smallest sense of shame," he wrote. "I told them that on my return I should have my family with me, and no one must come near us in that state."

sonal hunger." Scudder attributes the improvement to the belated decision of the colonial government to distribute food when crops failed, and to Tonga innovation. For generations they'd grown maize, the staple of southern Africa, in the sediment-enriched soil along the shores of the Zambezi twice a year—planting once at the onset of the rainy season in November and again in May, after the annual flood, in the banks exposed by the receding river. But population was beginning to outstrip supply of land. The Tongas' response was to cultivate inland garden crops for the first time, in plots as big as several acres and as far away from the river as a mile or more. The soil there wasn't rich, and it probably would have given out within a dozen years if the reservoir hadn't inundated it first.

The dam diverted the Gwembe Tonga from a disagreeable fate to a disastrous one. In the luster of the dam's projected production of 1,320 megawatts of electricity, nearly twice as much capacity as the federation possessed in 1955, the Tonga were entirely obscured. Of the two main entities in the short-lived and loosely linked federation, Southern Rhodesia wanted energy to fuel its budding industrial belt around Salisbury (now Harare), and Northern Rhodesia wanted energy for its booming copper mines, which generated four-fifths of its foreign exchange. The mines ran on coal, which was delivered from Southern Rhodesia on a single overburdened railroad track for most of a 450-mile journey, and wood, whose use denuded more than two hundred thousand acres of forest in less than a decade—enough to be deplored even in the preenvironmental era. By 1960, projections suggested, fuel shortages would hamper mine expansion: the only conceivable solution was hydroelectricity. So energy-starved were the mines that they paid a substantial share of Kariba's cost. Even so, if copper had been the only interest that the dam served, it wouldn't have been located at Kariba. The copper mines favored another proposed dam, which would have spanned the Kafue (kuh-FOO-ee) River, a Northern Rhodesian Zambezi tributary, and would have displaced a thousand people at most. The Kafue Dam would have been smaller and cheaper, it could have been constructed more quickly, and it was closer to the mines, but Kariba's telling advantage was political. In straddling the river that defined the border between the two Rhodesias, Kariba would give both Rhodesias shares in an invaluable asset. For four years, the battle between Kariba and Kafue went back and forth, until it threatened to undo the fledgling federation. Instead, Southern Rhodesia, which dominated the federation, won out. Not coincidentally, the federation's prime minister Sir Godfrey Huggins (later known as Lord Malvern) was a Southerner—indeed, his twenty-year term as Southern Rhodesia's prime minister set a British Commonwealth record as its longest-running prime minister. When Kariba was chosen, Huggins triumphantly declared that it would connect the two Rhodesias with an "arch of concrete." Though the federation desperately needed schools, hospitals, and roads, it chose to build a dam for £80 million (equivalent to about $225 million in 1955 and nearly $2 billion in 2002), double its annual budget. As David Howarth wrote in *The Shadow of the Dam*, a 1961 account of the Tonga resettlement, the dam was "a glorious castle in the sky" that would show the world that African colonies were capable of great technological feats. The resettlement of the Tonga was considered trivial and did not figure in the deliberations.

Dam construction turned into a race to reverse the faltering supply of energy for the copper mines. Construction began even before the federation secured the vast loans it would need to build the dam. The World Bank, whose crucial $80 million loan started the flow of international funding, did not commit itself until January 1956, seven months after the federation hired engineers to design the preliminary works. Instead of pausing six months to collect competitive bids, the federation selected a Frenchman, André Coyne, who'd already advanced the cause by writing a decisive report favoring the Kariba site over Kafue. Coyne, then considered the world's foremost designer of arch dams, declared at a press conference that Kariba would deliver the cheapest power in the world. The pursuit of electricity was so single-minded that the planners showed no interest in constructing a multipurpose dam, generating irrigation as well as electricity, conceivably mitigating some of the dam's social and environmental harm.

Work on the dam proper was the last step in a construction minuet. First, the site had to be become accessible. An airstrip was installed, and roads to Salisbury and Lusaka substantial enough to bear the weight of bulldozers and concrete mixers were built through jungle. On the day after the Salisbury road was completed, trucks began arriving with supplies. Within a month, African workers had erected a functioning quarry, a brick and block factory, a power station, and transmission lines. In May 1956, the federation's governor-general used golden scissors to cut a cord opening the world's longest suspension footbridge, which spanned the gorge. A British firm built a town in nineteen months; five months later, it was the sixth-biggest town in the federation. The town included a bank, a hospital, and a mortuary, with separate entrances for whites and blacks.

Haste probably contributed to the frequent accidents and design missteps that Kariba suffered. Out of the ten thousand men who built the dam, probably more than a hundred died. They were electrocuted, caught in machines, crushed by falling rock, and killed on the road; five died of malaria. When scaffolding collapsed on February 20, 1959, seventeen men fell two hundred feet into a shaft of hardening cement and were instantly entombed. A crew required half a day to recover the bodies with pneumatic drills. That incident spawned a walkout by sixteen hundred African workers, more than a third of the workforce; Impresit, the Italian contractor, had already realized that it could meet its remaining deadlines with far fewer workers, and cheerfully accepted the resignations.

A paucity of hydrological data about the Zambezi contributed to the design errors. Accurate measurement of the river's flow didn't start until 1924, which left designers with a meager three decades of records. They proved to offer a deceptively benign picture of the river's flood patterns. In March 1957, nineteen months after construction began, the Zambezi produced its highest flood on record, rising eighteen feet in twenty-four hours. It overtopped a cofferdam, washed away a derrick, and nearly consumed the only road bridge across the gorge. If the bridge had collapsed, construction could have been held up for as long as a year; as it was, the delay was six weeks. A dam's spillway is designed to protect the dam's downstream foundation by deflecting some of the energy of the water cascading down it. If water overwhelms the spillway, it can hurtle to the bottom of the gorge with enough force to damage the foundation. On the evidence of the unprecedented flood, Kariba's spillway was hurriedly redesigned.

A year later, the river flooded at twice the rate of its record 1957 flow—one journalist called the 1958 spate a thousand-year flood; another said it was a ten-thousand-year flood. It arrived at Kariba at the same time as a violent thunderstorm, which triggered landslides across the access roads and into the gorge. The river rose forty yards over its dry-season level, gobbling up the road bridge that had survived the 1957 flood, submerging the dam's foundation, and rising to within eight feet of the dam's partially completed upstream face. Water found a weakness in a corner of the upstream foundation and poured through it until the dam's interior was flooded. A cofferdam was overtopped; the footbridge collapsed; the chief engineer's office slid into the river. Workers toiled under arc lights at night to save some of the piers that once held up the road bridge and poured stones into the hillside gashes torn open by the landslide. Sightseers paid about $15 for plane rides from Salisbury that gave them a minute's look from the pilot cabin into the devastated gorge. The flood set the project back by months, and triggered a second spillway redesign. A World Commission on Dams study concluded that if the 1957 and 1958 floods had not occurred, Kariba's spillway "would have been completely under-designed," suggesting that a major flood could have undermined the dam's downstream foundation.

Resettlement was the project's unwanted child. The federation swiftly detached it from the rest of the project and passed off responsibility to the Northern and Southern Rhodesian territorial administrations. The British government, then in the process of shedding its colonies, was sufficiently concerned about the federation's indifference to the Tonga that it required as part of an £18 million loan the federation's promise to safeguard the resettled Tongas' well-being. The federation grudgingly accepted, then made clear to Northern and Southern Rhodesia that resettlement money could only compensate Tonga for their losses, not improve their living conditions. Since Tonga possessions were nearly without marketable value, this decision ensured that their payments would be meager.

In the mid-1950s, the idea of a dam resettlement "program" was largely hypothetical—Kariba's resettlement was the largest in Africa, and its administrators were ill prepared to provide what it would entail. The white colonial district commissioners and district officers who knew the Tonga sympathized with them and worked hard to produce a decent resettlement program, but they were overwhelmed by their lack of knowledge and the legal and financial constraints placed upon them. By 1956, the Northern Rhodesia officials had produced a workable plan in consultation with chiefs, headmen, and villagers, many of whom were taken to the proposed resettlement sites to make tentative choices of homesteads and field sites—all areas were in the valley, often up a tributary from a resettled village's original location, usually no more than twenty miles away. Then the plan was rendered obsolete by the federation's decision to increase the dam's height by twenty feet. A twenty-foot rise meant that the homes of an additional seventeen thousand Tonga would be inundated, along with much of the land being prepared as resettlement sites. Now the planners had to resettle more people and at new sites, some outside the valley and as much as a hundred miles away. Despite the 40-plus percent increase in resettlers, the planners' budget remained the same—and the dam would be sealed in thirty months.

If it is axiomatic that the suffering caused by resettlement begins at the moment when villagers learn their homes will go underwater, then a corollary must account for denial. In the Tongas' case, denial temporarily served to ward off comprehension of the incomprehensible. The Tonga, after all, had never seen a lake, and they had no words for "dam," "turbine," "electricity." How could they understand a reservoir? Howarth reports that when two government officials broke the news to the residents of a Tonga village called Chisamu, the villagers couldn't absorb it. The officials had to give them time to think about it, and promised to come back for another conversation. The idea of a wall that could obliterate the village seemed otherworldly, yet it instantly began undermining the foundations of village life. Howarth writes:

> The headman could not be certain whether he would still be headman, or whether the village, in some foreign land, would be split in pieces or merged with other villages; he foresaw that if he helped the government in the move and something went wrong, the people might turn against him. Men who had sweated to clear a patch of forest, and so been able to claim it for their own, saw all their labor wasted. Others who had inherited a winter garden by the river and hoed it and planted it every year since they were children could hardly bear to think of leaving it. A man who wanted to marry a girl would wonder whether he should build her a hut, as convention demanded, or whether she would come to him if he waited till the village was moved. A girl with a lover [across the river] would wonder whether she should marry him and go with him to new lands on the other side, and never see her family again.

Even more disturbing to the villagers was the expected impact on their ancestors, whose spirits they believed they'd inherited. Their ancestors' shrines could not be moved with them, for their meaning derived from their location, and connoted the connection to the land of past, present, and future generations. If the villagers moved and the shrines were inundated, then, inevitably, the villagers' crops would fail until they returned.

For months after the officials' visit, nothing happened, and the villagers figured they'd been right to doubt that a wall could drown the valley. Howarth writes, "The old men argued that if anyone thought he could stop the river by building a wall across it, it only showed he had no idea of how strong the river was. Let them try, the people of Chisamu began to say; the river will push the wall over, or run round the ends of it, and the flood will never come."

Of the four Northern Rhodesian chieftaincies facing inundation, the resettlement was most widely disbelieved in Chipepo (chuh-PEP-oh), the chieftaincy of Mazulu and Chisamu villages. This was not surprising, since Chipepo would suffer most from the resettlement. Two-thirds of the chieftaincy's nine thousand inhabitants were being asked to make the biggest move of any resettlers—a hundred miles downstream, thirty miles beyond the dam itself, to an arid, densely thicketed, tsetse-infested scrubland called Lusitu (loo-SEE-too). Chipepo's people had already suffered the disappearance of friends and family members who lived across the river in Southern Rhodesia: they'd vanished as part of the stern Southern Rhodesian resettlement. Now Chipepo would also be split internally, between the three thousand people sent to Gwembe foothills and the six thousand consigned to Lusitu. In place of the Zambezi's alluvial soil, the residents of Mazulu, Chisamu, and thirty-five other villages were being given infertile inland fields with a brackish or nonexistent water supply. Recession agriculture—the growing of dry-season crops on the banks of receding rivers and the key to Tonga subsistence—would be impossible: the Tonga would have to live on one crop a year instead of two. In their view just as bad, the people who already populated Lusitu were Goba, whom the Tonga accused of rampant witchcraft. As evidence, they pointed to the Gobas' deployment of cemeteries in groves, which contrasted with predominant southern African burial placements in individual plots close to home. To Colson and Scudder, the siting of Goba cemeteries suggested a different but equally ominous explanation: the Goba had experienced frequent flu, smallpox, or sleeping sickness epidemics.

By 1957, Alex Smith, the Northern Rhodesia district officer responsible for Chipepo, grew so frustrated with the chieftaincy's denial of the dam that he arranged truck tours of fifty or sixty people at a time to Kariba Gorge. Two roads now connected the dam site to the plateaus, and thousands of men toiled there. Smith took his sightseers (most of whom were in a moving vehicle for the first time in their life) to a vantage point from which they could see cement silos and sand mountains and cableways—but the dam wall had barely formed, and the villagers still could not see what would drown them. A reservoir was still unimaginable to them, but elsewhere in the valley, the Tonga were being moved out, and the Chipepo villagers could see that pairs of huge tractors linked by battleship anchor chains and eight-foot-high steel balls were toppling trees at the rate of thirty acres per hour. The purpose was to clear the reservoir floor so that fishing nets would not be snagged on the tips of inundated trees, but the villagers saw the land clearing and felt confirmed in the suspicion they'd had since they first heard about the dam: the Europeans were using it as a ruse to move them so that whites could take their precious farmland.

The 1958 flood caused as much chaos in Chipepo, where it lasted for several weeks, as it did at Kariba Gorge. It topped the Zambezi's banks and inundated villages for the first time in their known history. It swept away century-old trees and flowed up tributaries for miles, bringing crocodiles with it. Smith, the district officer, plied Chipepo in a dinghy, rescuing hundreds of villagers who'd retreated to higher ground only to discover that it wasn't high enough. Villagers who found themselves stranded in the forest ate berries and leaves to survive, and some died of disease or exposure. The colonial administrators hoped the villagers would interpret the flood as a harbinger of the *real* flood to come and would therefore submit to resettlement. Smith noted that the people he rescued all said they were willing to leave Chipepo when the rainy season ended, but if that was their true sentiment, it was not the predominant one. Chipepo villagers still believed that the white colonials were engineering a land grab, and a few representatives of the African National Congress, then nascent in Northern Rhodesia, encouraged them to resist.

As 1958 unfolded, disaster loomed closer. The gathering dam towered over the valley, and would be sealed in early December; with decent rain, half of Chipepo would be submerged by the end of the year. What occurred in Chipepo from May to September was an asymmetrical but lockstep escalation, as the two sides displayed their weaponry in intensifying shows of force: the Tonga brandished their stones, spears, clubs, and a few nineteenth-century muzzle-loaders against the automatics, rifles, shotguns, and tear gas of the police. The governor himself arrived in Chisamu with a military band; arrayed in plumed hat, sword, and white uniform, he held spirited conversations with the resistant headmen but accomplished nothing. At last, on September 10, 150 police tried to cordon off Chisamu and move its people to Lusitu, by force if necessary. The Tonga charged the police with spears and shields, and the police opened fire. Eight Tonga were killed, and at least thirty-two were injured. Tonga resistance in Chipepo melted away.

The same day the administrators' messengers went from hut to hut, taking pots, pans, blankets, and stools and loading them on flatbed trucks. Some of the women and children who'd been hiding in the forest asked permission to load their own goods. By late afternoon, fifteen five-ton flatbed trucks departed for Lusitu. So that no resettlers would be tempted to return, government messengers burned down the huts. Then bulldozers obliterated what was left.

Colson writes that the survivors of the "Chisamu War"

> were hurriedly loaded onto the lorries before they could recover from the shock of Chisamu, with the assurance that their goods would be brought later. They rode the swaying open lorries for a hundred miles, over rough roads, in the blazing sun of the hottest period of the year . . . to reach an unknown land they dreaded . . . Vomiting women and children hung over the sides of the lorries. Drinking water and water for cleansing was gone long before they reached their destination . . . They emerged exhausted and sick to find themselves in what they regarded as a wilderness . . . They struggled to cook and eat. Then they lay listening to the [unfamiliar] trumpeting of the elephant which still frequented thickets along the Lusitu River . . . Next day they rose to the task of turning a strange land into home.

Years later, an American Catholic priest named Michael Tremmel wrote down the words to the songs that some resettlers sang during their journey. One began:

Let the driver overturn the lorry;
we are perishing,
we who used to be very happy.
Hit the brake, driver,
we are perishing.

"On superficial inspection," Colson writes, "Gwembe villagers seem to have few possessions other than their stock, but the Kariba move demonstrated that they are thoroughly encumbered with possessions which they cherish." They had "coils of bark rope and string, grindstones and mortars, traps, ploughs, sledges, blacksmith equipment, stores of medicine, mats and stools, drums and bottles." Most objects were left behind. Some were stolen by messengers and lorry drivers, who were said to have feasted on chickens left behind in the old villages. The bigger animals—sheep, goats, some cattle—walked. Unaccustomed to long treks, many arrived with broken limbs, causing their owners to slaughter them on the spot—as Colson put it, "They at least provided meat in the first days." The only preparation performed at the resettlement site was the stocking by prison labor of firewood and thatching grass for windbreaks. In the three months before the rains came, the resettlers were expected to clear the fields of its scrub vegetation and build homesteads, all while sustaining themselves on their meager compensation. Most places lacked water.

The consequence was a two-year famine, coupled with assorted epidemics. Bacterial dysentery killed eighty people. An unidentified ailment that the Tonga attributed to Goba witchcraft killed fifty-six women and children, but no men. Old people and children died in large numbers. Fewer women than usual delivered children. "This is an area where people were not meant to live," a senior headman told Colson. "It would not be so bad if the adults died and children lived, for that would mean that life would go on. But when children die, as they do in Lusitu, this means an end to life."

While the Tongas' relocation went unnoticed, the prospect that the reservoir would drown the valley's animals provoked international concern. The drownings began as soon as the Zambezi overflowed its banks, and forced crickets by the billions to emerge into the unfamiliar daylight from their underground colonies; birds attracted by this unique feasting opportunity were said to blacken the sky. Officials had assumed that the valley's larger animals—a menagerie from antbears to zebras, from badgers, baboons, buffalo, and bushbabies to warthogs and waterbucks—would sidestep the reservoir by moving inland, instead of finding more temporary refuges, as so many did. In February 1959, as the reservoir began to fill, a feature writer at the *Rhodesian Sunday Mail* wrote a story describing how a few Southern Rhodesian game rangers were risking crocodile attacks and lethal snakebites to rescue drowning animals. The rangers found monkeys and baboons clinging to nearly submerged trees, bucks trapped on disappearing islands, and civet cats hissing as they hung onto floating logs—the rangers' work, the story said, amounted to the biggest animal rescue operation since Noah. Soon reporters, photographers, and television crews descended on the valley to report the story, and the idea of "Operation Noah" took shape. With exquisite tunnel vision, the Faunal Preservation Society of London raised £10,000. (The officials knew that elsewhere in Rhodesia, some of the same species being rescued in the valley were being trapped for food or shot to eliminate tsetse fly vectors.) The augmented staff of rescuers swept the islands with nets as long as two football fields, or used beaters to drive the animals into the water, where they could be easily captured. Rhinos and antelopes were immobilized with drugs and taken from the islands in jury-rigged rafts. Many animals were treated for shock. Operation Noah is still remembered as the forerunner of more sophisticated animal rescues, while its ineffectiveness has been overlooked. In Southern Rhodesia, it saved only about five thousand animals, and in the understated bureaucratese of the World Commission on Dams case study, did not yield "tangible benefits for the wildlife involved."

For a few years after resettlement, the Lusitu Tonga were fortunate to receive adequate rain. In 1960, they reaped a decent crop, and the 1962 harvest was phenomenal. In 1962 and 1963, as the reservoir drowned the last stretches of Old Chipepo land, the resettlers began to think of Lusitu as their home. Indeed, the shock of resettlement seemed to startle the Lusitu Tonga into a new, more energetic configuration. It invigorated the survivors, who took pride in having navigated their families through the first treacherous years after resettlement. Tradition seemed to fall away, and youth became an asset: chiefs and headmen lost authority, while men with imagination and education seemed to grasp it. For generations, land in Old Mazulu had been passed down according to rigorous matrilineal rules of inheritance, but in New Mazulu, Scudder enthusiastically writes, "the most enterprising men" got the land, regardless of their genealogy and the amount of land they owned before. (Women lost out in this arrangement, a fact Scudder acknowledges in passing.) With tsetse flies contained by DDT sprayings, villagers switched from hoe to ox-drawn plow. General stores flourished; ominously, so did beer halls. Some Tonga used their compensation payments to buy bicycles, paraffin lamps, and transistor radios for the first time. Funeral drumming recommenced; Tonga drummers and singers were even chosen to perform greetings for political leaders when they returned to Lusaka from some foreign capital. Secondary schools served the Tonga for the first time.

In 1962, Scudder returned to Zambia to settle in New Mazulu for eight months. He and Molly drove their Bedford Dormobile, an admired precursor of the Volkswagen minibus whose roof unfolded upward, accordion-style, all the way from the Kenyan coast to New Mazulu, some twelve hundred miles. He writes, "Returning to the Middle Zambezi Valley at this time was a joy," for he thought Lusitu was becoming enmeshed in the gears of development. By 1965, the suffering of resettlement seemed safely in the past. Amazingly, some plateau Tonga even moved to Lusitu. By the 1970s, the valley had produced its first government minister, its first university professor, and numerous civil servants and jet pilots.

Yet even during Lusitu's false dawn, most Tonga were aware of what they'd lost. The isolation in which they'd reveled in Old Chipepo no longer served them in Lusitu, because they'd become dependent on the government, on outsiders. If the boreholes or pipelines on which they relied for water broke down, the water department worker who could fix it was thirty miles away, and might take several weeks to pay a visit. Having switched to plows, farmers needed plow parts; dispensaries needed health workers; schools needed teachers; the cattle needed inoculations and DDT.

In the dimness of the slideshow, Scudder's images from the '60s, when Mazulu still had hope, commingled with his memories of the later grimness, during his last dozen trips there. His slides showed the village ambulance and the retail shop, "fashion store A," that briefly graced Mazulu and seemed to promise the burgeoning of local enterprise. In one photograph, a Mazulu resident happily played a toy piano as he sang about his good fortune in wife and children—he was "a wonderful guy,"

Scudder said, except that he was widely considered a witch, and Scudder once found him dancing naked around another man's house, acting witchlike; murders perpetrated in the name of witchcraft had exploded during Mazulu's bad years. Introducing a photo of his research assistants from the '80s, Scudder said, "This is Bishop, he died of AIDS. This is Paul, he died of AIDS. This is Zaka, he fears he's gotten AIDS and accuses his wife of giving it to him."

At the end of the slide carousel, Scudder paused. Kariba, his formative resettlement, was his first disillusionment, "one of the worst resettlement projects I know," and he didn't relish talking about its unraveling. The next carousel, he said, was called "Things Fall Apart."

7 LESOTHO

From Lusitu to Lesotho, Scudder spanned a career; the two places formed the brackets around his education in dams. Lesotho (luh-SOO-too) is a Belgium-sized smidge of a country, a geographic curiosity, entirely surrounded by South Africa and in perpetual danger of being subsumed into it, which nevertheless is the site of one of the world's largest dam projects. During the apartheid era, Lesotho maintained its independence by shrewdly playing South Africa and the antiapartheid international community against each other. To South Africa, it often presented itself as supportive of apartheid, while attracting abundant foreign aid by posing, as one academic observer put it, "as a small and weak but plucky and willing opponent" of its giant neighbor's white settler government. When the apartheid era in South Africa ended in 1994, Lesotho lost some of its rationale for a separate existence; now that role is played by water.

Lesotho is a poor, mountainous country of two million–plus people, many of whom lived off the money its male laborers earned in South African mines, but in recent years most have lost their jobs. Elevation is Lesotho's last asset worth selling, and the Lesotho Highlands Water Project marks the transaction. It's an $8 billion investment in outlandish plumbing, redirecting the country's pristine water from its natural southwestern course down the Orange River to a new northward path ending in the factories, mines, and houses of South Africa's heartland province, Gauteng, the region of Johannesburg and Pretoria. Only one of the project's four planned phases has been completed, but it alone will probably be enough to slake Gauteng's thirst for a couple of decades or longer. In engineering terms, the project is state of the art. Phase 1 includes Katse (KOT-say), Africa's tallest dam, a graceful arch of white cement commonly called an "engineering masterpiece"; the blunt, bottom-heavy, triangular wall of Mohale (mo-HALL-ee), one of Africa's most massive dams; and sixty miles of broad tunnels, which convey water through mountains, from reservoir to reservoir to electric power station to South

Africa. The tunnels nullify mountain fall lines and rearrange the effects of gravity; one tunnel's flow is even reversible, balancing water stored in two reservoirs so that neither overflows its dam. Scudder became the project's consultant on social impacts in 1989, and applied to its social and environmental plans everything he'd learned since Kariba. In the process, he hoped, the project would deliver his elusive good dam.

Scudder's plan had always been to work from within, as a consultant. Doing so nicely abetted his scholarly career, for his access to project documents as a consultant far outstripped what he could expect as an academic, but the approach also seemed to suit him. He was the outsider on the inside, who habitually picked fights with the Bank over its feeble resettlement and environmental policies. By the late '80s, he'd lost so many clashes that he was on the verge of deciding that being an insider didn't work. For such a broad and disheartening conclusion, he fashioned a narrow solution: instead of working for the World Bank as a consultant, he'd work for dam projects directly. That was one reason the Lesotho job delighted him: he'd be on a panel of four foreign experts working for the institution charged with administering the project, the Lesotho Highlands Development Authority. By working for the builders instead of the donors—by leaving out the middleman—Scudder hoped to increase his influence. One of the themes of his career was his insistence that the policy of restoring resettlers to a standard of living comparable to the one prevailing before resettlement inevitably fails. The change that resettlers experience is so drastic that as a practical matter, unless the resettlement objective is *improving* living conditions, the resettlers' living standards plummet. If, for example, they're shifted from an area with rich soil to a site with poor soil, it won't suffice to provide the same living conditions as before, because the resettlers will lack the means to maintain them. To have a chance of success, Scudder argues, the resettlers must be given some means of improving their lot; they must be given some tools of development. The solution for dam-affected people, Scudder thinks, is to give them a stake in the dam's bounty: they should be each project's first beneficiaries. Whether by reaping irrigation water or electricity or fishing rights in the reservoir, they should be awarded a stake as a matter of social justice. And resettlement was not all that the builders needed to do right. He wanted authorities to release some reservoir water to mimic the flows of the predam rivers, so as to minimize damage to downstream environments and the people who depend on them. He wanted an end to single-purpose dams such as Kariba, so that no opportunity is wasted—dams should generate electricity, irrigation, fisheries, and tourism all at once, in step with an amply funded development program.

The achievement of a huge, internationally funded project in a southern African backwater on behalf of an outcast regime attests to the creativity of bankers and politicians. South Africa had eyed Lesotho's water since the 1950s, and carried out feasibility studies through the 1970s and early '80s. South Africa grew so enamored of the project that by the mid-1980s it offered to pay all costs related to the diversion of water to Gauteng, including the dams, tunnels, and even the resettlement programs within Lesotho; Lesotho's only financial responsibility would be a modest power station. Of course, the project was too expensive to be financed strictly by South African banks. It required loans from international agencies, but they were hindered by the existing antiapartheid embargo on aid to South Africa. Despite the embargo, the World Bank wanted to support the project, at least in part because it thought the royalty payments that South Africa would pay Lesotho for its water could galvanize the moribund Lesotho economy. (Though the royalties are substantial—about $30 million in 2003—they have steadily declined as a percentage of Lesotho's export earnings as the nation's textile manufacturing has expanded. From a peak of 14 percent in 1996, the first year of dam operations, the royalties fell to 4 percent in 2003.) Bank lawyers helped Lesotho negotiate the treaty with South Africa that established the project, and the Bank became the chief coordinator of international donors, to whom it presented the project as a worthwhile investment opportunity. Even with the Bank's support, the project would not have happened without Lesotho's existence as a country apart from South Africa. Because of it, the Bank and other agencies were able to sidestep the embargo, by steering all project loans through the Lesotho government. Lesotho, in fact, received rate discounts because of its status as a poor country, and passed the savings on to South Africa. Even so, South Africa was not satisfied. Fred Pearce reports in a 1992 book, *The Dammed*, that by 1984, South Africa was prepared to sign the treaty, but held off because it did not trust Lesotho's prime minister, Chief Leabua Jonathon. Instead, in January 1986, South Africa imposed an economic blockade on Lesotho, whereupon an ally, General Metsing Lekhanya, staged a successful coup. Nine months later, the two countries signed the treaty.

In 1989, when Scudder joined the project, Nelson Mandela still moldered in a South African prison, and few people could imagine apartheid's dissolution in five years. Scudder nevertheless supported the project because he assumed that independence was coming. Sufficient water would be crucial to South Africa's stability, he argued, and South Africa's stability was crucial for the rest of southern Africa. The impacts on Lesotho would be a bonus. The water royalties could fund poverty alleviation and development, and the power station would make the nation self-sufficient in electricity. With the sealing of Katse Dam still six years off, there was enough time to do resettlement and environmental mitigation well. The good dam was a solid possibility.

In satellite photographs, Lesotho's Maluti Mountains look like abrasions on the smooth skin of the South African plateau. Mountains occupy three-quarters of Lesotho, and form arid South Africa's most substantial watershed. Its waters comprise most of the flow of the thirteen-hundred-mile-long Orange River, which falls westward across the width of South Africa and empties into the Atlantic Ocean. The Maluti Mountains are not particularly tall—Lesotho's highest point is a little more than eleven thousand feet—and only seasonally snow covered, but living in them is formidably difficult. The Basotho (bah-SOO-too) people, who comprise 95 percent of Lesotho's population, once occupied the mountains only in summer, when they brought their cattle to graze, but as the lowlands population increased, some took permanent refuge there. About two-thirds of Lesotho's people are considered poor, and of those, another two-thirds—half the total population—live in destitution. The highlanders are the poorest people in Lesotho, the people most lacking in schools, clinics, and roads, who felt their isolation as both blessing and curse. Elsewhere in Lesotho, the project was widely perceived as a windfall, promising limitless wealth and electricity. The authorities even promised rural electrification, a tricky task because of the mountainous and roadless terrain that an electric network would need to negotiate. The highlanders seemed no more than ambivalent about the project, and gave the impression of being dazed or indifferent. Even the few volunteer groups that sprung up to defend the resettlers' interests didn't dare oppose the project: doing so would have consigned them to oblivion. Among some of the younger resettlers offered the chance to relocate down the mountain, to the foothills or even the outskirts of Maseru, the capital, the project seemed to offer opportunity. The project's impact fell hardest on older people, who drew satisfaction from their hard-won knowledge of the mountains and the networks of mutual help that they'd relied on to survive. "We are dispersing like the chicks of a bird," an elderly highland woman told an oral history interviewer, and that was true. Highlanders can't swim—they feared the reservoir—and now a vast reservoir would insinuate itself between them and half the people they knew. Unlike at Kariba, many villages splintered, as some resettlers moved all the way to Maseru, and others went halfway. Of the roughly thirty thousand people in the Katse and Mohale basins, about a tenth would be resettled. Many thousands more would lose farming and grazing land enriched by the alluvial soil near the rivers, forcing them farther up the barren slopes. Scudder knew that cash compensation for lost land would eventually run out—only development programs could foster enough new productivity to make up for it.

Scudder paid his first visit to Lesotho as an expert panelist in 1989, and downgraded his estimation of the good dam's prospects almost immediately. From the beginning, he sensed project authorities' ambivalence about his expert panel. Unlike a comparable expert panel on dam safety, whose members received multiyear contracts, Scudder and his colleagues on the social and environmental panel were given only annual contracts, and had trouble procuring documents they needed to do their work. Construction on Katse wouldn't start for two more years, but Scudder's nemesis had already surfaced, in the form of the project's ultimate authority, the Joint Permanent Technical Commission. Its six members—three from South Africa, three from Lesotho—were predominantly engineers, who grew annoyed with the imprecision of social statistics. Unlike the sturdy numbers that they relied on, statistics on such topics as resettled families fluctuated in response to marriage, migration, birth, and death. The engineers were chiefly interested in finishing the construction projects on time and without cost overruns—Scudder says they disregarded most environmental and social issues, and misunderstood the remainder. As a condition of receiving the World Bank's first $110 million loan, Lesotho had agreed that all aspects of the project, including resettlement, would meet World Bank guidelines, but here the Bank got tangled in its own funding stratagems. The responsibility for restoring dam-affected people's living standards fell to South Africa, but since it had never signed on to the Bank's guidelines—unlike Lesotho—it had no intention of following them. Scudder reports that the South

African delegation showed interest "in the timely removal of local villagers from the reservoir basins rather than in their rehabilitation." The project's feasibility studies reflected the South Africans' views: they underestimated the number of people to be resettled at Katse and Mohale by a factor of six, and blithely dismissed the dam's environmental impacts as inconsequential. The South African delegation also vetoed proposals to fund fifteen rural development studies on such subjects as animal husbandry, community forestry, mountain horticulture, rural electrification, and village water supplies. All this frustrated Scudder so much that during a meeting he found himself yelling at an Afrikaans adviser to the South African delegation who showed neither familiarity with nor interest in the World Bank's resettlement guidelines.

Two years passed before the South African and Lesotho delegations agreed to split the cost of rural development programs, but that created a new problem, as Lesotho could not afford its share. A 1991 report by Scudder's panel blamed the Joint Permanent Technical Commission and the Lesotho Highlands Development Authority for "protracted arguments," "unnecessary delays," and "unwarranted interference, pressure, and criticism"; the commission demanded a retraction; the panel refused. Several officials told Scudder that the commission wanted to disband the panel. If not for the Bank, which strongly backed the panel, the commission might have succeeded. Not even the tectonic shift in South African politics—from apartheid to Mandela's African National Congress in May 1994—affected the behavior of the commission's South African delegation, which continued to be dominated by white engineers. The result of all the arguments and delays was that in mid-1995, when construction on Katse was nearly finished, most components of the resettlement and rural development plans were three years behind schedule, and affected people's morale was deteriorating. The commission needed the Bank's approval to seal the dam on schedule in October, but housing for some of the Katse resettlers still hadn't been built—and the Bank wouldn't permit sealing until all resettlers received housing. Faced with the prospect of missing a chance to fill the reservoir during the impending rainy season, thereby delaying by a year the generation of water to South Africa and royalties to Lesotho, the commission found a use for the panel. Scudder spent two weeks wandering from office to office in Lesotho's capital, Maseru, arranging for the construction of the homes. Though he managed to win temporary jobs for dam-affected people, he didn't feel good about his role. All he'd done was get people moved so that the dam could be sealed on time; five years later, a majority of Katse's resettled families still lacked water and sanitation. The commission's discovery—"oops, there's value in the panel," as Scudder put it—was bitter consolation. Scudder and his expert colleagues determined to prevent a repeat of this "lamentable" situation in the project's next phase.

A few months after Katse construction finished, it began at Mohale, with slightly improved consequences for dam-affected people. The experts argued that resettlement should be offered not just to people whose houses would be inundated (as at Katse), but also to those who would lose at least half their land, since without some kind of assistance, these people would face dire consequences. A reluctant commission agreed, but not until 2002, fifteen years into the project, when the resettlement effort was winding down. The commission's concession at least marked a precedent, which could be applied to the project's future phases. But in most ways the Mohale treatment of dam-affected people resembled the Katse treatment. At Mohale, resettlement continued to emphasize housing construction while neglecting compensation and development activities. Three years after moving to the principal Mohale resettlement community, resettlers still didn't have enough water, and development projects hadn't even started. The Lesotho government was so inexperienced and understaffed—short of "capacity," in the bureaucratic lingo—that issues awaiting decisions evanesced into the Maseru mist, while high-ranking officials devoted most of their energy to political survival.

Even Nelson Mandela seemed disingenuous on the subject of Lesotho villagers' rehabilitation. At the inauguration of 'Muela Dam in January 1998, he proclaimed that the project's "impact will be felt in the remote villages of the Highlands that have been opened by the roads, communications services and electricity supply that have been installed. It will go on feeding the local economy through the skills which were gained in construction and which are being expanded by the continued training offered to those directly affected by the project." But electricity never reached Highlands villages, and the training project ended in failure.

In a misguided attempt to quash a festering Lesotho army mutiny, South Africa saw fit to invade in 1998. It sent hundreds of troops to several Lesotho locations, notably including Maseru and the Katse Dam.

South Africa expected a warm reception, but instead triggered riots, arson, looting, and gun battles that left sixty people dead. At Katse, a training program intended to compensate about two thousand people who lost all their fields to the reservoir abruptly closed. The Lesotho Highlands Development Authority didn't even bother to station employees at Katse until the late '90s, after the dam was finished. Compensation was provided erratically. People given compensation in grain suffered hunger because they did not receive it at the same time of year as their harvests would have been reaped. The rural development plan, in which Scudder placed much of his hope, proceeded "with budgets unutilized, a few plans dropped or seriously delayed, and the majority inadequately compensated." Having promised rural electrification at treaty time, the Lesotho government dropped it. The project would still make Lesotho "self-sufficient" in electricity, but the sufficiency would be entirely urban. The most plausible hope that highlanders possessed of benefiting from the dam was extinguished.

When I first talked to Scudder, in October 2000, he said he was considering resigning from the panel because of the botched resettlement. "I'm not just frustrated, I'm bloody angry," he said. He did resign, a year and a half later, but protest wasn't the reason. Molly's accumulating maladies—breast cancer, arthritis, drug allergies, and an ankle fracture—finally were calling him home. He declared repeatedly (but not entirely convincingly) that his next trip to Lesotho, his eighteenth as expert panelist, would be his last trip to Africa. As we'd planned, I met him at his Maseru hotel room in the late afternoon of his last day as a Lesotho project expert. He certainly didn't look like a man about to sever his connection to a continent: he was practically jaunty, as if Africa had invigorated him. The needle on his good-dam-ometer was edging upward, too: he said resettlement for Mohale was going much better than at Katse.

At five-thirty, the doorbell rang, and Scudder welcomed in two young black men, Lenka and Mothusi, activists from the Transformation Resource Center, a tiny nonprofit that was fighting for decent treatment of the resettled people. Scudder included nonprofits among the sources he consulted, for their advocacy was useful, particularly when Scudder concurred with their project assessments. The two men gave Scudder the African three-part handshake (conventional grip/knuckle clasp/conventional grip again). Lenka, who smiled easily, spoke better English than Mothusi and did most of the talking. Scudder mentioned that the two men had recently spent time with Medha Patkar in India, thanks to travel grants from a London trust fund; Lenka said Medha was "very inspirational." Scudder began by clarifying his position on dams. "I'm anti most big dams," he told them, "but I'm pro some big dams, like this one, for reasons you know." Medha, he said, was antidam; Jan Veltrop, an engineer on the World Commission on Dams, was pro. "What we all agree on is that people affected by the dam should be the first beneficiaries of the dam." The speech served as a kind of meeting preface.

Lenka delivered his litany. The Development Authority was mismanaging money. Some people were unjustly disqualified from receiving compensation due them. And villagers who suffered losses from an earthquake caused by the filling of the Katse reservoir hadn't been reimbursed.

Scudder listened impatiently, for all this was old news—then he changed the subject. He wanted Lenka and Mothusi to know about a "huge issue": a plan to develop a large island in the middle of the Mohale reservoir as a combination resort and high-altitude athletic training facility. The island was already set aside as national parkland, and any development there was supposed to get approval from the Lesotho Highlands Development Authority—but Monyane Moleleki, Lesotho's minister of natural resources, simply ignored that requirement. Instead, he signed a contract with a British developer to erect a slapdash facility in the time remaining before the rising reservoir inundated a connecting road. Scudder suspected that several other ministers had joined Moleleki in investing in the island. Schemes like this should have been designed to benefit dam-affected people, but instead, Scudder said, "It's going to take care of the ministers' children, their grandchildren, and their great-grandchildren."

Scudder warned Lenka and Mothusi that the information was volatile. "This is very delicate. Go well and I expect to find you alive."

Another round of handshakes followed, and Lenka and Mothusi left. As Scudder closed the door, he said, "I get so bloody enraged about these kinds of things."

We adjourned to the ground-floor restaurant for Scudder's "last supper" in Africa. Two of the three other panelists were present: John Ledger, the director of a trust for endangered animals in South Africa, and Mike Mentis, a South African environmental consultant. The South Africans on the panel carried out their duties with more caution than Scudder, for they could not afford to offend their own government.

The conversation bounced from Lesotho to Washington to China. Scudder said that China had conducted three of the world's ten best resettlement projects, and their resettlement policy was the best of any government in the world; the trouble was that implementation didn't always follow the policy. In fact, that was why he couldn't sign off on Three Gorges Dam, the world's biggest dam project, to which the World Bank sent him as a consultant. Three Gorges would create one and a half million refugees, and Scudder was afraid they'd end up "in Tibet or Xinjiang or wherever the Chinese want more Hans."

When dinner was over, Ledger declared that a handshake was insufficient, and gave Scudder a bear hug, bellowing, "I love you, man." Scudder looked startled, and did his best to hug back.

Being large and close to Cape Town, the Lesotho project is on every dam expert's must-see list. To pass through South Africa for a World Commission on Dams meeting without taking a look around Lesotho was like touring San Francisco without seeing the Golden Gate Bridge. I'd visited Lesotho a year earlier, when I began looking into dams. On one notable weekend, in the company of Keketso Sefeane, a Lesotho Highlands Development Authority official, I rode in a 350-mile circle through the mountains, beginning and ending in the lowland capital of Maseru. Sefeane drove a Development Authority four-wheel drive, a vehicle that is both necessary for mountain administration and emblematic of the immense gap in wealth between officials and highlanders. (Indeed, infused with project capital, jobs in the Development Authority were so lucrative that the authority had no trouble tempting capable employees away from other government departments. Many of the government's most capable officials made the transfer, which slowed down the bureaucracy even more.) Sefeane himself was amply endowed in bureaucratic skills: immaculately dressed in casual Western clothes, he possessed an economics degree acquired in England, and looked as different from the highlanders as I did.

Maseru (mah-SAIR-oo) is one of the world's least distinguished capitals: not colorful, not surprising, not even bleak. It's a low-slung sprawl of highly variable construction that is in the process of absorbing the entire Basotho population—a few decades ago, only 5 percent of Basotho lived in Maseru; by 2020 or so, the number is projected to reach 60 percent. Even so, the evidence of bustling Maseru commerce is mixed. Foreign-owned textile factories on the outskirts of the city disgorge thousands of employees at five o'clock each day, but the louche city center still displays the remnants of stores torched during the 1998 riots.

It was fall in Africa, and in Maseru hot, but in the mountains, the weather turned an argentine, sunlit cool. The mountains were adorned in green, evidence of the recently completed rainy season. By geological standards, the Maluti Mountains are old and therefore smoothed, arranging themselves into a succession of rock-clad horseshoe-shaped canyons that rise to innumerable soft, looping crests silhouetted against powder-blue sky. On the arable slopes of the lower elevations, the hills are less aggressively terraced than in Asia, and they looked not staircased, but rippled. There, too, the rondavels were exquisitely round and precisely trimmed of thatch, seeming to confirm the Basotho belief in the superiority of their housing styles. Higher up, the rondavels seemed to lose their snap, as the struggle for survival intensified.

The dams looked heroic; the dam-affected people looked dazed. Katse is a cloud-gray curve—a double curve, in fact, bending into the reservoir both from abutment to abutment and from foundation to crest—whose grace obscures its height and power. Walking atop Katse was like strolling across the deck of an aircraft carrier, windswept, feeling something alive beneath my feet, faintly aware of occupying an illegitimate space, as if I were committing a transgression by being there. Just as an aircraft carrier's violence resides beneath the deck, evidence of the dam's violence is hidden underwater; maybe that is why zoologists Bryan Davies and Jenny Day compare the Katse reservoir to "a very long, very heavy axe head with the blade burying itself in the valley floor."

If Katse is a sculpture, Mohale in midconstruction was an assembly-line ingot, an elongated, flat-topped triangle stretched across a riverbed a third of a mile wide. From our hilltop vantage point above the dam, the project looked like an elaborate erector set construction, with rock quarry, tractors, the world's largest dump trucks (sporting tires taller than our minivan), and a zigzag road across the dam face like an African cicatrice; only the scale was too large, by a factor of thousands. Katse, the sleek arch dam, consumed two million cubic meters of cement; rock-filled Mohale, though barely three-fourths Kariba's height, would require seven million cubic meters of rock. In the words of an Irish engineer who gave us a tour, "A rock-fill dam is just an ugly pile of rock that you put a concrete slab on to make watertight." We drove on to the tunnel construction headquarters, a few miles upstream, and got a briefing from a chain-smoking engineer who was agonizing over the twelfth major delay in the drilling of the twenty-mile-long, thirteen-foot-diameter Mohale-Katse tunnel. Once more, the tunnel had started leaking, and the engineers tried in vain to stanch the leak with injections of tons of grout. They'd injected so much grout into the tunnel wall, in fact, that two springs used by villages above the tunnel had ceased flowing. By now the tunnel was a year behind schedule, which Scudder deemed a good thing: it would give the Mohale resettlement plan a chance to catch up to the rest of the project.

We stopped in a village where the pressure of the filling Katse reservoir had caused the ground to emit eerie moans that frightened villagers, and then, in January 1996, a 3.2-magnitude earthquake left a mile-long crack in the terrain. The crack extended from the reservoir straight up a slope, and when I inspected it four years later, the crack was still so wide that I could see through it and down into the earth for several feet. The filling of the Katse reservoir caused hundreds of small earthquakes; by the time the ground finished shifting, five village springs stopped flowing. On the other side of the reservoir, we talked to a village chief who stylishly sported white shin-high rubber boots and was wrapped in a bright red blanket depicting the four suits in a deck of cards. We stooped inside a stone-walled, dung-sealed, corrugated-tin-roofed house to talk to the chief while he ate potatoes and greens from a metal bowl with a spoon. It was cold and overcast now, and everything was damp. We were high on the mountain, looking down on the steel-blue reservoir. The village chief rattled off the dam's impacts, as if he were used to delivering the litany: the villagers had lost agricultural land, grazing land, and the source of their firewood. "Most of our resources have been inundated," he said in Sefeane's translation. "What is left is quite minimal."

At Mohale, we talked in the smoky hut of a thirty-six-year-old woman with six children and an absent husband as her three-year-old boy, an albino, sucked halfheartedly at her breast. She looked stunned. The project had brought nothing good to her. The fields on which she sharecropped were submerged, promised compensation hadn't been paid, no one in her family got a job on the project. "Maybe the nation has benefited," she said, "but my family hasn't."

A year later, I was riding the same roads, only now Scudder was driving. I'd asked him to show me whatever facets of the project he considered noteworthy, and now we were climbing in a rented four-wheel drive through the foothills of the Maluti Mountains. It may be indicative of Lesotho's lowly international standing that its sole geographical distinction amounts to a backhanded compliment. The lowest point in Lesotho, at an elevation of 4,600 feet, is higher than the lowest point of any other country in the world. We still had not driven high enough to feel the altitude's effect, even when I followed Scudder out of the car and straight up a hill. At the top, he sat on a rock. Over his right shoulder was a magnificent pink-tinged rock-ribbed Grand Canyon–esque gorge, and over the other shoulder, nestled into a slope, was a Mohale resettlement village. It looked to me like a fetal Levittown, a couple of dozen identical gray cinder-block houses, all fronted by identical green plastic water barrels, arranged in precise rows: a relentlessly linear village in place of traditional villages' preponderance of curves. The ambiguous gift of modernity now reached deep into the Malutis.

Scudder said this site was better than the first Mohale resettlement site, which lacked both a water supply and a development plan—equally grave sins in his view. "The houses were good, but people complained," he said, so officials started blaming the resettlers. "It's a complaining culture, a dependency culture," Scudder heard officials say. Given the absence of water and hope at the site, Scudder said, "Well, of course it is."

The resettlement village we surveyed was a slight improvement. The green barrels were connected to rooftop gutters, thus comprising simple water-harvesting devices. Every unit had an identically sized plot of garden, and every unit had an outhouse. "They're good, they're vented, but they're pit outhouses only," Scudder said. "In the old areas, the dogs were the sewage disposal areas, so this is not so bad!

"The Development Authority has begun to get its act together, but there's still no water supply here—God damn it all, you can't resettle people without a water supply!" We watched two girls carrying water canisters on their head as they arduously made their way up the steep gorge beneath the site: the village's water source was at the bottom. The daily round-trip probably took an hour, and according to villagers we talked to, sometimes entailed being accosted by men along the way.

Unlike the Katse resettlers, these villagers were forming a community, Scudder said. They were participating in the Development Authority's new wheat, maize, potato, and poultry co-ops, "So what does that say about the 'complaining culture'? The complaining culture doesn't exist! It happened only because the Development Authority didn't create opportunities."

It was early afternoon by the time we'd driven all the way to Scudder's destination, a rock outcropping near the road overlooking a steep canyon that would be mostly underwater in a year or two. The dam was miles away and out of sight, but it was already transforming this terrain. Scudder pointed across the canyon to a prominence rising star-shaped to a point, surrounded on three sides by higher crests. It was no more or less remarkable than its bucolic surroundings, but it had become exceedingly desirable land—this was the "island" that Scudder had told Lenka and Mothusi about. When the reservoir filled, it would turn the island into a spectacular site, a star-shaped "jewel" surrounded by the reservoir, itself surrounded by mountains.

"Two weeks ago we heard that they were going ahead with the plans before inundation," Scudder said, and truculently pointed out the evidence. Looking more carefully now, I could see a road winding partway up the hill, with trucks plying it—the road was under construction. Even the road's lower portion looked like a raw slice into the hillside, for the road was temporary, meant to exist just long enough to supply the hilltop with enough building materials to turn it into a resort; then the reservoir would consume the road.

The land had already been designated as a national park, not a commercial enterprise. Now, with the connivance of the minister of natural resources, it was becoming a tourist destination where, Scudder surmised, villagers would not only not be principal beneficiaries, but would get menial jobs as maids and bare-breasted dancers.

"This has got to be stopped," he said. "If this goes forward, it will be a disaster. The World Bank has to go to the minister of finance, and the cabinet has to say, 'No, no!' "

We sat on the rock and ate box lunches, and watched the trucks kick up dust across the canyon.

In resettlement studies—a field of no more than two hundred practitioners—Scudder is considered a founding father. In the much larger field of development anthropology, he is an eminence. Yet for all of Scudder's expertise, he can turn surprisingly reticent. When he was given the opportunity in his Bronislaw Malinowski Award speech in 1999 to summarize the lessons of his career or expand upon a favored cause, he backed away. He instead conducted a survey among eighty-nine of his anthropology colleagues and reported on the result. He has produced one well-known theory on resettlement, which can be summarized in one (long) sentence: Resettlement proceeds through four stages, from the "recruitment" stage (when the government decides to proceed with its project and makes fateful decisions about such issues as the location of resettlement sites), through the "transition" stage (when the villagers' problems begin, from the moment they first hear about the dam, and the government stops building schools and maintaining services for them, through the shock of their involuntary relocation), on to the "potential development" stage (having survived the resettlement, the survivors become energized and move toward a "risk-taking and . . . open-ended society"), all the way to the "handing over" stage, when resettlers feel themselves at home and are able to take their place among surrounding communities. It's pretty obvious stuff, and most dam resettlement projects never reach stage three or four anyway. So far, at least, it describes Scudder's vision of how resettlement ought to unfold much more than it does reality.

In resettlement studies, however, Scudder's four-stage theory is a milestone, a reflection of the newness of the realization among Western scholars that resettlement induced by development is rampant and agonizing. As a commissioner on the World Commission on Dams, Scudder produced enough evidence to convince his colleagues that dams had displaced forty to eighty million people, a number higher even than that proposed by Patrick McCully, the canny antidam activist whose book *Silenced Rivers: The Ecology and Politics of Large Dams* proposed the number of thirty to sixty million resettlers. Whatever the true number, Scudder is responsible for its magnitude being known at all.

Scudder's stature as an anthropologist rests on pursuits that by some academics' definitions aren't anthropological. He himself says he has become less and less an anthropologist, more and more a "river basin development specialist." He has effectively carried out two careers: an academic one that began spectacularly with Kariba and has been fueled by the connection ever since, and another one as a (well-paid) development consultant whose goal is changing the world. To many academic anthropologists, he is practicing the inferior discipline of "applied anthropology," dirtying his hands with policy, likely becoming entangled in webs of neoimperialism and corruption. Scudder hates the word "applied," with its implication that his work is not academic, not rigorous. Instead, he's a "development anthropologist," he says, employing every bit of the rigor he learned at the outset, from Colson and other admired mentors.

Anthropology is often described as a discipline in crisis, unsure of its purpose, reluctant to assert its understanding of other cultures. The overt reason for the crisis is the lack of university jobs. As a result, as many as half the people with anthropology doctorates are crossing over from academe into advocacy, taking jobs as "applied anthropologists" with the World Bank, the United States Agency for International Development, the United Nations Food and Agricultural Organization, the Ford Foundation, and on and on. Scudder believes that he was the first anthropologist the Bank ever hired; now the Bank employs at least sixty full-time. They struggle to influence the economists who run the place, but still are not widely appreciated, for their role is usually to slow projects down, to point out the complications in plans that builders assume are straightforward, to force long-term considerations on people whose vision is profoundly restricted.

During World War II, when applied anthropology got started, anthropologists unhesitatingly contributed their skills to the war effort. Most spectacularly, the U.S. Office of War Information hired Ruth Benedict, an anthropologist at Columbia University, to provide insight into enemy cultures, and even plied her with classified documents. The result, in popularized form, was *The Chrysanthemum and the Sword*, a classic (and bestselling) analysis of Japanese culture written by someone who'd never been in Japan. When the war ended, development took its place as motivation for mobilizing resources in the Third World; "war" became a metaphor, in the form of the international "war on poverty."

At first, anthropologists were called in to explain why projects failed.

In one notable case, officials administering food aid in West Africa gave powdered milk to villagers who refused to use it because they said it contained evil spirits. Anthropologists concluded that the villagers did not act irrationally; they were simply lactose-intolerant, and the milk hurt their stomachs. Anthropologists eventually grew tired of playing coroner for failed projects, and clamored for a place at the drafting table with project designers, in the name of averting calamities. What they got were frequent conversations with the designers, but marginal influence.

For a time, it was axiomatic that the biggest problem in the Third World was poverty, for which development was the obvious solution. The only questions that got asked were about how to do development right. It took a while before the evidence of development's failures began to accumulate. Some academic anthropologists turned their analytical tools on the development industry itself, in recognition that it, too, is an exotic culture. They argued that far from being altruistic, development projects typically served the donors' interests more than the recipients', facilitating the extraction of raw materials from Third World countries while developing markets for the First World's manufactured products— effectively maintaining the hegemony over Third World countries that First World powers previously enjoyed in the colonialist era. Some critiques stressed the connection between development and the cold war, an inescapable link in the case of American economic aid, which was concentrated among the United States' political allies as a way of bolstering anticommunism. Other critiques underlined the relentlessly "technicist" approach that development fostered, in which economists with little or no experience in recipient countries decided that economic growth should be the countries' major objective, even if it encouraged corruption and disenfranchised the poorest citizens in the process. What was constant in all the critiques was the notion that development, particularly development in Africa, had been a spectacular failure.

After lunch, Scudder and I drove on to the Mohale Dam, which was completed but not yet sealed. The distinction was crucial, for Scudder knew that once the reservoir began to fill, the influence of the experts and even the Bank would drop precipitously. In fact, the donors' leverage started to decrease the moment they agreed to fund a project. Now that Mohale was nearly completed, the Bank's last bit of leverage was to threaten not to recommend sealing as scheduled. In its rush to reap more South African water royalties, the Lesotho government could ignore the recommendation, but if it did, it risked losing support from dozens of international donors.

Scudder wanted me to see Mohale's downstream face, and drove back and forth on the roads above it. He was so intent on looking at the dam that it occurred to me that for all his dam-induced fury, his emotional makeup still includes a dab of engineer. We never did get more than a glimpse of Mohale. From our distant vantage point, it looked merely massive, like a blurred satellite image in a Pentagon briefing, which might or might not be missile silo, munitions plant, dam. We wandered through the construction village where the engineers once lived: this Levittown had moved past embryo to immaturity. Compared with the resettlement site down the mountain, houses here were much bigger and closer to one another, more blindingly synthetic. I'd seen the village a year ago, when most houses' lawns contained mushroomlike satellite dishes; now it was a tidy ghost town enclosed in chain-link fence. Scudder said he was proud that at least the place was built with permanent materials—as a result of the experts' recommendation—but there was no plan to reuse the village.

On the way back to Maseru, we drove over God Help Me Pass, which looked less treacherous than the name suggests. It was late afternoon by the time we reached Makhoakhoeng, a resettlement village on the city's edge. Makhoakhoeng's problems, Scudder said, "threaten the entire Mohale resettlement program." Given a choice between resettling elsewhere in the mountains or moving to the outskirts of Maseru, twenty-two families had chosen the city in hopes of making a living there. Human habitation is a new thing to greater Maseru. In Makhoakhoeng, tiny rectangular homes occupy neat plots on bald red earth. The only exception is dramatic: predating the resettlement houses, Lesotho foreign minister Motsoahae Thabane, known to Scudder as Tom, built a mansion in Makhoakhoeng, but arsonists burned it down during the 1998 riots. We walked around the barbed-wire perimeter, and all we could make out was a light fixture, a water tank, and some rosebushes that had grown back—the home itself had vanished. Thabane continued to exert his influence over the settlement, as he proved by resisting the arrival of the resettlers in 1999. He encouraged the existing residents—the "hosts," in resettlement parlance—to treat the resettlers with hostility and demanded benefits far out of proportion to the Development Authority's payments to other urban host communities. Resettled children were harassed, and when the first resettler died, the survivors found out they were not allowed to use the neighborhood burial ground. In November 2000, in an act that suggested an alliance with Thabane, the minister of natural resources, Monyane Moleleki, the mastermind of the island resort, declared that the twenty-two families would have to move from Makhoakhoeng. Harassment of the resettlers intensified. At that point the experts stepped in, and denounced a second resettlement as "unnecessary, unacceptable, and undesirable." The World Bank backed the experts. The Bank even sent a mission from Washington to have chats with the two ministers, but, according to Scudder, found both men "unavailable."

Scudder blamed the debacle on the Lesotho Highlands Development Authority, the one government agency that was bloated with foreign money, thanks to the dam project. "This is the stupidity of the Development Authority," Scudder said. "They bought this land next to the home of the minister of foreign affairs, and they never talked to him about it. The bad blood was such that even if the Development Authority had gone to him, he might not have listened." Scudder threatened to resign from the panel of experts if the resettlers were moved again. The International Rivers Network and local nonprofit groups including the Transformation Resource Center pressed for a halt in resettlement until the program's credibility was restored. The issue reached Lesotho's Cabinet, which decided that the twenty-two households could stay in Makhoakhoeng. The Development Authority instructed Thabane, the foreign minister, to inform the resettlers officially, but that didn't happen. Instead, in July 2002, Thabane was promoted to home minister, a more powerful position.

We walked around the resettlement site, with its tidy homes and fledgling gardens. "Look at all the onions!" Scudder said. "That's a good vegetable garden! And look at the fruit tree growing here!"

At the side of the street, four little girls were playing in an imaginary playhouse that they'd demarcated with bricks. One girl crouched inside a nonexistent room, as if hiding. "Aren't they adorable?" Scudder said. "I'd like to take a couple of them home. Hel-lo!"

He tried to talk to the girls in English, but they just stared back at him.

Scudder adopted an exaggeratedly stentorian voice. "Little do they know," he said, "that I played a role in their being able to stay here!"

One of the most provocative books about development written by an anthropologist is James Ferguson's *The Anti-Politics Machine*, a devastating critique of a World Bank project in Lesotho. The book was written in 1985, before the Lesotho dams were built, and it deals with a livestock development project, not a dam, but it's congruent with antidam thinking, as it calls all development projects into doubt. Ferguson describes prodevelopment authors as people who "either as insiders or as sympathetic outsiders, see 'development planning' and 'development' agencies as part of a great collective effort to fight poverty, raise standards of living, and promote one or another version of progress"—people including both Scudder and World Bank officials. Against them, he sets conventional neo-Marxist critics, "who seem satisfied to establish that the institutions of 'development' are part of a fundamentally imperialistic relation between center and the periphery and take the matter to be thus settled."

In reply to both groups, Ferguson looks closely at Lesotho's Thaba-Tseka project. It was launched in 1979 chiefly as a livestock project, evolved midway through its existence into an "integrated rural development project," and ended in 1984 as an abject failure in both efforts. The failure was hardly unusual. Lesotho attracted an inordinate amount of aid—in 1979, for example, it received $64 million in "development assistance," $50 for every human inhabitant, and more per capita than many much larger African countries. According to Ferguson, between 1975 and 1984, twenty-seven countries gave aid to Lesotho; so did seventy-two international agencies and organizations, from the Abu Dhabi Fund and the Danish Volunteer Service and the International Potato Production Centre and the Mennonite Central Committee and the South African Mohair Board and three United Nations agencies all the way to the World Food Programme, the World Health Organization, the World Rehabilitation Fund, and the World University Service. A 1977 Food and Agricultural Organization report listed two hundred–plus development projects in progress in Lesotho, yet, Ferguson writes, "If all observers of Lesotho's 'development' agree on one thing, it is that 'the history of development projects in Lesotho is one of almost unremitting failure to achieve their objectives.' " Given all that, Ferguson asks, why did the donors all clamor to give money to Lesotho, an inconsequential country whose population then was a mere 1.2 million people? Whether the aid was given to spread altruism or capitalism, it failed, so why did the donors keep giving? With understandable wonderment, Ferguson says, "One can hardly help posing the question: what is this all about?"

His approach is characteristically anthropological. He takes a step back, until the World Bank and the Canadian International Development Agency, a project cofunder, and Lesotho officials and the target population in the mountains all become part of an intricate system, which he proceeds to describe. For starters, donors are organized to donate: the one unforgivable sin that the World Bank could commit would be to fail to "move money," fail to propagate development. Projects are the institutional imperative, so the institution must find recipients that meet its requirements. This means, among other things, construing countries' problems so that they seem solvable by economists. To do this, the 1975 World Bank Country Report on Lesotho, which presented the country as a suitable site for a project like Thaba-Tseka, portrays Lesotho in a way that Ferguson finds "truly fantastical." For example, the report starts with a couple of whoppers. At the time of Lesotho's independence, in 1966, the report said, the country "was virtually untouched by modern economic development," and "still is, basically, a traditional subsistence peasant society." Yet Lesotho had been "touched" by development as far back as the beginning of the century, when it began producing cash crops for the South Africa market, and its economy was far from "subsistence." In fact, 60 percent of its male adults then worked in South African farms, mines, and factories, and the economy depended on their remittances home. Ferguson writes that the Basotho participated in a cash economy "since before the middle 1800s at least."

The Country Report isn't just riddled with mistakes; it's also oblivious to Lesotho's politics, from its corrupt leadership to its encirclement by a powerful apartheid regime in South Africa. As Ferguson explains, "An analysis which suggests that the causes of poverty in Lesotho are political and structural (not technical and geographical), that the national government is part of the problem (not a neutral instrument for its solution), and that meaningful change can only come through revolutionary social transformation in South Africa has no place in 'development' discourse simply because 'development' agencies are not in the business of promoting political realignments or supporting revolutionary struggles." In fact, the World Bank's aversion to addressing political reality stems directly from its constitution, the Articles of Agreement, which forbids it from making political interventions.

The Country Report is full of statistical errors and internal contradictions, but the errors all point in the same direction: they suggest that Lesotho's "subsistence economy" is collapsing. For this condition, the solution can't be political, so, Ferguson says, the Country Report "must make Lesotho out to be an enormously promising candidate for the only sort of intervention a 'development' agency is capable of launching: the apolitical, technical 'development' intervention."

Not surprisingly, the Bank's technical solution didn't work. Despite the report's claims, the mountains of Lesotho were not ripe for any sort of production increase. Only 10 percent of Lesotho's land is arable, and that percentage drops in the eroded soil and steep, rocky inclines of the mountains. Many households maintain livestock, but not out of a Western spirit of entrepreneurship; rather, they store their wealth in the form of cattle, sheep, and goats, and are reluctant to sell them even when the animals are dying. While mountain villagers grew modest crops of maize, wheat, sorghum, and marijuana (a cash crop, by the way, not a subsistence one), most still depended on remittances from relatives in the South African mines. Ferguson cites a 1981 anthropological study of migrant labor in Lesotho that aptly calls the country's population "a rural proletariat which scratches about on the land."

In this unpromising arena, the World Bank and the Canadian International Development Agency tried to commercialize and "rationalize" livestock production. "Livestock," declared the project mission statement, "must play the major role in the development of the mountain areas." At the core of the plan was a scheme to double livestock production and increase the animals' weight by 20 percent, by setting aside three-eighths of grazing land in the Thaba-Tseka district solely for improved stock. Unfortunately, this plan violated Lesotho law, which prevented its implementation for two and a half years; then project officials got permission to start an "experimental" project on a single fenced pasture comprising less than 5 percent of district grazing land. As Ferguson explains it, the plan was to allow farmers with better-quality animals to use the rangeland within the fence. The farmers would have to keep pure-bred livestock and follow good management practices; their success would show the wisdom of commercialization to other farmers. But excluded farmers did not appreciate the project's appropriation of six square miles of what had been communal grazing land. Even most of the farmers with livestock good enough for inclusion in the program resisted it. Within a year and a half, Ferguson reports, "the fence had been cut or knocked down in many places, the gates had been stolen, and the association area was being freely grazed by all. The office of the association manager had been burned down, and the Canadian officer in charge of the program was said to fear for his life." By 1982, project officials were calling it "a complete disaster."

Yet even though the project did not change livestock practices, it had a powerful, seemingly unintended impact: it extended the reach of Lesotho's central government. Thanks to the project, a tarmac road now connected Thaba-Tseka to the capital, and Thaba-Tseka had a new administrative center that housed everything from a post office to an army paramilitary unit. Thaba-Tseka had been a stronghold of the political opposition in Lesotho. Now, to counter the opposition, the central government had its own command posts and supply lines, all supplied by foreign development agencies. As Ferguson puts it, "The 'development' apparatus in Lesotho is not a machine for eliminating poverty that is incidentally involved with the state bureaucracy; it is a machine for reinforcing and expanding the exercise of bureaucratic state power, which incidentally takes 'poverty' as its point of entry."

Over lunch, I once asked Scudder what he thought of The Anti-Politics Machine, and he brushed it off as "postmodernist." I asked him again a couple of months later, and he replied: "Ferguson is a bright guy and his analysis of the Thaba-Tseka project is not only the best available but very perceptive. Problem is he pushes his arguments about 'development' too hard and too far, as if there was a single 'development' discourse. He neatly divides experts and academics into two classes, as I recall, 'prodevelopment' and 'antidevelopment.' Well, where does that leave people like me, . . . various NGOs, the Institute for Development Anthropology [which Scudder cofounded], etc.? Our starting point is that the huge majority of the world's population *wants* development—I have example after example. But they want development on their own terms, not those of Ferguson's pro- and antidevelopment discourse factions. And those of us who struggle, not very successfully, to help them achieve their goals, see development as a very complicated process;

hence, we do not take kindly to those who try to simplify it for their own ideological reasons."

In refuting Ferguson, Scudder was defending his entire career. Even so, the questions Ferguson raises hover over Scudder's work like whispered reproaches. Has his work chiefly served the poor people he has striven to aid, or has it inadvertently served central governments and officials more? Has his function been to defend poor people and the environment, or to supply a respected imprimatur to projects in need of legitimacy? Is World Bank–style development intended to lift people from abject poverty, or is it a ruse to promote First World power?

Scudder allotted just enough time for showers at the hotel, and then we were back in our rental car. (A Budget rental car, to be precise. The reach of the Budget Group, the 642nd largest company in the United States, extends even to Maseru.) It was dark by now, and Scudder was nervous. He kept the doors locked and windows up, to ward off thieves — never mind the heat. Our destination was the Lesotho Sun, a brassy hotel on a hill and cousin to our humbler digs, the Maseru Sun, in the city flatlands. The Lesotho Sun was the temporary residence of a visiting World Bank team, whose two anthropologists were to be our dinner companions: Dan Aronson, lead social scientist of the Bank's Africa region, whose bland appearance (pear shape, short-sleeved white shirt, plastic pens poking out of his breast pocket) disguised his bonhomous character and acuity; and Roxanne Hakim, a raspy-voiced, endearing young Indian who'd written her doctoral thesis on a tribal village in the Narmada Valley.

The fact of this meal, twenty-four hours after Scudder's hotel room meeting with the activists from the Transformation Resource Center, pointed to his distinction. No other figure in the field of large dams managed to maintain good relations with both international donors and the NGOs that opposed them. Over dinner, the three anthropologists' differences seemed unimportant. Now they were coconspirators, comparing notes on contending with the prickly, frequently corrupt and incompetent host government.

"We have a real anthropological problem that we hit on today, so we come to the master," Aronson declared with senatorial deference. He explained that the Development Authority compensated resettlers at a higher rate for inundated gardens than inundated fields. But in Sesoto, the Basotho language, the word "garden" refers to any land used for growing, so the resettlers, no fools, were claiming the bulk of their agricultural land as garden. In the first stage of the Mohale resettlement, the Development Authority had tried to enforce an informal policy of compensating no more than four hundred square meters of garden per household, but eased to a thousand square meters when it realized that only six households' claims went over the four-hundred-square-meter limit. Now, in the second stage, many households claimed more than four hundred square meters of garden, and Bank officials were worried that in the third stage, every household would.

Scudder's eyebrows assumed the up position. He thought a moment, then declared that since the Joint Permanent Technical Commission had agreed to the thousand-square-meter compromise, it ought to keep its promise. "This is where you can hoist them on their own petard," he said.

Disdain for the commission was general. Aronson said, "They have some very pretty petards, too."

Later, Scudder put the cap of his ballpoint in his mouth and on his napkin drew a map of the island-to-be. He proposed that if the road to the envisioned resort was completed, the Bank should refuse to recommend sealing the dam. Aronson pondered the effect this might have on Monyane Moleleki, the minister of natural resources. Moleleki, Aronson said, had stood up the Bank for seven straight scheduled meetings, including once when a Bank official flew in from Pretoria and two others came from Washington. He'd heard people say of Moleleki, "He's so oily he'd slide uphill."

Just before dinner was over, Scudder explained his retirement from the panel of experts. Though he was leaving Africa for good, Asia was still in his plans—he was continuing as an expert panelist on Nam Theun 2, a proposed dam on a tributary of the Mekong River in Laos that the Bank was still deciding whether to support. Nam Theun 2, he said, "may finally be the one project which does resettlement well."

I once told Scudder that I had no trouble summarizing the views of the International Rivers Network, or, for that matter, the International Commission on Large Dams, the prodam group representing dam engineers, but I found it much harder to describe his views. "How would you summarize them?" I asked.

Scudder answered without hesitation, as if he'd already thought this out. "Yes but, no but," he said—not exactly a rallying call.

I told him his answer was too formulaic, as if all he stood for was complexity. This time his answer was longer but not much more illuminating. Some dams were necessary, he said, but each requires analysis.

I tried asking him what advice he'd give to World Bank officials if he enjoyed complete access to them. Again, he seemed to hedge. If the access was only short-term, he said, his chances of making impact would be small, but given long-term access, he'd try to get officials to understand that they were falling increasingly short of their professed intention to alleviate poverty. If he still held their attention after that, he'd recommend that they embrace the World Commission on Dams's final report, and that they require that projects be reviewed by outside peers before final approval. He'd urge them to insist on making poor people the beneficiaries of dam projects, and he'd press them to require expatriate contractors to allocate a share of their subcontracts to local firms.

What struck me, when I pondered Scudder's answer, was that it was so technical, so bureaucratic—it was mostly about procedure, not politics. He didn't call out for attention to the injustice and suffering he'd seen, he didn't try to fill holes in the officials' grasp of dams' impacts. His agenda was relentlessly apolitical, as if the good dam were a matter of fine-tuning, as if honest, well-informed leaders reinforced by steadfast bureaucrats and altruistic donors would produce the good dam as a matter of course, simply because it was rational. The good dam would reward vast numbers of poor people with electricity and water, and then its administrators would devise programs to guide them down the path of economic development. The good dam would spread its uses around, having not just one objective but many: drinking water, irrigation, electricity, flood control, fishery growth, tourism, ecosystem survival—it would sacrifice some of each objective for the good of them all. The good dam would make the people it displaces its first beneficiaries, who'd use the opportunity to find roles in the modern world. He thought he'd caught a glimpse of the process during the '60s in Mazulu, when villagers briefly prospered. The good dam would be designed with enough flexibility to allow timed water releases of appropriate temperature, volume, and sediment load to sustain the downstream environment. By its mere existence, the good dam would validate the assumptions that dams represent. Then it would be clear that what warps dams isn't the concept of them but their implementation, that the problem isn't science, but politics. But dams are too big to be apolitical; indeed, they're the most political of public projects. The huge sums of money involved, the potential for bribes and patronage, and dams' capacity to reward and punish entire regions (one region gets water and power, another is inundated) assures major political involvement. Even though Scudder said that 70 percent of the world's big dams shouldn't have been built—in casual conversation, he placed the number at 80 percent—he never seemed to wonder whether a proportion so one-sided suggested that the concept of dams itself might be flawed. In downplaying politics, he risked being a cog in the antipolitics machine.

We left early the next morning and drove toward Katse, knowing we'd have to turn around short of the dam to get back before dark. Maseru's outskirts extended for miles. Again, Scudder kept the car doors locked. He was explaining that the Bank could put more pressure on a small country like Lesotho than on big ones like India, China, and Brazil, its biggest clients. "I feel the same way about the Bank that I do about dams," he said. "The Bank as it exists has too many flaws, and dams as they exist have too many flaws. But I don't say they shouldn't exist. I say we should eliminate the flaws."

In front of us, the rear of a public bus bore a painted illustration of the Katse reservoir, an idealized portrait in brilliant blues and greens that shows a shimmering lake spanned by a graceful bridge as a contented peasant looks down from flourishing fields. The message, though wordless, was obvious: put your hopes in the dam.

Scudder and his colleagues on the panel of experts had just finished devising a list of six points the Bank ought to require the Development Authority to achieve before recommending sealing the dam. Scudder recited the list. Three of the points involved completion of promised resettlement tasks, such as construction of roads and fair compensation for dam-affected people's losses. The fourth point pressed for a policy on adjusting downstream flows from Mohale to benefit downstream communities, which meant reducing the flow to South Africa by devoting some water to the estimated 150,000 villagers who lived within five miles of the river downstream from the dam. The fifth point argued for the creation of a Lesotho Biodiversity Trust to redress the expected dam-hastened extinction of the Maluti minnow, an inch-long fish whose population has already decreased by 90 percent. "I think this shows the unethical behavior of some NGOs," Scudder said. "They say the project will cause the extinction of the minnow. No, the trout that were introduced by the colonialists for fly-fishing are causing the extinction!"

It was the sixth point that stood to catch the attention of dam opponents outside Lesotho. It said the Bank should refuse to support Mohale's sealing unless the government agreed to follow its own regulations and drop the planned construction of the island resort. "The Katse impoundment couldn't be stopped because South Africa was undergoing a drought, but now South African reservoirs are full." That meant impoundment of Mohale could responsibly wait a year, so the Bank risked little by carrying out this threat. Scudder's anger went in two directions: he castigated the cabinet ministers who would profit from the resort, but then, unexpectedly, he also denounced the Lesotho project's critics. "God damn it all, decisions aren't easy—it's very easy to sit on your can back in Berkeley," home office of the International Rivers Network. "I feel much better about the way we dealt with Mohale than we did with Katse.

"Well, for Christ's sakes, turn!" he said to the car in front of us.

We were driving through rolling hills. As we ascended, the views improved, while the road, a two-lane highway, remained first rate. Before Katse was built, "this was a dirt road," Scudder said. "These people are better off because it's a paved road now. One hundred thousand people are better off because of this road. The NGOs like to forget that!"

We were surrounded by range after range of jagged peaks, alternately pink-rocked and shimmering green with grass. Deep canyons, the eroded remains of a lava plateau that covered most of Lesotho millions of years ago, opened in succession in front of us. Thumbnail-shaped humps lined the horizon. "Look at that vista," Scudder said. "This is the kind of thing I'm going to miss." We passed a man riding a donkey and a procession of walkers clad in rubber boots.

I asked Scudder whether poor people such as those forced to move because of the Katse and Mohale dams wanted development. "They want development, but on their own terms," he said. "The Nuer are classic pastoralists who lived along the banks of the Nile. When I was doing work in Sudan—this was in the seventies, between the two civil wars—we arrived at a Nuer cattle camp, and the young men were stark naked. They were doing hundred-yard dashes singing the praise names of their oxen, simulating their oxen's horns." Yet for all their seeming bucolicism, "they wanted better medicine for their cattle. They wanted education. Then they brought out their Uzis, and they said, 'We're having trouble getting parts.' They knew damn well what they wanted—that's how they defined development. It's not necessarily how the World Bank would define it. That's the starting point: we have to do more participatory development. If that's not possible, it leads to environmental degradation and fundamentalism. It's a process I don't think Bush and Colin Powell understand."

We were driving up a long series of switchbacks, with spectacular peaks behind us. "But when a dam comes," I said, "it forces people into development."

Scudder agreed. "All of it is involuntary. They're developmental refugees."

Then if his role wasn't to prevent involuntary resettlement, what was it?

"I think we have helped make a lot of people less impoverished than they would otherwise be."

Considering that humans have been building large dams for nearly two hundred years,* the understanding that they cause extravagant damage downstream is shockingly recent—so recent that in 1986, when Lesotho and South Africa signed the treaty that launched the project, the negotiators assumed that most of the water flowing through the affected rivers served no purpose, and therefore could be redirected to Gauteng. The treaty posited that at least 95 percent of the water in the two soon-to-be-dammed Orange River tributaries could be diverted without seriously harming the plants, animals, and 150,000 predominantly

---

*Nineteenth-century dams were earthen embankments built without scientific understanding of soil, rock, and rivers, and therefore tended to collapse. A tenth of American embankment dams built before 1930 collapsed. In 1889, the most notorious of them, a dam above Johnstown, Pennsylvania, gave way, and twenty-two hundred people drowned.

poverty-stricken Basotho living downstream. The assumption was built into Katse Dam, and justified the omission of a preconstruction assessment of downstream environmental impacts. On one level, it's easy to understand how such a mistaken idea took hold. Unlike dams' impact on upstream residents, which is sudden and dramatic—their homes and fields are flooded—downstream effects take years or even decades to surface, and often occur so far downstream that the connection to a dam isn't obvious. Through the 1970s, dam planners assumed that they could expropriate 90 percent of a river's flow without causing serious harm. Even after they began conducting environmental impact assessments, they typically looked only upstream. The omission of downstream impacts significantly skewed cost-benefit analyses in favor of dams.

The size, timing, and contents of a river's flow determine the character of a river's ecosystem. Change any of those variables, and the ecosystem is forced to adjust; substantially change all of them, as dams do, and the ecosystem declines. Small floods trigger fish and insect migrations; big floods scour riverbeds to create fish habitats and carry nutrients across floodplains. Changes in a natural river's water temperature cue fish reproduction, while the water's chemistry supports animals already adapted to the river environment. All this, dams disastrously disrupt. Hydroelectric dams, for example, typically release water through their turbines in response to electricity demand, which varies enormously from season to season and from hour to hour. As McCully puts it in *Silenced Rivers*, "The link between water releases and power demand means that river levels downstream of Glen Canyon now change not according to rainfall in the Colorado Basin but because of factors like the drop in electricity use on Sundays and public holidays." Not surprisingly, cycles generated by electricity demand don't suit downstream plants and animals. They're adapted to cope with entirely different cycles that include, most significantly, floods. The relentless hourly variations in energy-driven flows wash away plants and animals downstream and accelerate erosion. The loss of vegetation causes a decline in animals that eat it, while animals that don't depend on it may flourish. Rare and specialized species that have evolved over millions of years go extinct; hardy animals, the ones humans consider pests, often prosper. The consequences of the altered flow extend all the way to the river's mouth, where saltwater intrusion may be accelerated, or a wetland or marine fishery may be lost.

The science of environmental flow assessment emerged as a way of ameliorating the damage, of coexisting with dams. The assessments are management tools, enabling decision makers to understand the impacts of the choices they make about such crucial issues as the amount of water they'll take from dammed rivers. An assessment commonly depicts several scenarios, including one with a flow regime sufficient to sustain the river's predam ecosystem. Environmental flow assessment got its first push in the 1950s in the United States. Here, fishermen noticed that favored game fish were dying out for reasons that had nothing to do with the usual culprit, water pollution; the problem was that dams had altered the rivers' flows. Since then, at least twenty-nine countries have incorporated flow releases into their water-management plans, and, as of late 2001, another eleven countries were considering the plans. In the United States, the Federal Energy Regulatory Commission requires operators of many hydroelectric dams to release water according to an environmental flow plan. But environmental flow assessments in the developed world typically omit human impacts, since relatively few people in countries like the United States depend on rivers for subsistence. The opposite is true in the developing world. No one has kept track of how many downstream people have been harmed by dams worldwide, but the number is surely in the hundreds of millions, at least three or four times the forty to eighty million upstream dam-affected people estimated by the World Commission on Dams.

Scudder was an early backer of environmental flow assessments, but he didn't push hard for one in Lesotho because he was "lulled" into accepting the conventional view among Lesotho consultants that downstream impacts would be minor. "I should have known better," he writes. By the time he wised up, in the mid-1990s, Katse was completed, and preparatory construction work on Mohale was beginning. Nevertheless, in 1997, he and his expert colleagues persuaded dam authorities to launch an environmental flow assessment that for the first time included estimates of losses to downstream people. The new "socioeconomic component" then predicted the expense of justly compensating them for their losses.

Published in September 2002, the pioneering study involved twenty-six research specialists spanning fifteen disciplines, including not just scientists but also anthropologists and economists. It identifies thousands of dam-induced changes, from the abrupt separation of upstream and downstream fish to the loss of species and the spreading of disease-carrying snails. It shows that the assumptions codified in the treaty were wrong: people downstream from the dams are vitally tied to the river, and the impact on them of restricting flow to 4 or 5 percent of its predam rate would be disastrous. It shows that distribution of the water is as important as its gross flow: even substantial environmental flows won't be effective unless they mimic natural rhythms of large and small floods. And that's not all. To be effective, a managed flow regime must release water of the right quality, at the right temperature, with the right sediment load.

The study also implies the likely inadequacy of paying downstream inhabitants for their economic losses, as the Development Authority planned to do. The problem arises from the low market value of many of the items—river-dependent shrubs, say, or herbs—that nevertheless play vital functions in downstream inhabitants' lives. Even if the inhabitants are compensated for the items' disappearance, the items' market prices don't reflect their importance, and certainly don't approach the market value of the water to industrial users in Gauteng. As a result, wrote Jackie King, the study's leader and a freshwater ecologist at the University of Cape Town, "In any water development, developers could be happy to pay the highest compensation costs for the most river-damaging scenario, and walk away with all the water. So, we need ways of assessing scenarios that do not just hinge on straight economics, but bring in quality-of-life issues, the intangibles such as rare species, level of health, and the existence values of beautiful rivers."

The Lesotho study envisions four scenarios of water flow, ranging from best to worst case. The "minimum degradation" scenario concluded that if the rivers lost a third of their flows, environmental impacts would be "low," and impacts on downstream people would be "negligible." The remaining scenarios described progressively grimmer outcomes, ending in the "critically severe" environmental and social impacts of the treaty assumptions. In the end, in 2003, South African and Lesotho officials adopted a plan much closer to the treaty scenario than the sustainable "minimum degradation" one. While the planned releases are several times those called for in the treaty, most downstream social and environmental impacts are still likely to be severe. Instead of trying to substantially limit downstream harm, the authorities chose to cash in, by diverting most water to Gauteng and paying the inhabitants modest compensation for their losses. Ferguson could have written the conclusion: the villagers suffered, while the central government prospered.

The international response was predictably mixed. On one side, the World Bank was content. Andrew Macoun, the Bank's forthright Lesotho project task manager, called the decision "a major concession to environmental and social needs" and declared the Bank "fully satisfied" with its logic. On the other side, dam opponents, who accept the scientific validity of environmental flow assessments but doubt they will ever be implemented appropriately, point to Lesotho as confirmation of their views. "Mitigation," McCully writes in *Silenced Rivers*, "is especially dangerous when it misleads the public into believing that dam builders can recreate the characteristics of wild rivers and fisheries and so allows more dams to be built." In most countries, he says, environmental flow levels are chosen arbitrarily, without ecological basis. Flow volumes are usually set too low to retain the ecological vitality of the predam river, and they give little attention to the importance of natural seasonal flow variations, not to mention large floods. The debate put Scudder in his accustomed place, beset on both flanks and ambivalent. The decision did not set aside nearly enough water for downstream villagers and environment, he thought, but it marked progress in gaining legitimacy for environmental flows.

We drove over a magnificent ten-thousand-foot crest and headed down switchbacks on the other side. As we drove, we caught glimpses of the silvery tail of the Katse reservoir, disclosed behind grass-green ridges. Katse forms one of the highest, deepest, and most sinuous reservoirs in the world; its surface altitude is more than sixty-five hundred feet. We were at about ten thousand feet when Scudder turned into a paved driveway, the entrance to the Bokong Wildlife Observation Center. The center is part of a Development Authority program to establish ecotourism facilities in the highlands, and Scudder took a nearly proprietary interest in it. At its core is a large split-level room with a magnificent view through full-length glass windows down a steeply pitched valley to a narrow strand of the reservoir. On one side of the center was a wispy waterfall that disappeared into a mostly dry streambed, which in turn snaked its way down to the reservoir. The bed's two banks were a shade greener than the surrounding hills, an illustration of the life-giving nature of water. In the other direction, atop a gentle crest, a vulture feeding station would soon be established for the delectation of tourists. I asked Scudder what food would be placed there.

"Road kill. Sheep. Goats . . . People." Scudderian humor.

The brisk wind up the mountain had tousled Scudder's hair until some of it was tangled in his eyebrows. We entered the main room, which was still unfurnished—our voices echoed around it. Scudder, a proficient yodeler, gave examples of Swiss and American yodels.

Before we left, Scudder instructed me to buy a T-shirt at the gift counter. For $10, I got a green one, in extra-large (the only size available), which said across the chest, slightly off-center, "Bokong Nature Reserve." Above the words appeared a drawing of a Cape vulture.

As we hurtled down the mountain toward Katse, Scudder was gathering momentum. He was eager to show me some of the development projects he supported as an expert, projects that gave him hope. We drove by a newly designated wetlands reserve just down the mountain from the observation center. We drove through overcast Ha Lejone (HA-la-JO-nee), the biggest town in the Katse basin, a town that nevertheless looks only a few years old, and raw. The town has little in it besides a marketplace, with few marketers and fewer patrons, and on its fringes a desultory work camp, a relic of dam construction. On this misty, sunless day, when the clouds seemed to have extracted every glimmer of enthusiasm from the sun's rays, men walked the dusty street wrapped in their blankets, with heads covered in baggy wool caps. At the far end of town, along the reservoir's edge, Scudder delivered us to Chocks' Centre, an incipient tourist resort. The potential patrons of Ha Lejone hotels, now that Ha Lejone finds itself near the edge of a large body of water, are, according to Scudder, lower- and middle-class white South African recreational fishermen, for whom the reservoir is being stocked. Chocks, the enterprising proprietor, was just the sort of small entrepreneur Scudder wanted to protect, in this case from ruination by the big hotels that Scudder envisioned taking over Ha Lejone.

Looking around, I thought that idea required considerable imagination. Chocks' Centre is a touchingly unpretentious hostelry, with handsome brick-walled, thatch-roofed guest rooms that possess much charm but no plumbed toilets. Scudder fears that the absence of plumbing will scare away the fishermen's wives (fishermen themselves presumably being immune to such concerns), so he'd like to see Chocks get a loan to finance indoor plumbing. In Scudder's ideal development plan, such a loan would be a given, but dam projects are rarely ideal. We drove by a field where a reservoir irrigation system had broken down, and the farmers declined to repair the broken parts. They actually preferred relying on rainwater over irrigation to avoid dependence on mechanization and available parts. This didn't sound like a development victory to me, but Scudder, with a logic I did not quite follow, said it exemplified "participatory development."

We stopped at 'Muela (moo-ELL-uh), the smallest dam of the project's completed triad, like Katse an arch dam but at 180 feet less than a third of Katse's height, and tending to disappear inside a narrow gorge. We drove around the dam, trying to get close, but the curved concrete slab had a way of eluding our view, as if its impact were inconsequential, as if a collector of six million cubic meters of water could ever be considered minor. The dam is the forward salient of an otherwise predominantly hidden installation. It receives water tunneled from Mohale and Katse, passes it through the turbines of an underground power station (which transforms its energy into Lesotho's power supply), then enables it to gather behind 'Muela Dam, which regulates its rate of flow through yet another tunnel into South Africa. The dam was completed in 1997, five years before our visit, and the surrounding village seemed to have weathered the shock. Rondavels in sparse arced clusters overlooked the reservoir, and in their neatly trimmed thatched roofs exuded sufficiency. But above them ominously loomed the project's operations building, possessing a scale so unlike anything in its human-sized surroundings as to suggest that it had been deposited there by aliens, which, in a sense, it had. Massive and bland, it announced the arrival in 'Muela of modernity, three stories high and twenty-five horizontal windows per story, as linear as a milk carton, bristling with antennae and surrounded, at a generous distance, by chain-link fence. Some six hundred feet beneath the operations building, embedded in a sandstone cavern, three turbines perpetually churned, lighting up Maseru and 'Muela alike. "These people got a road, a water supply, electricity," Scudder said. "It's hard for me to see how they are worse off."

As we began our drive back to Maseru, a policeman flagged down Scudder for speeding, yet he avoided the usual payment of an instant "fine." I asked him how he did that. "I flashed the old Scudder charm," he said. "I gave him a big smile and asked him what the problem was, and he said, 'You may proceed, sir.' I seem to be able to get along with everybody—NGOs, the World Bank, and relocated people."

We passed a soccer field that sloped dramatically from sideline to sideline, indicative of the scarcity of flat land. As we went by, a game in progress was temporarily halted when the ball careened down the hill. We saw children playing with rolling tires, people riding donkeys, ox-carts, tractors, people tending to sheep and goats and cattle, people guiding wheelbarrows down the road. An easy majority of the middle-aged and old women were obese, and waddled. We passed at least half a dozen funeral homes, signs of the impact of AIDS in Lesotho, where a third of adults are believed to carry HIV. As we got closer to Maseru, we passed rusted car chassis, the Rose Pillars Bar, the Prince of Peace High School. Scudder's sociability was apparently scant comfort to him, for now he was ticking off the reasons for his discouragement about the state of the world—the widening gap between rich and poor, the rise of fundamentalism, the global undermining of ecosystems. If reason does not ultimately prevail, he said, "I think we'll have failed in our role as the dominant species."

The Lesotho Highlands Water Project has been noteworthy for one other reason, and it is not one that dam advocates like to trumpet: corruption. Not that corruption is unknown to dam building—indeed, in *Silenced Rivers*, McCully finds a connection between dams' immensity and the corruption they inspire: "The corporate beneficiaries of dam building—including environmental consultancies and electricity-intensive industries like aluminum—are not just passive recipients of government largesse, but actively persuade politicians and bureaucrats to build more dams. Such 'lobbying' routinely involves bribery: the massive costs of large dams means that they are an almost unique effective channel for kick-backs, greatly increasing their attractiveness for business executives, aid bureaucrats and politicians. In recent years dams have been at the centre of major corruption scandals in Britain, Malaysia, Kenya, Japan, Italy, Brazil, Paraguay and Argentina."

On this list of countries, Lesotho is now at the top. The conventional understanding of corruption in international projects lays the principal blame with poor countries, presumably festering dens of lawlessness into which morally superior multinationals are lured—corruption has even been called "the African problem," as if developed countries don't figure in it at all. The Lesotho project has placed this understanding on its head. Here, the international dam industry's leading companies went to extravagant lengths to pass bribes to the Lesotho Highlands Development Authority's first chief executive, Masupha Sole, for the apparent purpose of winning contracts. According to L. F. Maema, the Lesotho attorney general who brought charges against the companies, the multinationals entered into formal agreements with intermediaries who opened Swiss bank accounts into which the multinationals deposited bribes for Sole. Over ten years, twelve multinationals paid Sole $2 million—by one estimate, the equivalent of twenty-six hundred years of work for the average Lesotho citizen. After indicting Sole, six intermediaries, and twelve multinationals, Maema has won his first five cases, in what are generally regarded as David vs. Goliath–type victories. Acres International, a Canadian engineering consultancy, was found guilty on two counts of bribery and fined $2.3 million; Lahmeyer International, a German engineering consultancy, was found guilty on seven counts and fined $1.6 million; Schneider Electric, a giant French-based electrical manufacturer, pleaded guilty to sixteen counts and was fined $1.5 million; Sole was convicted on thirteen counts for accepting bribes and is serving a fifteen-year jail sentence.

By multinationals' standards, of course, the sums of the fines are trifling, but the impugning of their integrity is potentially disastrous. Some of the bribes took place even as the World Bank and the International Monetary Fund launched an anticorruption drive. World Bank president James Wolfensohn declared in October 1996 that because "corruption is detrimental to development," the Bank would curtail loans to countries undermined by it. "If we find evidence of corruption, we will cancel the project," he said. Nevertheless, the Bank's response to the Lesotho case has been decidedly ambivalent. For its anticorruption constituency, the Bank turned over useful information to Maema as he faced what he called "a barrage of lawyers not seen before (or since) in the High Court of Lesotho." And Bank officials held a meeting in Pretoria with Maema and project donors at which the attorney general received praise and promises of financial support—but, according to Maema, none of the promises were kept. He concluded, "When you prosecute international bribery, you are on your own."

In its newfound enthusiasm for fighting corruption, the Bank adopted policies specifying that firms found to have engaged in corruption will be declared ineligible for future Bank projects and has disbarred about a hundred companies. But none of the companies are leading ones in their fields, or instrumental in dam building. Most of the indicted companies in Lesotho are both large and instrumental, and the Bank has shown a reluctance to punish them. At the outset of the trials, Bank officials declared that since the Bank financed only a small portion of the Lesotho Highlands Water Project, it had no contracts with most of the indicted companies, and that meant that they were exempt from disbarment. And though Acres and Lahmeyer, the first two convicted companies, both *were* financed with Bank loans, the Bank at first declined to disbar them, then reopened an investigation of Acres based on trial transcripts in March 2004. The ambivalence of the Bank's response goes unacknowledged in Bank declarations by its president. "The one thing I'm proudest of is our work on corruption," Wolfensohn declared in 2003. "It is central to what we do."

8 THE OKAVANGO DELTA

The Okavango Delta is the center of the center of the world. Surrounded by the Kalahari Desert, the world's largest continuous expanse of sand, it arises out of nothing, like Renaissance Florence on the moon. On the oldest continent, where three pairs of three-million-year-old footprints of humanity's ancestors have been found in petrified mud, the Kalahari is the oldest terrain, from which rocks nearly as old as the earth—3.8 billion years—have been extracted. On its foundation of inhospitable sand, the Okavango Delta hosts an improbable biological jamboree, life in its most effulgent display. The delta is slightly bigger than Connecticut, yet it sustains 450 bird species, half as many as in all of the United States and Canada. Among the species that reside in the Okavango are 164 mammals, more than 150 reptiles, nearly 100 fish, and more than 5,000 insects. Nearly all the members of the great African menagerie are represented: elephants, lions, giraffes, rhinos, hippos, and zebras; crocodiles and black mamba snakes; hornbills, eagles, and vultures; tiger fish, tsetse flies, malarial mosquitoes. This effusion of nature red in tooth and claw is enabled by a delicately poised hydrological regime, which spreads across the terrain a sheet of water barely more substantial than a mirage: brutality, beauty, and vulnerability are intertwined. The creation so exhausts the Okavango River, which feeds the delta, that just beyond it, the river dies in the sand. The Okavango, says writer Adrian Bailey, "is the greatest of Africa's wetland wildernesses, and among its last."

The Kalahari's lineage dates back 135 million years, when the African supercontinent, then the world's biggest landmass by far, began to break up. Over the next thirty or forty million years, the continent shucked off India, Madagascar, then Antarctica and Australia, finally South America, and began drifting southward, drying along the way. As the continent assumed its modern shape, its southern edge lifted above the Kalahari, which became a giant drainage basin. The Okavango River formed on the Angolan plateau northwest of the delta, and once joined with the Upper Zambezi to become a kind of megariver, descending in a southeastward direction across the Kalahari all the way to the Orange River and the Atlantic Ocean. About three million years ago, the river reversed direction after traversing the Kalahari and joined with the Limpopo River on an eastward course to the Indian Ocean. Still later, the earth's crust delivered up a natural dam that stopped the giant river and created vast Lake Makgadikgadi in what is now northern Botswana. As biologist Karen Ross writes in *Okavango: Jewel of the Kalahari*, the delta "is the last remnant of this ancient lake."

Earthquakes made the delta, and may one day cause it to cease to exist (if humans don't beat them to the job). Over eons, they redirected some of the rivers away from the lake, launching its drying phase. As it shrunk, two sets of parallel fault lines, perpendicular to one another, emerged upstream from the lake, giving shape to the delta. The first two faults, only ten miles apart, formed a fifty-mile-long channel, through which the river's meandering delta "panhandle" flows; the third and fourth fault lines, a hundred miles apart, mark the boundaries of the delta proper, the delta's spoon, which filled with river- and windborne sediment, which is to say, alluvium. The delta is defined by its flatness: over its 150 miles, it loses only two hundred feet of elevation, a slope too gentle to be visible to the eye. The river spreads out and loses its momentum, accelerating evaporation. The sediment blocks some river channels and creates others, spreading water from the top of the delta as if the river were a hose with a thumb on its nozzle; from one wide channel in the panhandle, the river splits into many narrow ones. Despite its name, the Okavango is not, strictly speaking, a delta, since it doesn't empty into the sea; it is, rather, an alluvial fan. It's an apt word, "fan," conjuring up rippled, elegantly embroidered skeins of silk. Water embroiders the Okavango's fan, making a constantly shifting filigree through its fifty thousand islands.

Now the Okavango is an adornment of Botswana, a Texas-sized expanse of savannah woodland, grassland, and desert that is both the world's fifth most sparsely populated nation (1.6 million people, 8 people per square mile) and the nation with the highest HIV adult infection rate (39 percent). Since it was granted independence in 1966, the former British protectorate of Bechuanaland also has enjoyed a reputation as Africa's best-governed country, partly a consequence of its relative ethnic homogeneity—the Tswana people comprise 79 percent of the population—and partly because of a lack of competition for the distinction. Nevertheless, nearly half of Botswana's people live in profound poverty, and official corruption is not unknown. In this country of extreme aridity, where 1 percent of the land is arable, water is so valuable that the national currency, the pula, is the Setswana word for rain. It is therefore not entirely surprising that until recently, the Botswana government considered the chief value of the Okavango River not the wildlife it causes to flourish, but its water. Proposals to divert the river began cropping up in the first quarter of the last century. In the mid-1970s, the United Nations Development Program and the Food and Agriculture Organization produced a joint study, aptly called "The Okavango Delta as a Primary Water Resource," which promoted the notion of building a dam on the Boteti River nine miles downstream from the delta, in conjunction with two smaller dams that would seal a reservoir. By the time the government approved the project, in 1988, it was grandly called the Southern Okavango Integrated Water Development Project, and was enlarged to include the dredging of the Boro River, the delta's central channel. By hurrying water down the delta, dredging would drastically reduce its annual flooding, thereby reducing the animal populations that depend on the flood, but it would increase the flow of water into the reservoir and on to human uses. The government said those uses included supplying the residents of Maun, the district capital at the bottom of the delta, and promoting commercial irrigation and flood-recession agriculture. According to Scudder, the government declined to reveal one other reason for the dam, and that was probably the most important one: the water would supply the thirsty diamond mines at Orapa, 175 miles southeast of the delta. Diamond mines earn four-fifths of Botswana's foreign exchange, and Orapa produces half of that.

The Okavango's human residents include the twenty-five thousand townspeople of Maun, plus Bushmen who in small numbers inhabit the flood lands. They didn't find out about the project until two years after the government approved it, when bulldozers arrived in Maun to start work. Local residents held what Scudder describes as the biggest antigovernment demonstration in the history of Botswana, and told the minister of water to commit suicide, as his predecessor did. The local residents' fury was fueled by their experience with Okavango dredging. As Rick Lyman wrote in a January 1991 article in *The Philadelphia Inquirer*,

In 1938, laborers hand-dredged the smaller Gomoti River and then sat back to await the blessed flow. Instead, the Gomoti dried up.

In 1942, the Thaoge River was hand-dredged. The next year, it dried up.

And in 1973, a few miles of the lower Boro were dredged. It did not dry up. Instead, its neighbor, the Santantadibe River, did.

Now the government intended to dredge not just a few miles of the Boro, but twenty-five.

The Orapa mines are jointly owned by De Beers, the South African diamond consortium, and the Botswana government. When Greenpeace, the Amsterdam-based environmental group, learned of the planned dredging, it threatened to counter De Beers's renowned advertising slogan, "Diamonds are forever," with a boycott and a slogan of its own: "Diamonds are for death." Astoundingly, government officials were so confident of the project's merits that they invited Greenpeace to send a fact-finding team to review it. Greenpeace was delighted to oblige. Predictably, the Greenpeace delegation soon issued a denunciation of the project.

Government officials still did not lose hope. This time they called on the World Conservation Union, a sober Swiss-based environmental alliance (and, later, cosponsor of the World Commission on Dams), to conduct a full-fledged scientific study of the project. The nearly thousand members of IUCN (as the World Conservation Union is commonly called, after its French initials) come in three forms—national governments, government agencies, and NGOs—out of which IUCN tries to forge a partnership to promote environmental sustainability. It also draws upon a large network of experts to produce scientifically grounded environmental studies. This was the first time a sovereign government invited IUCN to conduct a major study on its soil, a fact Scudder considers noteworthy. "Can you imagine," he said, "the U.S. Corps of Engineers or the Bureau of Reclamation asking South Africa to come over and evaluate Grand Coulee?" IUCN signed on to conduct a yearlong study of the Okavango, and chose Scudder to lead its team. He'd always wanted to visit the Okavango, and now he'd have a chance to do research that would complement his Zambezi work. If the project was bad, he might play a part in saving one of the world's natural wonders; if it was good, he'd finally be involved in a dam success story.

Scudder left Molly and Caltech in October 1991 for what turned out to be an eleven-month stay in Maun. He and IUCN assembled a team of thirteen experienced scientists with expertise in wetlands, water resources development, and southern Africa. The deputy team leader, Robert Manley, is considered one of England's leading hydrologists; Gilbert White, the senior adviser, is a much-honored geographer and former president of Haverford College. Scudder traveled through the delta by road, water, and air; a local couple gave him rides in a microlight aircraft so that he could observe and photograph delta settlement patterns, land and water use, and wildlife. He traveled in and around the Okavango, and satisfied his desire to cover the terrain that millions of years ago linked the Okavango and Zambezi rivers. When IUCN bureaucrats tried to wrest control of the project from his team, he warded them off by threatening to resign.

The team studied the Okavango for seven months, then concluded that the project was a terrible idea. The project would improve living standards in only one of nine zones of the delta, while at least six would be harmed; and the judicious drilling of wells could satisfy Maun's water needs just as well. Scudder and Manley intended to announce the findings at an afternoon lecture, but the government of Botswana preempted them. A half hour before the talk was scheduled to begin, officials announced on the radio that the project had been suspended because of the strength of local opposition. That opposition, galvanized by foreign nationals and local tribal leaders, was vital, but the study tipped the verdict. For a man without religious convictions, Scudder derives notable pleasure from telling acquaintances that he's been called "God, the Guardian of the Okavango," but the phrase errs only in its hyperbole: as much as anyone else, Scudder helped save the Okavango. Scudder's ambition was turned inside out. The greatest achievement of the man looking for a good dam has been to stop a bad one.

Christopher Scholz, a Columbia University seismologist who published *Fieldwork: A Geologist's Memoir of the Kalahari* in 1997, writes of the Okavango, "It looked as though an enormous glass of water had been spilled over the desert, and every point that had been touched by the water had turned bright green." The Okavango River gathers momentum in the Angolan highlands, then rambles into the Kalahari and crosses Namibia's Caprivi Strip—the 275-mile-long sliver of territory acquired by German chancellor Georg Leo Caprivi in 1893 in the mistaken belief that it would give German South West Africa a water route along the Zambezi River to the Indian Ocean. (Alas, the strip's portion of the Zambezi wasn't navigable.) Across the Botswana border, the river enters the delta's fifty-mile-long panhandle, a swamp valley of crocodiles and papyrus. Here, sand deposits have raised the Okavango River's bed, so that each year the full-throated river overflows its banks and makes of the valley a floodplain. The panhandle is the breeding ground for most of the delta's crocodiles, including many startling specimens sixteen feet or longer. Crocodile mothers bury their eggs in sandbanks suitably flattened by strolling hippos. During the ninety-day incubation, the mothers linger nearby, warding off crocodile egg devotees such as the Nile monitor lizard. When the eggs hatch, croc mothers delicately carry the young in their mouths to placid backwaters. The young dine at first on dragonflies, beetles, water bugs, and crabs; then, as they grow older, catfish, squeakers (fish named for the sound they emit when alarmed), and bream; and finally mammals such as sitatungas (a kind of antelope), goats, and cows.

The river is an organized channel until it runs into the Gomare Fault, at the northeast border of the delta proper, where the river spreads from a single arm to numerous fingers. The biggest channels—the Thaoge, Boro, Santantadibe, and Maunachira—are no more than fifteen feet deep, and they are exceptional: vast areas of the delta are covered by no more than an inch or two of water. This sheet of water is broken by countless tiny islands, some rimmed by sycamore fig trees and dotted with a single palm tree in the center. The islands are themselves engines of purification, absorbing the salt that would otherwise poison the delta, just as it has the Dead Sea and the Great Salt Lake. Some islands start as little more than termite mounds. Trees on their perimeters take in water through their roots, leaving behind the water's salts in the groundwater beneath. As the salinity of the groundwater intensifies, the islands' centers become toxic, and vegetation congregates on their fringes, causing the islands' expansion. As a result, even small islands have distinct vegetation bands, ranging from salt-vulnerable on their fringes to salt-tolerant at their centers. The swamps at the top of the delta are perennial, but farther down, to the southeast, they turn seasonal, as water becomes more precious, and the flood more variable. Rainfall in Botswana is erratic, but five hundred miles upstream, in the Angolan highlands, it teems: the annual flood supplies more than twice as much water to the delta as the December-to-March rainy season. It's the timing of these two principal sources of water that enables the delta to flourish. The flood doesn't reach the upper end of the delta until March or April, just as the rainy season is subsiding, and then it takes five months to traverse the delta's hundred miles. Americans are used to thinking of floods as sudden cataclysmic inundations, but that's because we've channeled our rivers, cut down our forests, and paved our city surfaces, all of which make bigger floods. But in the thickly vegetated African tropics, rainfall percolates into the soils and is slowly released into rivers. Even so, U.S.-style floods are increasing in Africa because of deforestation, heavy rainfalls possibly induced by climate change, and poor management of reservoir releases. The forests of the Angolan highlands are still largely intact, and the flood is therefore a magnificent but gradual thing: once it reaches the delta, it is only a few inches high, and moves at the stately rate of half a mile a day or less. By the time the flood crosses the entire delta, in July or August, the dry season is a few months old. For parched animals, the delta becomes a haven. For the delta, the flood provides an extra season of water, a kind of second life. As the watery sheet retreats from the floodplain, fast-growing but nutrient-poor grasses appear, and are consumed by the migratory herds. Birds patrol the shrinking ponds looking for trapped fish; when they find the fish, as many as thirty bird species may participate in a feeding frenzy. Humans grow crops in the drying banks of the major channels, in textbook illustrations of recession agriculture. So delicately poised is the delta that if it were closer to the Okavango River's source (so that the rainy season and flood overlapped), or if it were less flat (so that water flowed through it more rapidly), or if the Angolan highlands received less rain (reducing the delta's flow), then the delta would not get its proper quotient of water and would be ordinary.

The delta ends as abruptly as it begins, when its channels run into another perpendicular fault line—this one faces the northwest, and acts as a natural dam, containing the delta. The seasonal Thamalakane (TOM-la-KAH-nee) River follows the path of the fault to the southwest. Only 2 or 3 percent of the Okavango River water that enters the top of the delta leaves it at the bottom—the rest is absorbed by the air or plants or seeps through the sand and becomes groundwater—but the two outlets that remain are enough to flush the delta of a last significant increment of salt. Both outlets extend from the Thamalakane. One is a salt-poisoned lake, Lake Ngami, which filled with water from exceptional rains as recently as the 1970s, and the other is a modest river, the Boteti (bo-TET-ee), the downstream extension of the Okavango River. The Boteti in summer and fall consists of mere intermittent pools; energized by its vestige of the annual flood, it flows to the southeast for another few hundred miles. Its bed is much too wide for it, and memorializes the more powerful river that once flowed through it. It ends in Lake Xau, a desiccated bed that contained freshwater as recently as the mid-nineteenth century, when Livingstone hunted sitatungas there. Now entirely dry, it is rimmed with the skeletons of animals it failed to sustain.

The Kalahari's reclaiming of Lake Xau may presage the delta's fate, or it may simply underline how much the delta differs from its surroundings. Indeed, thirty miles from the panhandle lie the Tsodilo Hills, the most sacred locale in Bushmen lore, where Bushmen as long ago as a hundred thousand years painted pictures on rocks of the animals they hunted and observed. The thirty-five hundred paintings that survive depict giraffe, zebras, rhinos, elands, even fish (as well as Bushmen with erect penises). As recently as three hundred years ago, vast herds roamed the hills, but now the animals and the grasslands and lakes that sustained them are gone, replaced by a lifeless wilderness, and the few Bushmen who still live in the area know nothing of their painter ancestors.

When David Livingstone reached Lake Ngami in 1849, its circumference was more than sixty miles. He was close enough to the delta to hear from inhabitants of "a country full of rivers, so many that no one can tell their number," but he never got there. The distinction of being the first European to set foot in the Okavango fell to Charles Andersson, a Swedish explorer who delighted in describing how he shot elephants, lions, and giraffe, and described delta women as "perfectly hideous."

In fact, the Bushmen who inhabited the delta were artful hunter-gatherers who had enlarged their skills to include fishing. For centuries, they fished from reed rafts; after other immigrants brought metal tools, they carved dugout canoes that facilitated travel. The delta's environment is too harsh to support many humans, but over the last two or three centuries its population expanded and contracted as it absorbed a succession of ethnic groups who brought farming, trap fishing, and even cattle. The presence of tsetse flies (and, therefore, animal sleeping sickness) kept cattle out of the delta until 1896, when an epidemic of rinderpest, a disease fatal to livestock and many wildlife species, devastated the delta's wild herds. Since the tsetse fly lives on the blood of many of these animals, its population also plummeted, creating an opening for cattle. If not for the eventual recovery of the herds and the tsetse fly, the delta might have succumbed to unchecked cattle grazing and fencing; instead, cattle were forced out of the delta. Now, ominously, cattle ranches merely ring the delta.

In 1915, the Batawana, then the Okavango's dominant ethnic group, set up their capital in Maun, at the base of the delta. Maun means "place of reeds," and the town was not much more than that until the advent of commercial tourism. Now, Maun's airport, with its fleet of small aircraft for charter, is one of the busiest in southern Africa.

Now, as Scudder and I drove around Maun in our newly rented four-wheel drive, he said, "Everything is changed." Since his last visit a decade earlier, the town's population had grown by half, to forty thousand people. Even so, the business end of town, with its low-slung white bungalows and rondavels, was minuscule. While Scudder waited in the car, I changed money at a bank—a narrow, heavily guarded room, where the clientele stood on one side, in a line that snaked out the door, and the clerks sat behind floor-to-ceiling partitions on the other side.

We were headed for Croc Camp, the pleasant tourist lodge Scudder had chosen for our stay, but it was twenty minutes out of town, and he wanted to see the Thamalakane River first. If you think of the Okavango Delta as a handle and rake, then the Boro River is the rake's central tine. Maun is at the midpoint of the delta's bottom rim, near where the Boro runs into the Thamalakane. Maun straddles the Thamalakane, a fact I discovered when Scudder drove across a short bridge spanning a seeming marsh of reeds and occasional pools, and announced that we were over the river. I wondered whether he'd somehow erred, for the Thamalakane was disappointingly unemeraldlike. The dry-season river was little more than a creek, meekly winding its way around reedy obstacles and occupying a fraction of the shallow bed that enclosed it. Ten years earlier, when Scudder last saw it, the Thamalakane had not been nearly this dry.

We drove to the camp on a deserted paved highway. We occasionally passed men riding donkeys and a few women wearing outrageous multi-colored petticoats and horned bonnets at least a foot wide—outfits that astonished me both because of their riotous beauty and their wearers' disregard for the heat of the day. These women are Hereros, Scudder said. The Hereros were latecomers to Botswana, descendants of people who revolted against the Germans in South West Africa in 1904, then fled the repression that followed. But their clothing, a kind of amped-up and Africanized Victorian style borrowed from the dresses of early European missionaries' wives in South West Africa, reflected their origins. The Hereros' livelihoods revolved around cattle, but in fleeing South West Africa (now Namibia), they abandoned their herds. In Botswana, they worked as servants while patiently building new herds with their wages. Now, many were prosperous cattle ranchers.

It was four o'clock by the time we'd stowed our bags at Croc Camp, which left time for a late-afternoon outing. Beneath his familiar plaid flannel shirt, which Scudder wore Herero-like, in spite of the heat, he was wearing a T-shirt with black lettering on the front and back. In his room, he showed me the wording. The front displayed an acronym for the IUCN study and its dates of operation, 1991–1992. The back said

I'M NOT A TOURIST HERE
I WORK HERE

Scudder said he once owned twenty of the T-shirts, but now he was running out.

I was beginning to understand that the Botswana leg of our trip addressed Scudder's delta nostalgia. We drove near the Thamalakane, and he spotted the first of many red-billed hornbills we'd see in the Okavango—they're dramatic black-and-white-feathered birds with long, intensely red beaks that curve downward in an eternal semicomical scowl. He drove by the house where he briefly lived before he got "mugged," as he put it, when his wallet and glasses were stolen. He walked up to the barbed-wire-topped gate and called out for the present residents, but was greeted only by four growling dogs.

Our last stop was at the house of Pete Smith, who before his death in 1999 had been Scudder's colleague. Smith was a Rhodesian accountant who in midlife threw over his job, moved from Harare to Maun—from Zimbabwe to Botswana—and embraced his passion, botany. Despite lacking an academic degree, he became an unrivaled authority on delta plants, and also befriended some of the last delta Bushmen. Scudder hired Smith as one of the study's six local consultants and considered him "far and above the most valuable" one.

Smith's house sits on two and a half acres of Okavango ground, facing the Thamalakane. The property is enclosed by a chain-link fence, inside which Smith grew indigenous species of trees and plants; he turned the plot into a woodland botanical garden, so thick with foliage that a month into the dry season, the house was still invisible from all four sides of the fence. Scudder said Smith intended to leave the land to some scientific institution that could protect it, but that hadn't happened, and now nobody knew for sure who owned it or what would come of it. Sometimes, Scudder said, when he grew tired of the motel room from which he ended up directing the IUCN study, he pitched a tent in Smith's garden and spent the night. He pointed out the treetop where an eagle used to nest.

We walked down an unpaved alley parallel to the fence, until we reached the Thamalakane. At the moment, the Thamalakane was a series of ponds, connected to one another by trickles narrow enough to hop across. Most of the wide, empty riverbed was green with vegetation, as if sustained by the memory of water until the flood arrived. For the last thirty years, the Okavango has been growing dryer—no one knows whether this phenomenon is part of some natural cycle or a response to human activity. Explanations range from global warming to the unethical but common tourist hunting practice of starting fires to trap big game. "When I lived here," Scudder said, "it was never this dry. Now it's just reeds and mud."

Smith and Scudder used to begin late-afternoon Thamalakane walks from a gate in the southeast fence, but now the gate was replaced with more fence. "Pete and I would come down and look at all the hippos," Scudder said. "But the human population grew, and the hippos became dangerous, so the authorities removed them. Before the droughts of the '90s, there were deep pools here, and the hippos would live in them. In August, during the flood, the river was full of hippos."

By now, Scudder's shirttail was hanging out, and his floppy green hat was pulled low on his head to ward off the russet sun. A blacksmith plover flew over us, and Scudder, an avid bird-watcher, whistled to it in his best rendition of plover language. "See this beautiful bird here?" Scudder said, and pointed to a pond in the middle of the riverbed. "It walks on floating vegetation." The bird was a jacana, also called a lily-trotter in honor of its ambulatory skill, which the specimen in front of us proceeded to demonstrate, before flying off. "See the way they fly, with their legs hanging down? . . . There aren't many water lilies left here because in hunger years they're food for people."

Scudder pointed out a yellow-vented bulbul. Then: "Look, look, look! Egyptian geese, probably—no, they're storks, black storks." We'd walked along the Thamalakane to a point where the river broadened, and a man punted himself across it in a dugout canoe. It took him no more than ten or fifteen seconds. Scudder lay in the long, damp grass of the riverbed and crossed his feet. Tall reeds framed his outline. "You're at the very edge of where the Okavango meets the Kalahari sand veldt," he said, "and you can see how important water is. Just since we've been here, we've seen about seven different kinds of birds, and there's two black storks staring at you." He pointed to the storks, uncommon Okavango residents that were now perched on the top limb of an acacia tree just across the river, enveloped in powder-blue sky. Long-billed and long-legged, they faced in opposite directions, as if posing for photographs of their exquisite silhouettes. For the first time in our acquaintance, Scudder looked something like relaxed. A yellow-billed duck flew by. "Look," he said, "the storks are going. They're graceful flyers." They flew in an upward spiral over the river before choosing their direction. In a voice infused with wonder, Scudder said, "Look at that."

As we started to wander back to the jeep, Scudder stopped in his tracks and pointed to the riverbed some thirty yards away, where a monstrous black bird strutted. "Oh, my heavens, there's a ground hornbill." It had a red face and red patches on its throat, and it was as big as a turkey. The species is so threatened that it is more or less confined to reserves like the one we were in, which encloses Maun. "You're really lucky to see him," Scudder said.

A second ground hornbill appeared. We stared for a couple of minutes. Scudder caught sight of a babbler, a drongo, a couple of starlings. We walked past two rusted car chassis, half buried in the Thamalakane sand. The ground hornbills took flight and spread their massive black-and-white wings over the river.

"Pretty spectacular, huh?"

By the time we returned to Croc Camp, a full moon hung over the Thamalakane.

The Okavango was the site of Scudder's greatest accomplishment, wherein he helped preserve an international wonder, yet something about the experience hadn't taken, for he was still looking for recognition for his deed. We careened from touring the delta to his quest for validation back to touring the delta again. In the case of the Harry Oppenheimer Okavango Research Centre, an outpost of Western science outside Maun, the two motives comfortably overlapped. The building was empty but unlocked when we arrived, so we walked in. The center is refreshingly understated; its compactness and lack of ostentation nicely camouflage the First World scientific gear and computers inside. Scudder conducted an instant inspection: he strode purposefully into the center's bathroom, turned a water tap and tried the hand dryer. "I always check these things," he said. "It's impressive that the hot water and hand dryer work. In Africa, usually the hand dryer doesn't work and there's no hot water, and sometimes there's no water at all."

The center arose directly from the collapse of the Southern Okavango Integrated Water Development Project. As Botswana government officials stirred through the ashes, they realized both that they lacked sufficient understanding of the Okavango to manage it, and that some nonpartisan institution was needed to oversee the required research. The University of Botswana eventually proposed to launch a research center focusing on the delta's sustainable use, and the government approved the plan. It is indicative of how completely the public perception of the Okavango changed—from swampy wasteland to invaluable wetland—that De Beers, owner of the Orapa diamond mine, is one of the center's major donors. The center, in fact, is named for a former De Beers chief executive officer who contributed its start-up funds.

Lars Ramberg, the Swedish director of the center, was so pleased to meet Scudder that he'd sacrificed part of his day off to show us around. He arrived in his Saturday garb—a polo shirt, shorts, and sandals—and welcomed us into his office. He's an amiable man, whose thinning, brushed-back hair is just long enough to suggest nonconformity and whose Swedish-cadenced English is disarmingly precise. At fifty-nine,

Ramberg was displaying a tendency to work in Scudder's old stomping grounds. For three years, he was director of the Lake Kariba Research Station in Zimbabwe, and would have happily stayed another three years to the end of his contract if Robert Mugabe's nationalization drive hadn't forced him out of his job. Now he told Scudder that he'd read Scudder's book *The Ecology of the Gwembe Tonga*—"I was very impressed by it, to be honest." Nine years later, Ramberg applied to become director of the center. He knew nothing about the Okavango, he said, so he prepared for the job interview by reading the IUCN study. That, he told Scudder, "was enough to get me the job."

Once the introductions were over, Scudder began questioning Ramberg. Occasionally, he leafed through his notebook for an empty page and recorded the answer in tiny letters. The longer the conversation continued, the more concerned I became about prospects for the delta's long-term survival. True, the dam project had been averted, and was not likely to be revived; the Botswana government's 1996 designation of the delta as a "Ramsar site"—a place afforded protection by an IUCN-sponsored international convention on wetlands—helped ensure that. But if the delta had eluded a single, devastating blow, it was still vulnerable to a death by a thousand cuts, most human-inflicted. The dredging of the Boro River in 1972 had already left a mark. Ramberg said its main impact was diverting water from the floodplain to the river, thereby "killing" the floodplain. Acacias, trees common in arid areas because of their ability to withstand drought, started growing on the floodplain, Ramberg said, and the area became a woodlands. Dredging even impeded the customary recharging of the Okavango aquifer beneath the delta, because the floodplain never received water to begin with. "The way we understand the delta," Ramberg said, "flooding is the absolute key."

Fencing perpetrates another kind of damage. More and more fences are being erected around the Okavango, as a consequence of the burgeoning growth of Botswana's export beef industry. Some fences simply confine cows; the bigger ones, the "veterinary cordons," are designed to separate buffalo, which carry hoof-and-mouth disease, from nonresistant cattle. By 1987, when Karen Ross wrote *Okavango: Jewel of the Kalahari*, nearly two thousand miles of fences were erected in Botswana, and hundreds of miles more have been installed since then. The most notorious fence, the Kuke Fence, runs for nearly two hundred miles along an east-west line some sixty miles south of the delta. When it was built, in the

1950s, southern African animal migrations were neither understood nor taken into account; the result was that in 1963, at the climax of a long drought, about three hundred thousand wildebeests died of thirst or starvation because the fence blocked their way to the delta. Many wildebeests died gruesomely, entangled in barbed wire; so did many other grazing animals. The cycle repeated itself during a drought two decades later, when animal populations suffered another dramatic decline. This in turn made life much harder for the Bushmen and other people who subsist on hunting. According to Ross, until the 1980s, more than 60 percent of the people in western Botswana depended on hunting for a share of their livelihood; now they had to travel farther and farther to find animals to hunt, and more and more ended up on drought relief. The tourism industry, which is led by white foreigners, dislikes fences, since wildlife is Botswana's main tourist attraction, but cattle entrepreneurs, who include many government officials, appreciate them. A 1998 study sponsored by an international environmental group called the WILD Foundation concluded that the fences have seriously damaged wildlife in the delta and surrounding areas, compromised the viability of the Okavango ecosystem, jeopardized the ability of local residents to live sustainably, and threatened the Okavango tourism industry. As Ross told me later, "Fences and wildlife do not mix."

Cattle have grazed in the Kalahari for only fifty years; the tsetse fly, a vector for animal (and, less commonly, human) sleeping sickness, kept them away. The first attempts to control the tsetse fly endeavored to deprive it of its natural cover by cutting down thousands of trees and to starve it by killing some fifty thousand animals that provide its food source. Nevertheless, as Ross notes in *Okavango*, "These drastic control methods had little impact on the fly population." In the 1960s, dieldrin, a carcinogenic insecticide spray that was eventually banned in most developed countries, became the weapon of choice, but its applications also failed to stanch the tsetse population. Success was achieved only after officials switched from dieldrin to endosulphan, which breaks down in air and sunlight; replaced hand sprays with aerial sprays; and developed a technique called "ultra-low-volume spraying" that spreads a diluted chemical mist over targeted areas. Even this mixture kills some fish and many insects, including a weevil introduced into the delta to consume a rapidly spreading alien water weed, salvinia—in other words, one kind of intervention, the tsetse spraying, nullifies another, the salvinia weevil. The latest tsetse fly control effort, Ramberg said, entails aerial spraying of a chemical called deltamethrin, a pyrothide considered environmentally friendly, over a two-year period, while simultaneously introducing millions of male tsetse flies sterilized with radiation. A facility for tsetse fly irradiation will soon be built in nearby Chobe.

Efforts to control the tsetse fly will certainly continue. So will the encroachment of cattle, even though overgrazing in Botswana has already destroyed large swaths of Kalahari grassland. The lushness of the Okavango is deceptive; it is nutrient poor, and its grasses therefore make poor fodder, but the number of cattle will probably grow there, too. Add to these encroachments on the Okavango the unknown impact of frequent fires started by safari lodges to flush out game for tourists, and the demographically significant number of large mammals they've killed. And take into account the mysterious three-decades-long drying-out phase that the delta has been experiencing.

Scudder eventually asked how the Botswana government now regards the IUCN report. Government officials were pleased with the outcome it helped produce, "but they won't admit it," Ramberg said. Even so, the government's probable solution to Maun's water problems broadly resembled the IUCN study's recommendations.

"So, okay, we were right!" Scudder said, exulting.

True enough, Ramberg conceded, a misguided consultant had recently advised the government to tap Okavango groundwater from a spot near the delta's center, but this wasn't likely to happen. For one thing, the center opposed the idea because it would deplete the Okavango aquifer, and, for another, it would give Namibia an excuse to divert its own share of the Okavango River. Instead, Ramberg proposed to pump groundwater from a much larger aquifer on the delta's periphery.

Scudder quoted Peter Rogers, an IUCN team member and Harvard environmental engineer, as saying that dam projects are like vampires in that they often rise from the dead. Might the Southern Okavango Integrated Water Development Project ever be resurrected?

"I don't think so," Ramberg said. "We'd oppose it."

With its shabby fence and nearly empty parking lot, the Maun Game Reserve looks like a glorified municipal zoo, so I needed a few minutes inside it to realize it was better than any zoo. It's a slice of protected wildness, one of whose long borders is defined by a fence down the middle of the wide but nearly empty Thamalakane River. As we walked, Scudder pointed out a tsetse fly trap, a synthetic screen steeped in pyrethrin, elevated a few feet off the ground and able to pivot in the wind, thereby simulating animal movements to attract flies. The trap is less harmful than aerial spraying, since it directly poisons no part of the delta other than its target. For us, the threat of sleeping sickness was in any case remote, since tsetse flies are scarce during the dry season.

From a decrepit elevated blind, Scudder and I could see the fence that bisects the Thamalakane: it's a homely affair of wood and wire rising out of the river's preflood trickle. On the Maun side, a few donkeys grazed, and then some cattle joined them; on the reserve side, where domestic animals were forbidden, warthogs, wildebeests, and impala dined on grassy stubble near the river, while baboons looked on. Eagles and storks flew overhead. Two impala with elegantly twisted horns pranced by, snorting loudly.

One day we went looking for the site of the proposed dam that Scudder's study helped defeat. We drove to the southeast out of the delta, into the Kalahari, and in the process rediscovered the elusiveness of freshwater. The road should have crossed the Boteti River a few miles out of Maun, but we saw no river, and drove on. The road turned to dirt, and narrowed to one rutted lane that sent my pen skittering across the notebook page whenever I tried to write. The only vehicle we saw was a cart pulled by two donkeys.

We drove on for nearly half an hour, as Scudder looked for Boteti clues. Had we simply overlooked it, or did the increasing density of foliage indicate that a river was near? At last we spotted a pickup truck coming at us from the opposite direction; like our van, it raised a veil of thick dust as it negotiated the ruts. Scudder managed to get the driver's attention, and both cars stopped. The driver was a burly blond South African; a black man and a dog rode in back. The driver cheerfully informed us that we'd passed the Boteti many miles back. And where was the driver coming from? Scudder asked. If we continued down the road, the driver answered, we'd run into his new cattle ranch—the government was parceling out ten-thousand-acre lots to ranchers like him. This meant the continued encroachment of fences.

"That's fantastic!" Scudder said as we turned around. "We went roaring across the Boteti without even seeing it!"

We soon discovered that the main reason we didn't see the Boteti is that it was extremely hard to distinguish from the woodlands surrounding it. Where the road crosses it, it was filled not with water, but mud. Donkeys grazed in it. Beer cans were half submerged in it. We marched through the mud and found a gauge, a kind of oversized ruler, sticking out of it, an apparent project relic. It was hard to imagine how the fifty-foot dam envisioned in the project could fit inside this shallow depression. "To make the dam," Scudder said, "they would have had to dig down quite a bit."

We found eight-inch-deep holes where cattle's legs had sunk into the mud. With all that mud, Scudder said, "I would have thought there would be a sedimentation problem as well as an evaporation problem . . . It was a bad project, a bad project. That was our conclusion: even if it was built, it wouldn't have operated as it was supposed to. My guess is that there must have been corruption. We had no evidence of that, but one reason the government was so angry is that it had to pay the contractors 25 percent of their total fees because they'd already arrived in Maun. It caused a big loss of confidence in the government."

Unlike in India, where life's vibrancy seems to have penetrated every surface, interior and exterior, the glory of southern Africa is entirely alfresco: it's as if the outdoors overwhelms the need for a permanent indoors. If I did not grasp the magnificence of the African outdoors at five-thirty one May morning, when we started our drive into the Okavango, I understood it by the evening, when we got back. Our destination was the Moremi Wildlife Reserve, a Delaware-sized slab of sand and silt that comprises a third of the delta, the first southern Africa wildlife sanctuary voluntarily given up by its human inhabitants. The reserve was created in 1963 in an attempt to protect at least a portion of Okavango wildlife so that the delta's traditional fishing and hunting communities could continue to subsist outside it. They gave up Moremi, where some inhabitants had lived, because they realized that protecting animals there would enable them to continue hunting them elsewhere, and because the hunting ban applied to professional hunting companies as well as them. Many of those companies have set up operations in Maun and even deep inside the delta. They envision Botswana as the third mecca of African wildlife tourism, emerging as wildlife habitats in Kenya and Tanzania are becoming overrun with tourists. The reserve may at least limit the damage commercial hunters cause.

Moremi includes Chief's Island, by far the biggest island in the delta: it is the shape of Manhattan and forty times Manhattan's size. It is wide and dry enough to contain in its interior a pocket of Kalahari desert, with vegetation and animal life to match—it's a dot of Kalahari yin inside the swampy Okavango yang, itself surrounded by Kalahari yin. Land, water, land, in all combinations and gradations: in the Okavango, water is blood. On a sandy spit of road before we even reached the reserve, a few dozen elephants stampeded in front of us, raising a dusty cloud that signified the absence of the flood. At a pan—one of the thousands of shallow pools that fill with water during the flood and sustain a formidable share of the delta's wildlife during the dry season—an array of birds (an Egyptian goose, a spoonbill, four kinds of ibises) sifted through the sandy fringes for prey, and a water turkey stood just beyond the water's edge with wings unfurled, sunning them. In another pan, crocodiles basked, and in numerous others, hippos, submerged to their ears and eye sockets, soaked. At a marsh, Scudder picked up a handful of soil. "It's mainly sand," he said. "You have this wetland of tremendous biodiversity on top of Kalahari sand." He stood on the top of a small mound and gestured, first, a few yards away to a brilliant green pool adorned with lilies— "permanent swamp"; then, to the waist-high reeds in whose midst we stood—"seasonal swamp"; and behind us, to stubbly land—"permanent grassland." Three distinct habitats within a few yards of one another, each overcoming its sandy constitution to produce an effusion of wildlife, each zone determined by the availability of water.

On another day we flew over the delta in a chartered Cessna at three hundred feet, and for the first time I understood the flood. For three-fourths of our flight up the delta from Maun, following the Boro toward the panhandle, the land looked distressingly dry, more gold and brown landscape than green, a substantially tarnished gem. In all the dryness, fires stood out. Smoke twisted up from a couple of blackened meadows, and fields bristled with new growth, evidence of numerous fires in the immediate past. Some of the elephants, buffalo, and impala that we flew over broke into nervous gallops at the sound of the plane. For much of the way, the Boro looked so narrow I could have walked across it. Perpendicular to this trickle, I saw a tiny dam of sticks made for traditional fishing: the impulse to dam is apparently universal. Then, near a safari camp called Mombo, I could see where the water had spilled out of its metaphorical glass and now crept along the floodplain, obliterating the Boro, spawning swamp, more like seepage than wave. From there to the panhandle, the browns and golds of the lower delta gave way to the blues of omnipresent water and the greens of island vegetation, from the faint grasses on the islands' edges to the intensely leaved palms in their center. The flood now was practically galloping, the Australian pilot said. Galloping, of course, is relative; even a smitten couple on a meandering stroll could overtake the flood. Yet the flood is fundamental, inexorable, life-giving. The Okavango exposes the fallacy of speed.

At a breakfast at the Croc Camp bar overlooking the Thamalakane, we had another conversation with Ramberg. He said he knew about ranchers like the one we'd encountered in our search for the Boteti, and he wasn't happy about them. They've drilled boreholes, thereby depleting the aquifer, so everyone's boreholes may dry up in the next drought. During the rainy season, they move their herds across the part of the Kalahari that remains unfenced, and in the dry season retreat behind their fences. "They can use the commons," Ramberg said, "but the commons can't use their land."

Part of the problem, Ramberg continued, is that the European Community subsidizes Botswana beef, thereby encouraging water depletion with artificially high beef prices.* He added that the World Bank had done the same thing in Brazil. World Bank officials, he said, should "have a lot on their conscience, but I don't think they *have* any conscience."

Of all of Scudder's students, Jonathan Habarad, an aspiring anthropologist who spent two years as a researcher in the Gwembe Valley, might

---

*In fact, beef exports to Europe from the Okavango region are severely limited because of problems with foot-and-mouth disease. Most livestock products exported from the area are sold as frozen meat in South Africa and other neighboring countries.

have been the one he was closest to. Then Habarad dropped out of anthropology and sight, and Scudder lost track of him—until this trip, when he heard that Habarad was running a nonprofit in Maun. We tracked Habarad down at his house. Habarad is a couple of inches shorter than Scudder, but his shoulders are much broader. He looked like an ex-high-school-wrestler-turned-reformer-rogue, with a stubbly full beard and modestly protruding gut, which widened the gap between the dangling bottom edge of his untucked sky blue polo shirt (collar turned up) and the waist of his blue jeans. Over the polo, he wore a collared shirt with long sleeves, rolled nearly to the elbow, unbuttoned and hanging loosely but with a couple of pens neatly attached to his breast pocket—he was swank and slovenly at the same time. In the shade, he wore his dark sunglasses over his head, like a tiara. For all that, it was impossible not to be disarmed by his earnestness. He'd started a nonprofit dedicated to defending, supporting, and advising rural people in Botswana—in his words, a nonprofit "dedicated to decentralization of opportunities and rural empowerment." Habarad embraced the goals that Scudder advocated throughout his career but rarely saw reach fruition, and worked directly on attaining them. Stopping the dam might have saved the Okavango, but the delta could still succumb to any number of other menaces. Habarad's nonprofit stood for the kind of development that Scudder considered not just beneficial but necessary to counter these threats; it picked up where the IUCN study left off.

Habarad's nonprofit focused on the indigenous people who lived in and around the delta. To counter the tendency of urban officials and entrepreneurs to impose their wills on rural people, he helped villagers start businesses of their own. To counter the incalculable impact of AIDS, he started therapeutic and educational workshops for AIDS orphans. In two-week sessions, he gathered as many as forty orphans from a single village and encouraged them to talk about their experiences. The tales are commonly grim. In Botswana, most AIDS orphans end up being passed to relatives, who treat them as inferiors, sometimes requiring that they sleep in cattle kraals, sometimes requiring sex from them. Habarad thinks that once the orphans of a village discover the similarities of their stories, they'll band together for support. He found the term he uses to describe this process—"creative kinship"—on the Internet. In fact, I was somewhat disconcerted to learn that he found the entire therapeutic approach he uses on an Internet search. His nonprofit, the People and Nature Trust, is improvisational and experimental, Habarad said. If an approach doesn't work, he drops it.

One of Habarad's most interesting gambits was to take advantage of a provision in the legislation establishing the Moremi Wildlife Reserve. In each of the delta's nine zones, inhabitants were allowed to form a community organization to develop a tourism business. Habarad provided funds for a village of five or six hundred people on the edge of the delta to build a restaurant and a cultural center for passing tourists. When the restaurant opened in June 2001, it was an immediate success. Community trusts got half the profits; the other half was divided among the restaurant's fifteen workers. But this infusion of cash disturbed the village's social order. Upset to find their status weakened, village elders tried wielding their power by blocking the firing of a couple of irresponsible restaurant employees; community members overruled the elders; then the restaurant was burned down, apparently by elders. Eventually, community members banded together to build a new restaurant and established reasonable policies on disciplining employees. "These dynamics are played out in every community, whether it has enterprises or not," Habarad said. In most communities, the elites control local enterprises in secrecy, but here, with Habarad's help, the elite's machinations were exposed, and a locally owned business took root.

But Habarad was working with only one of nine Okavango communities with plans for local involvement in tourism enterprises. "What concerns me," I said, "is that without your kind of oversight, other communities will squander this opportunity." His work was the exception, and so were the private funders who kept it going.

I expected Habarad to explain how his success might prove infectious, but instead he said, "You're right. The system needs to be changed." That didn't sound encouraging.

Habarad's workshop was impressive for the upbeat faces of both orphans and counselors, achieved despite a dearth of funds and supplies. Yet the project's smallness and thriftiness surely were factors in its success: the attention that the orphans received from the camp's counselors was personal, and the skills taught, including traditional ones such as animal tracking and trapping, were appropriate to the delta. Habarad's nonprofit worked because it started at the bottom, with a few precious resources, and closely supervised their use—its scale was human. Huge dam projects were just the opposite, involving huge sums, far-reaching adverse impacts, and decisions made by officials in distant capitals. From there, the people and ecosystems hurt by dams seemed so far away they might as well have been observed through the wrong end of a telescope.

One night we had dinner with Desmond Green, one of Croc Camp's owners, and Yssel du Plessis, a white South African who was captivated by the delta as a boy, thirty years earlier. Back then, he said, "The delta was a jungle, a canopy of trees. Not anymore. It was just one big flower garden, and it's all gone."

Du Plessis was overstating the delta's decline, but he was right about the trend. Scientists like Ramberg lay most of the blame on fences and the long drought, but du Plessis reserved his animus for hunters. It's the safari lodges that start most of the delta's fires, he said, beginning his rant. The fires don't just lay waste to foliage by burning, du Plessis said, but by concentrating animals in safe areas, such as Moremi, where they overgraze, or, in the case of elephants, topple too many trees. Numerous antelope species have disappeared from parts of the delta. And for what? Hunters get the thrill of watching a contraction of a finger muscle cause a six-ton elephant to crumble, at a price of $50,000 per crumble. (For this sum, the safari lodges throw in a few shootings of lesser beasts.) The lodges employ pilots who spot game from above; trackers on the ground use cell phones. Until a moratorium on hunting lions was put into effect, they cost a mere $30,000 a pop. Some hunters lured lions from customary reclusivity by playing audiotaped cries of impalas in distress. They drive up to elephants and shoot them from the car. For souvenirs, they cut elephant ears into the shape of the African continent, and take home elephant penises, nipples, scrotums, and eyelashes.

"I might be an alarmist," du Plessis said, "but I get depressed when I go into the delta."

Scudder stayed up late into his last night in Africa, copying by hand passages from an October 1997 Botswana Department of Water Affairs report that Ramberg had loaned him. The next morning, he showed me its conclusion: it calls on planners to tap both surface and groundwater to meet Maun's water needs. The idea was that when the Thamalakane was flowing, some of its water could be used for municipal consumption, and groundwater could be used when it wasn't flowing. In choosing this path, the department was following the recommendations of the IUCN study so precisely that it had even borrowed the report's phrase, "conjunctive use." The Botswana government never acknowledged that the IUCN study influenced its decision to stop the dam project, so Scudder took this as vindication: "Those were our words, and that was our conclusion!"

I was going on to Victoria Falls, where I'd meet Ben Clark, a former Gwembe Valley researcher, and travel with him to Mazulu; Scudder was flying home. Our last stop before the Maun airport was the Maun Department of Water Affairs, the unit that published the report specifying "conjunctive use." Scudder wanted a hydrological briefing on developments in the five years since the report had been published, but, to my amazement, he was skittish about asking for one. He wanted me to request the briefing, and, if we got one, to pose the questions he wanted answered. When I asked him afterward why he found the briefing so unsettling, he answered that he feared that bad feelings remained in the department because of the IUCN's rejection of its project.

As it turned out, Scudder was neither abused nor even recognized, and our request for a briefing was promptly met. T. G. Badirwang, a department technical officer, unhesitatingly showed us an April 2001 report on the "Maun Groundwater Project Phase 2." Phase 1, the drilling of seven boreholes near Maun, had already been carried out. Phases 2 and 3 would involve the installation of new well fields.

Once we were out the door, I asked, "They're doing exactly what you recommended?"

"Exactly," Scudder said.

9 MAZULU

From Maun, I flew to Victoria Falls in a six-seat Cessna low over the hinterlands of Chobe National Park, where once the Upper Zambezi and Okavango joined. Now Kalahari shrubland stretched to the horizon out both sides of the plane. It betrayed few signs of human habitation, for it is some of the last land in Africa that is still chiefly the domain of lions, elephants, and wildebeests. The world's largest surviving population of elephants migrates across this land between the Chobe and Linyanti rivers and the Okavango Delta. For a while I could see the meandering Chobe River out the left window. It's a Zambezi backwater that flows in both directions, depending on the height of water in the Zambezi.

At the town of Victoria Falls, a Zimbabwean recreational outcropping bereft of tourists since Robert Mugabe's "nationalism" drive intensified, a taxi driver offered to slow down as we crossed a bridge over the Zambezi River so that I could glimpse the magnificent cataract upstream. Alas, all I could see was mist, from which a wispy rainbow seemed to rise: the rainbow is a nearly permanent feature of the falls that in the Tonga view embodies the spirit of the creator. Victoria Falls forms the world's largest sheet of falling water, dropping four hundred feet over its mile-wide rim. The mist created by the pounding collision of water against basalt at Victoria Falls's base raises a mist that usually entirely obscures the cataract at this time of year. At its peak, the mist rises far enough above the upstream river to be visible twenty-five miles away: in recognition of the mist, the Tonga call the falls "the smoke that thunders." On the Zambian side, a salient extends out from the cliffs just downstream from the waterfall, and offers, at least theoretically, a vantage point. I was a drought beneficiary: because of the Zambezi's reduced volume, I could see a section of the falls, silvery in the late-afternoon light, and behind it, the broad expanse of water, so flat and wide that it looked less like a river than a rapidly moving lake flinging itself over a precipice. Upriver, low trees along the distant shore formed the horizon line; above it was a single long, horizontal smear of sun-light—blue at the bottom, then purple, red, gold, yellow, finally bleeding into the slate white sky. Here was one of nature's most awe-inspiring creations, the antithesis of a dam, the epitome of release instead of containment, but now also the upstream counterpart to Kariba Dam, together enclosing the Gwembe Valley in disharmonious juxtaposition.

With his cowboy shirt, jeans, and sparse red-tinged beard that made up in enthusiasm what it lacked in outright hair, Ben Clark, Scudder's former research assistant, looked both swashbuckling and droll. He needed humor to navigate the unmarked and mostly unpaved roads over which we traveled first to the town of Monze, where we spent the night, then to the valley the next morning. Before we left Monze, Clark spent an hour loading his Pajero (a kind of ancestral Mitsubishi SUV) with our supplies, and now they took up most of the car's storage area: among them were our cooking utensils, a tent, and three huge sacks of maize. Each of the three Tonga villagers whom Clark had chosen to help us in Mazulu—Emmy and Richard to translate for us, Delly to act as our watchman—would receive a sack in return for two days of intermittent work. This was not an insignificant payment: one sixty-five-kilo sack of maize could feed Emmy's family of six for more than three weeks.

At a supermarket ("Zambia's own hypermarket, Zambia's pride") in the town of Mazabuka, we bought Dettol, a mild concentrated disinfectant we would add to our bucket of bathwater), biscuits, bread, and yes, bottled water. As we descended into the valley, the temperature rose into the nineties, and conversation turned to the Tonga. In recent years, they rented some Goba land in return for plowing Goba fields, but with the drought, the Goba, the earlier inhabitants of the area, were asking for the return of their land. At the same time, Clark said, the Tonga were abandoning some fields near the Zambezi because hippos and elephants too often trampled them. All this was deepening the hunger of Lusitu's inhabitants.

We stopped at a stand selling baobab fruit, where Clark made a purchase. The fruit are velvety foot-long ovals that Tongas believe are aphrodisiacal, and are therefore the subject of many jokes, but their trees play an important role in the region's ecosystem. Most of a baobab's mass is in its astonishingly thick trunk, which holds water that it draws upon during droughts, while its numerous hollows provide havens for owls and other birds. I found the fruit pasty and slightly sour.

Kariba Dam, where we spent the afternoon, is as utilitarian as an execution chamber. It is massive but entirely plain, verging on Stalinist. Its most striking features are the disturbing black, white, and brown streaks that run down its downstream face from cement joints and sluice gates, and the gray and brown stains that traverse the face like blurred stripes. From the rim, it's impossible to get a sense of the immensity of the lake behind the dam, but the meager flow circling the turbines and making its way farther down the Zambezi suggested a void. After extended discussion with dam authorities, we were granted a tour of the power station on the Zambian side, where four giant turbines convert falling water into 600 megawatts of electricity. In the absence of sufficient Zambian electricity demand, which is to say, since the mid-1970s, the Zambian River Authority has exported some of Kariba's power south to Zimbabwe and South Africa, west to Namibia, and, on occasion, north to the Congo. Tanzania may get electricity from Kariba before Mazulu does, even though Mazulu is only thirty miles downstream: the dam's patrons are far-flung, just not nearby. In the control room, we were shown charts and maps and panels with fluctuating needles and flashing red lights and signs painted bright red that said DANGER, and we were led up and down narrow spiral staircases through the echoing, metallic power station maze. Our tour guides, Zambian River Authority officials, evinced engineers' pride in the dam, partly, I think, out of relief that they'd found employment at a Zambian facility that works. Sort of: with two of its turbines disabled, the power station was running at half its capacity. The officials graciously shouted explanations of the heavy equipment we toured over the din of the functioning turbines, and were largely inaudible. What stayed with me was the intense heat and humidity of the building's bowels. Considering the drought, it may have been the only humid place in Zambia.

It was in the mid-1970s that things began to fall apart in Mazulu. The international market in copper, which earned 80 percent of Zambia's foreign exchange, collapsed, and the price of petroleum imports quadrupled. A war across the Zambezi border, pitting black rebels against a racist white regime in the former Southern Rhodesia (renamed Rhodesia after Northern Rhodesia became Zambia), disrupted trade. Most fundamentally, the government of Kenneth Kaunda, Zambia's only president in its first twenty-seven years, was unelected, mismanaged, and corrupt. For all these reasons, a 1994 World Bank report said, "From being one of the most prosperous countries in Sub-Saharan Africa in the early seventies, Zambia became a country of low economic development, declining incomes and deteriorating social indicators."

One feature of Zambian mismanagement was that the government lavished its favors on industries and cities, while the rural poor were ignored. A working paper for Scudder's 2005 book, *The Future of Large Dams: Dealing with Social, Environmental, Institutional and Political Costs*, cites a report by economist Ann Seidman that claims that even after the mid-1970s, the government possessed a sufficient surplus to foster development in the countryside, where a majority of Zambians live. Instead, the surplus was spent on subsidies and increased salaries for public officials. A 1981 International Labor Organization analysis reported that the subsidies for high-paid officials included "car allowances, entertainment allowances, payment of electricity and water bills, the provision of one or two servants and security guards, and generous subsistence allowances while traveling abroad." The cost was substantial: the ILO analysis said it "exceeds the sum we estimate to be required to meet basic needs in water, health, education and housing over the next five years." Cities and the industrial "Copper Belt" were favored, while rural villages suffered.

Of all of Zambia's regions, the Gwembe Valley felt the effect of these developments most acutely. For one thing, the war's impact was most immediate there; this was not surprising, since the valley shared its border with Rhodesia, the seat of battle. Some guerillas established camps on the Zambian side of the valley, which led Rhodesian military units to mount periodic border raids, blowing up Zambian bridges and mining the valley's main road. As a result, the road was not maintained, and the valley was cut off from government services. Among those services were tsetse fly control measures, which caused a resurgence of animal sleeping sickness and the death of so many cattle that some farmers had to abandon animal-drawn plows for hoes. Mines and Rhodesian ambushes killed civilians, including subjects in both Scudder's and Elizabeth Colson's studies. Even lake fishing, which rewarded Tonga fishermen with increasing catches as the reservoir filled, peaked in 1963, when filling was completed, and then plummeted from four thousand tons to less than a thousand four years later. Though Kariba's planners didn't understand the phenomenon, this boom-and-bust phenomenon is typical of fish populations in new reservoirs. As Scudder explains in *The Future of Large Dams*, the reservoir's inundation releases nutrients from the flooded soil and vegetation. The nutrients foster an explosion in the food fish population, which lasts until predators extend their range and the nutrients dwindle.

At the same time, the river's twenty-eight native fish species faced an invasion of migrating and introduced exotic species, some of which enjoy the crucial advantage of being suited to lakes. Kapenta, a tasty sardine-sized lake fish, was the first foreign species introduced to Lake Kariba, in 1963. A central African research institute funded by the Food and Agriculture Organization and the United Nations Development Program released the kapenta on the Zambian side of the lake, and then the fish seemed to disappear; studies suggested that the introduction "failed." It took a while to determine what really happened. The kapenta abandoned the comparatively nutrient-poor Zambian side of the lake for the Rhodesian side, where the water was enriched with fertilizer runoff from the large commercial farms in the Rhodesian interior. When that fact surfaced, Zambian Parliament members "raged about the loss of Zambian fish," Scudder writes, "with one MP urging that a net divide the waters between the two countries in order to keep Zambian fish where they belonged."

In time, kapenta production flourished again, not just in Lake Kariba but in the downstream reservoir behind Cahora Bassa Dam in Mozambique, to which the fish migrated. By 1985, Kariba fishermen landed twenty thousand tons of kapenta, towering over the Tonga's subsistence fishing volume by a factor of four or five — the trouble was that the Tonga were excluded from the kapenta bounty. For one thing, Zimbabwe fishermen took two-thirds of the catch, as they continued to reap the consequence of agricultural fertilizer. On the Zambian side, the Tonga could not overcome their lack of capital. Kapenta fishing is performed at night by fleets of at least three pontoon boats wielding huge nets; a viable kapenta operation required an initial outlay of $100,000. As a result, the only beneficiaries have been large commercial operators who can afford all the equipment. Scudder cites an estimate that in recent years, 380 kapenta boats have plied Lake Kariba, but Tonga own only a few of them; what Scudder calls "better capitalized immigrants"—both Zambians and white expatriates—own the others. Instead of running the businesses, the Tonga end up as laborers in them: men work on the rigs, and women dry the catch onshore. Given the lack of jobs elsewhere, the lake has become, in Scudder's words, "probably the most important single source of employment outside of village agriculture in the Kariba lake basin today." But the Tonga laborers earn as little as $20 a month, which is insufficient to meet a family's minimum needs. Scudder thinks they make up the difference by smuggling Gwembe Valley marijuana across the river to Zimbabwe and by stealing fresh kapenta from the rigs: a 2003 doctoral dissertation estimates that between 30 and 60 percent of kapenta harvests are lost to theft. Even so, others reaped the preponderance of the lake's wealth, while Tonga laborers merely subsisted.

The Tonga suffered other indignities. Lake tourism should have favored them but didn't—even though they occupied much lakefront land, outsiders bribed their chiefs to take possession of the choicest properties. The most notorious case of this kind involved the most spectacular site, at the Kota Kota peninsula, where the valley's tallest mountain descends into the lake. There, an Italian expatriate used a paramilitary force to drive out local Tonga farmers and fishermen, killing at least one in the process. Then he fenced off the peninsula, stocked it with elephants and other big game, and turned it into a private hunting reserve. Scudder thinks the area should have been designated a national park, where "careful planning as a joint venture" between local Tonga and the National Park Service "could have provided important benefits." Instead, the elephants occasionally swam across the inlet to forage and trample on Tonga agricultural land.

The Tonga received none of the electricity that Kariba generates. And when the government built a foot-diameter pipeline to carry water from Lake Kariba to a mine, it didn't bother to provide offtakes to Tonga villages along the way, never mind that the villages were already suffering serious potable water supply problems, and that the government had promised to provide all resettlers adequate clean water in perpetuity. As if this weren't sufficient insult, the mining operations that the pipeline served further polluted the water supply to some of the villages. Water shortages still cause cholera and dysentery outbreaks in the Lusitu area that includes New Mazulu.

The Tonga might have ridden out all their calamities if the soil on which they were resettled had been fertile, or if their number had been small enough to match the carrying capacity of the resettlement area— they were, after all, farmers, with experience in surviving hard times. But in Lusitu, Scudder estimates that at the time of resettlement, "six thousand people were densely packed into an area that should have no more than two thousand." Lusitu's population has since grown to twenty thousand.

Under the pressure of population, Lusitu farmers abandoned their traditional practice of letting agricultural land lie fallow for several years at a time. Predictably, yields dropped. In the densest areas, erosion stripped off topsoil and shrank fields. Deforestation led to increased flash floods in the rainy season and the drying out of once-perennial streams in the dry season. Scudder writes, "The Lusitu resembled areas in the West African Sahel," where "wind-swept barren land expanded as the years went by."

Evidence of the decline was abundant. Broken-down bicycles hung from trees, yards contained abandoned beds with rusting springs. In his slide show, Scudder showed me a broken check dam, a broken windmill, a bridge that ended halfway across the dry Lusitu riverbed, and a school without desks or chairs or benches or blackboards—"Some of them have no roofs," Scudder said.

The villagers' beer consumption changed so dramatically that Scudder and Colson wrote a book about it: *For Prayer and Profit: The Ritual, Economic, and Social Importance of Beer in Gwembe District, Zambia, 1950–1982*. In the old days, possibly for centuries, the Tonga brewed beer in small quantities for use in rituals and as a means of attracting volunteers to work parties. As *For Prayer and Profit* explains, "Illness was commonly attributed to the anger of ancestors who had not had beer poured in their name," so beer offerings to ancestors were a staple of rituals. Another ritual, marking the full establishment of marriage (several years after betrothal), gives the woman the right to make beer for both her own and her husband's ancestors; "household (symbolized by hearth) and field (symbolized by grain) were brought together in the synthesis of beer, made by the wife and poured out over the doorway of the couple's house by representatives of the husband's and the wife's lineages." Ritual drinking was moderate: senior men did most of it, and men under thirty rarely drank at all. And even though women brewed the beer, few of any age drank more than a pint in the course of a year. Colson and Scudder wrote that even more than food, beer "represented the basic reciprocities of social life. It derived its full meaning from its association with a way of life dependent on agriculture, in which each household had fields and raised its own food; and it had multiplex associations with water, that agricultural necessity, and fire, that essence of human life associated with hearth, food, and the heat of sexuality that leads to new life. It was thus a 'key symbol,' linking almost everything that Gwembe people thought important."

All this changed after resettlement. Young couples inherited no fields, and the fields that existed often failed, so many men lived idle lives. And they could obtain beer more and more easily, as beer manufacturers extended their fragile supply lines all the way to Mazulu. What Colson and Scudder call "enterprising businessmen" imported manufactured beer called *chibuku* into Lusitu in fifty-five-gallon containers, and taverns sprang up to purvey it. Chibuku's urban manufacture contributed to its high prestige: it was considered cleaner and more sophisticated than village beer, and was therefore more popular. It shifted beer profits from the Tonga women, for whom it had been a singularly profitable enterprise since the '50s, to investors in the city and a small number of local male tavern owners. In the '70s, another, stronger drink appeared, called *salopi*: it could be made in the afternoon for an all-day drinking party starting the next morning, by adding sugar and yeast to boiling water. Salopi is a virulent concoction that often induces both vomiting and violence, yet for some resettlers it became the inebriant of choice. Fewer and fewer rituals were enacted, and when they were, neither chibuku nor salopi was used. Even beer ritual was undermined. Before resettlement, men drank only in groups, passing a single cup from hand to hand, reinforcing the act of sharing; after resettlement, many men harbored beer for their private use, and refused to offer it to others. Chibuku and salopi were commodities; their purpose was intoxication unadorned. Women and young men began drinking. For the first time, the problem drinker became a familiar village character. "The majority of men started drinking early in the morning," Scudder said during the slide show, "and they drank all through the day." Some men sold their stock and crops to buy beer. Others stole. Children expropriated and sold their parents' crops. Drunken men beat their wives and children. Colson and Scudder learned of two beaten women hospitalized with broken arms, and many others sported bruises. Instead of spending money on their children, alcoholic mothers bought beer. The ancestral belief system collapsed, and was replaced by nothing, or witchcraft. By 1995, Scudder wrote, suspicions reigned throughout the valley "that any misfortune is caused by the witchcraft of jealous neighbors or, in the case of illness leading to death, the witchcraft of kin who wish to inherit one's wealth." The mortality rate among infants and the elderly jumped, and a drought in 1995, the fourth and most severe rain failure in six years, brought starvation.

Eventually even the beer halls succumbed. Scudder showed me a "before" photo of a tidy beer hall and the "after" photo, when paint and plaster had fallen off the walls and the ceiling vanished. "One of my indicators of when a country is going to pot," Scudder said, "is when the breweries break down." Of course, if not for Kariba, the beer halls might never have been built at all.

Near the end of our visit to the Okavango, Scudder had said, "This is a much better place for you to look at than Kariba, because at Kariba you'd be looking at a disaster." Now Scudder had returned home, and I was hurtling down a road to the disaster. We were still making our way down the switchbacked escarpment road when we caught sight of the reservoir. Disclosed by a gap in the hills ahead of us, it receded into the distance as far as the horizon and looked as massive and featureless as the ocean. At the bottom of our descent was a resort town, Siavonga (SEE-uh-VONG-uh), a creation of the lake. By Zambian standards, Siavonga is posh. We drove by an assortment of resorts and ostentatious homes—one had Greek pretensions, another one a giant satellite dish. The homes had once hugged the reservoir's rim, but now the reservoir was doing the hugging. Deceived by the low reservoir level caused by the mid-1990s drought years, a dozen or so landowners built their homes beneath the reservoir's rim, ignoring warnings that the homes would be inundated when the rains returned—which is precisely what happened. Clark showed me a house with what appeared to be a grandiosely colonnaded carport, which now opened onto Lake Kariba.

The office of the Gwembe Tonga Development Project, a World

Bank–supported effort to alleviate the worst of the disaster the dam visited upon Kariba-resettled people, sits upon a hill overlooking the reservoir in Siavonga, and is therefore not in danger of being inundated. Siavonga was chosen as project headquarters because of its functioning telecommunications facilities and what the project planners called its "good access to the major cities," but the office's location and accoutrements both suggest a level of wealth and comfort unknown to valley resettlers. It's as if the office is turned to face Lusaka and Washington, not Lusitu. Each room has its complement of computers and air conditioners, and recent-vintage four-wheel drives with Zambian power company insignias on the doors are parked outside. Even so, after a year and a half of labor, its accomplishments were scant.

The project is Scudder's baby. "I created it," he told me, and in a way, he did. The failure of the Lusitu resettlement dismayed Scudder, and in 1995 he saw a chance to do something about it. He was on one of his Gwembe Valley visits when he heard that a mission of World Bank officials had arrived in Zambia to discuss a planned $215 million project to modernize the Zambian power industry. Scudder knew the head of the mission, Donal O'Leary, an Irishman whom he'd befriended when O'Leary was the World Bank's deputy task manager for the Lesotho Highlands Water Project. Scudder invited O'Leary to see how bad the Tonga resettlement was. On this point, the World Bank remained in official denial. As late as 1996, a Bank document said of Zambian resettlers, "There is evidence that there was an improvement in the standard of living." Scudder thinks the "improvement" happened in the 1960s; the Bank apparently hadn't done any surveys since then that would have documented the Tongas' decline. O'Leary, a Bank senior power engineer, responded to Scudder's offer by sending a team including his environmental and sociological consultants. According to Scudder, the consultants were appalled. Scudder asked the environmental consultant how long it would take to rehabilitate Lusitu's gullied, barren land, and the consultant answered, "Forty years—and that's if there are no people living in it."

The consultants' report got O'Leary's attention. Soon afterward, he and the chief executive of Zambia's power company, the Zambia Electricity Supply Corporation, showed up unannounced in the valley. They'd driven a Land Cruiser from Lusaka that morning, and would drive it back that night—in between, Scudder gave them a tour. At the end of it, O'Leary asked Scudder to make recommendations for how to rehabilitate the valley, and those recommendations formed the foundation of the rehabilitation project. Scudder proposed rebuilding the mined Bottom Road that connected the three Gwembe districts to one another and facilitated commerce and travel in and out of the Tonga region; providing water, electricity, and health care to the villages; promoting appropriate land use by teaching the Tonga how to grow different crops, improve grazing, develop small-scale irrigation, and store grains; and giving forecasts of reservoir water levels to the Tonga so that they could at least plant crops on the reservoir's shores as the level dropped—that wouldn't even cost anything, and it would instantly improve the Tongas' living standard. But, disturbingly, instead of administering the program directly, the Bank tacked the project onto an existing plan to rehabilitate the Zambian electricity industry, and it passed administration of the plan's most important component, rebuilding the Bottom Road, to the Development Bank of Southern Africa.

Scudder said he was offered a chance to give a speech at the launch of the project in Siavonga in December 1998, but by then he was limiting his travel. Instead, at Scudder's suggestion, the speech was given by Mwindaace N. Siamwiza, a Gwembe Tonga resettler who became a chemist at the University of Zambia and, in Scudder's view, the nation's leading scientist. Siamwiza said he'd grown up in a now-inundated village where he looked after cattle, goats, and sheep; played and hunted; made clay toys of favored cattle; and swam in the Zambezi, all "in synergy and symbiosis with the land."

When we were told that we should abandon our ancestral lands and move to lands we knew were barren hills, we were incredulous and contemptuous. First came bewilderment followed by anger which gave way to utter despair. It is not easy to describe this despair but sufficient to say that my paternal grandmother Buyuni wondered aloud how she could die in peace knowing fully well that she would be buried in the strange and hilly land of Mulungwa where our village was to be relocated. With our culture, in which, on death, one becomes a citizen of the interconnected and interdependent worlds of both the living and the dead; and the strong belief that on death one joins the ancestors who protected you in life while remaining, unseparated, an integral part of the living, relocation was like walking out on one's spiritual protectors; it was insanity personified. My grandmother died in August 1958, just before my village moved. Buyuni now lies tens of meters below the surface of the waters of Lake Kariba; certainly most content, I hope, because she believed that lying beside her ancestors, her soul would not be restless, condemned to eternal wandering. As for me, I have yet to fully come to terms with the uprooting because I have yet to accept that Mulungwa is my home, in spite of the forty years that have gone by!

Siamwiza said he'd managed to get an education because he had a firm cultural foundation, something the generation after him lacked—its members were neither of the old society nor of the "modern" one. Women were particularly hard hit because they were routinely denied education.

Siamwiza welcomed the rehabilitation project, but he called for going much further, by preventing developers from completing their land grab around the lake, giving the Tonga land rights that resettlement deprived them of, and building a girls' secondary school. He saved mention of Colson and Scudder for his last paragraph. "We owe them a debt that cannot be repaid," he said. "I would like to propose that we honor them as outstanding Gwembe Tonga, whose contributions to the understanding and knowledge of the Gwembe Tonga are unsurpassed."

For all that, the project bogged down. Starting in 1996, it took six months to define the terms of reference for a social survey in the resettled region. The survey itself took a year. The review process at the World Bank and the Zambia Electricity Supply Corporation, the plan's administrator, took another six months. Recruitment of the project's staff required more than a year. Funding arrived late, if at all: South Africa, which pledged $4.5 million of the project's $5 million "seed money," delayed its contribution as its own economy withered. Five months after the launch, the project's centerpiece, the reconstruction of the Bottom Road, was derailed when a land mine killed a Norwegian consultant. A road-building project became a mine-clearing one, and money for mine clearing was not readily available. As late as November 2002, six months after my trip and three years after the project's launch, about thirty miles of the road remained to be cleared of mines. Engineering designs for more than four hundred miles of road were nearly completed, but, thanks to lax oversight by the Development Bank of Southern Africa, they were ludicrously overambitious: they planned for an all-weather gravel road, which would have cost more than $100 million, when the project's total budget was $12.6 million. At about the same time, the South African rand was devalued, so that no more funds were available for road building. A limited electrification program started in 1999; most other facets of the project didn't start until mid- or late 2001. The project included plans to build nine small dams, to provide irrigation infrastructure, to teach agricultural techniques, and to start an HIV awareness program, but as of November 2002, only borehole drilling was near completion, and many facets of the project were pared back. A World Bank decision in February 2003 to extend the project for two years gave Scudder hope that its major goals would be realized, but as of November 2004, the Bottom Road remained uncompleted. At least for the time being, the project touted as an effort to redress the injustices dealt resettled Tonga became one more disappointment to them.

Scudder originally hoped for a binational project, which would help Gwembe Tonga in both Zambia and Zimbabwe and would show the Bank's willingness to repair its early dam debacles, but that idea was abandoned when the Bank opted to limit its support to Zambia. In a working paper for *The Future of Large Dams*, Scudder berates the World Bank for "losing a major opportunity to show its commitment to environment and resettlement issues by addressing serious defects relating to its first major dam project." Yet the Bank's involvement meant that major jobs in the project had to be internationally bid. That dictated huge salaries, which ate a significant chunk of the project budget and inevitably were meted out to political appointees. "They haven't even hired any Gwembe Tonga," Scudder told me. When Scudder visited Siavonga, he questioned a project liaison officer on her knowledge of key Tonga figures, then declared her "a complete flop." When Clark and I arrived at the project office, the manager was absent, so we were introduced to Chrispin Nchobezyi, a public relations officer. Nchobezyi was enthusiastic and largely uninformative, declining even to tell us the project's revised cost. He apparently forgot that a pamphlet he gave us cited the project's $12 million budget. On the other hand, he knew all about

Scudder: he called him "the father of the Gwembe Tonga people." Stationed behind his Toshiba laptop, Nchobezyi told us sweetly that by the end of the project, the resettlers would be on a path to sustainability. His blithe use of "sustainability" seemed to hasten the word's slide into meaninglessness.

After the interview, Clark and I had lunch at a resort overlooking the reservoir. Aside from us, the place was empty. It took a few minutes to get the waiter's attention because he was in the next room, playing pool. Clark had spent six months in Mazulu as a research assistant, and he'd acted as the operational link between Scudder and Tonga villagers for another three years—his experience, therefore, was not as extensive as Scudder's, but fresher. Neither Nchobezyi's briefing nor the project had impressed him. Under project auspices, Tonga villagers were given fertilizer and seed, with which they were expected to grow sustainable crops, but the drought nullified the gift. "So how are they supposed to do it for themselves next year?" Clark asked. Plus, in the absence of a functioning Bottom Road, "how are they supposed to get their products to market?

"How are the boreholes being placed? Will they be close to arable land? What arable land—is there any? And is it owned by commercial interests? And how will the community maintain the boreholes? The Tonga might face the choice of walking one kilometer to the clean water at the borehole, or walking thirty meters to some dirty, muddy water. Which are they going to do? Particularly if the water is being carried by a twelve-year-old girl, she's going to get the dirty water from the mudhole so that she can play." Before the dam, of course, the villagers didn't need a Bottom Road as acutely as now, for they were largely self-sufficient, and they drew water from the river. The dam created need where before none existed, and then proposed development as the answer to the need. The dam forced development officials to consider a myriad of issues for which they lacked sufficient knowledge, never mind will, to solve. Even their victories were mysterious. "When I've seen a development scheme succeed, it wasn't what the donors wanted," Clark said. "You keep simplifying and simplifying until you find a project that can work, and then the donors wonder if this is still development."

The conversation returned to Mazulu, where we'd be spending the next two nights. Clark said, "There are basically three things to do in the village at night—go to bed, get drunk, or have sex." If the villagers were

exercising option two during our stay—and that was likely, Clark said—
we'd probably get little sleep.

Darkness falls quickly over tropical Africa; even twilight is meted out
sparingly. By the time we reached Mazulu, around seven, it was already
dark. Mazulu is at the end of a long, dusty red dirt road, so rutted that the
Pajero constantly bucked and lurched. Clark knew Richard and Emmy
well, for they were Scudder's two male village informants. Using the
English they'd learned in Zambian schools, they chronicled developments in the village and sent the accounts on to Scudder. For the last
three years, their intermediary was Clark, who passed the accounts in
one direction and cash payments in the other, and in the process remained as entangled in village life as when he'd conducted research
there. We still had to set up camp, so we didn't have much time that
night. I could barely make out Richard and Emmy as they gave us the
village headlines. The drought was terrible, they said. No one was working in the fields because there was nothing to do. A couple of weeks ago,
a hippo trampled the fields around the village.

The collapse of Lusitu society has coincided with the rise of witchcraft. Clark knew that in this era of scarcity, villagers who experience
good fortune become targets of jealousy. A windfall as commonplace as
a sack of maize might inspire accusations that the recipient is possessed,
which can be used as a pretext for murdering him. Clark chose to pay
the men in maize because it is nutritious and, unlike money, not easily
converted into beer, but it is also more conspicuous than money. We
therefore gave the three men their maize now, in the dark.

We camped in a vacant field and slept well—what drumming we heard
was barely audible. The next morning I saw that we'd set up in the
shadow of a magnificent baobab tree, leafless but multilimbed, each
branch seeming to end in a squiggle of switches. Near it, I conducted my
first interviews, with Richard and Emmy themselves. Richard seemed
precisely the sort of man Siamwiza alluded to in his speech, "not of the
old order but not fully assimilated into the modern way of life," for he'd
had the double misfortune of being born in 1958. He has no memory of
Old Mazulu, the place that has ascended in Tonga memory to the status

of Eden, where he'd lived for the first few months of his life, yet he has lived through the resettlement and all its aftermath. In these circumstances, alcoholism was not an entirely irrational response. For all that, he had served more years as Scudder's research assistant than any other villager.

Richard still possessed a sense of style. He wore a not quite opaque long-sleeved shirt with a fancy geometric print on its chest; its cuffs were tucked beneath the sleeves at midforearm. He wore pleated pants, and, of all things, wingtip shoes. He was slight and angular, except for his smooth oval face. The hair on his head was clipped short, and it was even shorter in a near circle enclosing his chin and lips, where the suggestion of a goatee appeared. His eyes were bloodshot, hidden behind what looked like layers of eyelids. He was black haired except where a tiny vertical white salient had found purchase at the forward edge of his hairline and pointed like a dagger down his forehead to the only age-altered portion of his face, his thoroughly furrowed brow.

It wasn't until I took in Emmy that I realized that for this brief stint of employment, both men were wearing their best clothes. Even more than Richard, Emmy displayed a fastidiousness that defied his surroundings. On the top of his head, he allowed the hint of curly black stubble to form a perfect oval. He wore a black button-down shirt and green denim pants that bore the label "Killer" and the slogan "For Those who are going places." When he sat down, I could see on his socks the word "LOVE," spelled out in inch-high letters encircling his lower calves. His manner was dignified and composed.

Richard and Emmy had only bad news. The crop that should have come in a month earlier was a 100 percent failure. There were no jobs. Everyone was hungry. Some village women walked ten miles to the Zambezi with five-gallon plastic containers that they filled in the river, then carried another half mile so they could sell the water in a market, and then walked home. Elephants killed four villagers in the last two years and have "destroyed almost the whole area." People had heard of AIDS, but no one knew who had it—those carrying the virus rarely admitted it. Animal sleeping sickness was making a comeback: in fact, Emmy had lost six of his twenty-two cows in the last year. Every couple of months or so, someone got "witched"—that is, anonymously murdered, often with poison, on grounds that the victim was possessed. Indeed, just a month earlier, Richard's wife died suddenly, apparently witched. Richard nodded slightly at the mention of this event: his response betrayed no regret, no sorrow, no emotion at all. Emmy, on the contrary, laughed uneasily when he talked about witchcraft. He said some people had left Mazulu for the plateau or Lusaka, but plateau land required fertilizer, which no one could afford, and there were no jobs in Lusaka. Many people came back.

The Tonga seemed to understand little of the politics that led to their debacle. Outsiders had turned their lives upside down, so now they overestimated outsiders' power. In the world as they conceived it, I discovered, Scudder was a hugely important figure, but their curiosity rarely extended beyond Scudder. When I asked what the villagers thought of him, I was surprised to learn that he'd been expected instead of me—to some people, my arrival was a distinct disappointment. "People want to see Scudder face-to-face," Richard said. "They think he's the person who surveyed the dam and decided where it would be."

"Scudder is acting as an intermediary," Emmy clarified.

Delly, the camp guard, joined in. Before the dam was built, he said, Scudder showed their chief this resettlement area—which seemed to suggest that Delly held Scudder responsible for it. I asked if that was so.

"The villagers just want to ask him questions," Emmy said. "They think he can take information where it has to be taken."

Richard said, "Scudder knows all the information about 1956 to 1958, so they are blaming him and asking how he can put them in this area."

Were the villagers angry?

"No, they just want to ask him questions," Richard said. "They want to ask how we came here. People say it's better to destroy the dam and go back."

That morning I met the oldest man in the village, a man who didn't know how old he was, just that when he moved to New Mazulu he was already old. His name was Nine Sikalimbo; Scudder Sikalimbo is his son. His right hand was gnarled, and he had trouble walking. He wore a faded baseball cap low on his head, and during most of our conversation he sat with his back to the sun—all the shadows made it harder to see his glassy eyes and the chasm beneath each cheekbone. He wore a blue-and-white-striped cotton shirt that once might have looked festive but now was thoroughly stained and muddy, and was missing half its buttons. He was barefoot.

Mr. Sikalimbo did not like New Mazulu. In the "reign" of President Kenneth Kaunda, he said, life was "a bit fair," but under Frederick Chiluba, Kaunda's successor, it got much worse. At least Kaunda allowed the killing of hippos and elephants when they destroyed crops, he said, but Chiluba took the game officials' guns away. (The idea that environmentalists pressed for the protection of elephants and hippos in the interests of preserving their species had clearly not penetrated Mazulu. If it had, it would have landed on the Tonga with the same disheartening impact as Operation Noah, alerting them that certain animals' survival was valued more than their own.) The man said crime had increased, as had infidelity, until the fear of AIDS knocked it back down. In Old Mazulu, people had enough food to last the year; here, they were hungry more years than not. "I have no hope," he said. "Things are always getting worse and worse."

Clark had warned me that many villagers would ask for money, but under no circumstances should I give it to them, as most would spend it on beer. The old man was unusual in that he didn't bother to conceal the fact that he wanted beer money. "I am always drinking," he said. "It's the same as food. I drink in the rainy season and I drink in the dry season. Drinking is just life—it's to enjoy life."

I thanked him for his time but said I couldn't give him money.

"If Scudder had come," the old man answered, "he would have given us money for beer."

Later that morning, we talked to a veteran of "the war"—the confrontation in 1958 that left eight Tonga dead and shattered the feeble resistance to the dam. The war lasted two hours, the man said. He and some other villagers were forced to remove the clothes from the bodies of the dead Tonga, then display them in the home villages of the slain, to reinforce the lesson that resistance to the dam resulted in death. Under order, he helped bury the dead in their nakedness and washed away the blood from the vehicles that carried the bodies to their single grave. After that, he said, the government moved the villagers to New Mazulu in trucks, even though some of their crops in Old Mazulu were already two or three inches high, even though some crops were ripe! All they brought with them were mortar and pestle stones, some clothes, some pots and pans—they lacked even poles for putting up houses and had no tools to prepare the infertile land.

Forty-five years later, we were sitting outside his brick-and-wattle rondavel in rickety wooden chairs, as if all those years were brackets enclosing destitution. He never stood up, but he sat with a melancholic dignity that was reinforced by his full head of short gray hair and his full beard. He, too, was wearing a pin-striped suit jacket, and beneath it, an off-yellow sweatshirt that said PURDUE UNIVERSITY. This wasn't the man's only set of clothes, but it was close: a few yards away, two pairs of hopelessly tattered trousers hung on a strand of thatch.

A few years earlier, the old man had possessed twenty cattle—a substantial number in Mazulu—but thieves took all but four of them. He had goats, too, but they died of disease. Over the next few months, he'd have to sell his last four cows so that he and his four wives and ten children could eat. Once they were sold, he said, "death follows."

His son, who wore a polo shirt, shorts, and thongs, joined the conversation. He spoke enough English to make his point. "There is nothing to do," he said. "Just sitting, listening to the growls of my stomach."

A few minutes later, his father echoed him. "There is nothing to do. Just sit and wait. Maybe die. It's soon a matter of waiting for death. That's all."

We had lunch at Emmy's compound. By now it was past one, and hot, and I was glad for a break. Emmy, Richard, Clark, and I found shade beneath a spreading tree whose bark is used to treat malaria. Across the way we saw, but did not talk to, one of Emmy's two wives, who was cooking the meal. She was dramatically tall and statuesque—enough of a wife, I would have thought, to dampen the wish for more. She was cooking in two rooms at once, constantly running between them, while looking after a small boy. After half an hour, the meal was ready, and we men dipped our hands in a bowl of warm water. Then, with our hands, we ate fried kapenta, greens, and groundnuts, and *nsima*, the boiled cornmeal paste that is a staple throughout southern Africa. It was a pleasant meal. Emmy looked like a survivor, capable of handling the hard months that were coming. The conversation eventually came around to his wives, who, he said, sometimes argue. "I say, 'No, this is bad,' because they are fighting for me."

After lunch, we talked with three women considerably less well-off than Emmy. One said she sometimes went three days without eating. The women were all in their thirties, and had collectively given birth to nineteen children, of whom thirteen were still alive; one woman, Renet Siasabe, had lost five of her six children. Molly Mazulu, named for Scudder's wife, said she hadn't eaten since yesterday morning. Her two-year-old suckled her as she said, "I really believe children will die." I tried to change the subject for a moment, asking about the difficulties of being a woman in Mazulu, but only one subject interested them. "All these years we have had no difficulties but the drought."

At the end of the conversation, they asked for money. When I said I had none to give them, one of the women felt my pockets to be sure.

We talked to a man who killed his stepson out of jealousy and served a two-year prison sentence. We talked to the headman of a neighboring village who by Lusitu standards flourished: for a while he farmed on the plateau, until a spreading tickborne disease there caused him to move his cattle back to Lusitu. He considered himself fortunate to have lost only two of his twenty cattle, and his house displayed confirmation—on the roof was an antenna connected to a television, which was powered by a car battery.

We drove across the scoured land to the bank of the Zambezi and looked across it to Zimbabwe. A national park lines the Zimbabwe shore, so the trees, which are native and protected, still thrive there. They blanketed the shore and the interior without a gap, while on our side, trees were the exception, barren soil the rule. "This side once looked like that," Clark said.

The next morning, just as we were sitting down with Heavy, the Mazulu village chief, someone told him that a man married to a woman from Mazulu had just hanged himself in a nearby village. The news did not appear to startle Heavy, as if he'd already heard too much news like it. "From my observation," he said in serviceable English, "there is nothing good about being a headman." Though headmen usually inherit their positions, in Lusitu the assumption of office failed to empower them; instead, the office holders seemed to shrink. Heavy's job, as he explained it, consisted of welcoming visitors and finding them food; meting out just punishment for villagers found guilty of theft; and handling quarrels, particularly drunken quarrels. If a husband and wife were arguing, he'd sit with them together to help them talk out their differences; if one villager accused another of witchcraft, he might send them both to a witch doctor. But Heavy could do nothing about hunger, he said, and hunger was causing more and more arguments.

Conditions were already terrible, he said, and the animals hadn't started dying yet. That would happen in August, still three months away, and a new crop wouldn't come in for six or seven months after that. For several years, some villagers had earned modest sums of cash by growing vegetables in small garden plots near the once perennial, now seasonal Lusitu River, but this year so many people were growing vegetables that they would glut the market, driving prices catastrophically down.

"If the water were let out of the reservoir," Heavy said, "people would go back" to Old Mazulu. "Everyone hates this place." I did not try to tell him how fanciful the wish for a drained Lake Kariba was. He probably knew, anyway.

Of all Mazulans, Jeffra took the deepest fall. He was Mazulu's first university student, first university graduate, first native to gain employment as a scientist. He was also the most proficient English speaker in the village, who spoke to me in slow, unbroken grammatical sentences, without an interpreter. Yet his home, which he'd built while living with his five children under a thatched roof held up with poles, was insubstantial, nearly nonexistent. He was gaunt. His eyes looked glazed. His short-sleeved dress shirt was missing some of its buttons, and the thong on his left foot was held together with what looked like a rubber band.

When he left Mazulu, he became a chemist at the Maamba Coal Mine, the only coal mine in Zambia. He even went to Japan for three months to study "core science and technology." As a company marketer, he traveled throughout Zambia and in Zimbabwe. "At Maamba, my house was clean," he said. "I watched TV. I played CDs and cassettes. My house was air-conditioned."

Then something happened. Jeffra said his company's "operations" weren't good, so he resigned in 1996. He tried to find another job but grew impatient and returned to Mazulu. There, he found that only three of the seventeen cattle he'd entrusted to his brother were left—his brother had sold the rest and spent the proceeds. Jeffra tried raising chickens in Lusitu and selling them in Lusaka, and he tried a similar tactic with fish, but the businesses failed. Then his health deteriorated, and his brother died of a mysterious disease, perhaps AIDS. His wife, the daughter of the district chief, left him.

Now he was trying to grow maize. After all his entrepreneurial failures, he might still have succeeded, if not for the deficiencies of his resettlement surroundings. This year he lost some of his crop to birds, hippos, and elephants, and the drought took care of most of the rest. He thought of going back to Lusaka to look for a job, but jobs there were almost impossible to find, and he couldn't leave his children. He sounded bewildered. "I have gone to developed nations, so I'm able to compare the kind of lives people lead there with the life I lead here," he said. "I see myself here, and I think to myself, I'm not supposed to be working like this."

We were sitting on two solid wooden chairs with upholstered padding, the only noteworthy objects in Jeffra's compound. They were the last surviving relics of his former life. Pointing to them, he said, "I still have good chairs, but the living conditions are no good. I lived in the country's best hotels. I used to cook on a stove. All that comfort is gone."

By now, I'd formed an impression of Scudder's frugality, which congealed around a comment of his: "I hate spending money, even though we're not hard up." My ears therefore perked up when Jeffra mentioned that Scudder had used his own money to promote development and education in Mazulu. In conversations with Richard and Jeffra, I learned that Scudder had put up $1,500 so that five of his former and current native research assistants could start a business together—the choice of business was up to them. For a couple of years the assistants made a go of growing and delivering vegetables to schools and hospitals, but then the Zambian government stopped paying for the service, and the business foundered. In the end, the research assistant in charge of managing the money put the last $500 into an extension on his house. Undeterred, Scudder gave two village headmen money to start a business, but all they did was hand out the money to friends. That bred resentment among villagers who didn't receive handouts, and the ones who did used most of the money to buy beer. Scudder still didn't give up. He paid to send

Mazulu students to secondary school on the plateau, but the quality of the schools declined and the graduates failed to get jobs. That caused Scudder to switch to supporting college-level students only: in 2002, he provided more than $2,000 as partial tuition payments for five college students.

When I asked Scudder if his gifts were bad ideas, he said no. "They were experiments. They provided data, very interesting data. I'm glad I did every one of them because I can analyze why they didn't work. And then I can bring that analysis to bear on other projects that have failed." Besides, he said, not all the secondary students were failures—in fact, he was planning to ask one of them to do some research work for him. He had a new plan involving two brothers. The older one would get $500 toward enrollment at the University of Zambia on condition that he tried to persuade the younger one to become a teacher. "Now, will it happen?" Scudder asked. "It's like [development plans for] dams—will they be implemented?"

Clark thought the problem with Scudder's projects was more fundamental than a failure of implementation. The subject riled him, and his customary smile was gone. "Scudder's charity is of the very worst kind," he said. "You can't hand the Tonga money without administering it. He's too willing to give people money without giving them an idea of what they need to do with it. Then he gets discouraged when it doesn't work."

For a man whose career was founded on an understanding of the Tonga, Scudder's philanthropy sounded naïve, as if he misread the Tonga on a personal level. He once told me, "I think I understand what makes societies tick, but I don't understand what makes individuals tick. I don't understand what makes me tick, what makes my daughters tick, what makes you tick." Scudder seems to function best in the middle distance, where he can take the social scientist's wide view; at a closer level, he's less prescient. His altruism is unquestionable, but it is sometimes tainted with self-interest and sabotaged by a disregard for inconvenient consequences. To that extent, his gifts resemble the World Bank loans he criticizes.

Throughout my stay in Mazulu, villagers mistook me for Scudder. While walking through the village, I was stopped by a man of infinite politeness and weariness, who wore, of all things, a tie. He handed me a carefully torn scrap of lined school paper, on which he'd written a note:

3-5-2002

Mr. Scudder,

I am hereby, on behaffe of the Voice of resettled people of Chipepo, making any appointment on Saturday or Sundy 4th and 5th May respectively, to meet you and have discussion ~~facing the~~ affecting the resettled people of Chipepo; Kindly accommodate our request.

By L. Syantumbu,
Chairman

The "L," it turned out, stands for Little, and he looked not just small, but constricted. Like almost everybody else in Mazulu, he was too thin. His mustache looked like a pasted-on smile, a determination to be optimistic. He wore a proper white button-down long-sleeved shirt, and his tie was perfectly knotted; his disarray showed only beneath the knee, with the mud caked on a trouser cuff, the turquoise-and-white-striped socks nearly gathered around his ankles, one shoe held together with tape.

It was a saddening letter, and not just because Mr. Syantumbu had mistaken me for a much older man. (Hadn't he *met* Scudder?) It took me a while to convince him that I not only wasn't Scudder, but had no connection to the World Bank or any other agency. The news clearly disappointed him. Out of consolation as much as curiosity, I told him that in my capacity as a writer, I'd be glad to listen to whatever he wanted to say. I take it as a sign of Mr. Syantumbu's desperation that he accepted immediately.

We met the next morning at Emmy's compound, beneath the spreading tree. Eight men representing a group called the Voice of the Resettled People of Chipepo, plus Emmy, Richard, Delly, and I sat in a circle on rudimentary low-slung wooden chairs, some stools, a log, a rock. A passel of children in brightly printed but torn and dust-dimmed clothes watched the proceedings from across the compound; they were huddled together with a density rarely seen in America, six out of seven of them virtually or actually touching, occupying no more than two or three square yards, all standing, sitting, or kneeling in the shade of a thatched house. No women were in sight.

Three of the men did the talking, one by one, in numbered points, until I realized I was experiencing a political harangue, delivered to the wrong audience. Mr. Syantumbu recited the ill effects of the Kariba Dam, including the two-hour war in 1958. Sibbuyu Abeshy, the organization's secretary and author of its newsletter, pronounced the land "not fit for human habitation." Abeshy, in his twenties and probably the youngest of the men, looked the most up-to-date, with his shaved head, crisp white T-shirt, and pants displaying a stylized American flag design down each leg—yet he was the angriest and most bombastic. "We the victimized people . . ." was how he began one declamation. "That dam was once our sweet home," he said, "but now we are brought to this dry land, which was once the home of ornery animals." The resettlement was forty-four years old, much older in fact than Mr. Abeshy, yet in his telling it seemed to have just happened.

In its rhetorical structure, the Voice of the Resettled People's harangue reminded me of a few North Vietnamese press conferences I'd covered during the Vietnam War. I finally grew tired of it, and asked the Voice leaders what they wanted from the government and the Bank. Considering the inflated language, their demands were disarmingly modest: repair of the Bottom Road, compensation, electricity. "At least there could be one light pole, with one bulb," young Mr. Abeshy said, with what I took as a dollop of sarcasm.

Before I left, Mr. Syantumbu asked me to pass on the group's message to Scudder. I should tell Scudder, he said, that the situation was unbearable.

10 ALTADENA

Kariba and Lesotho enclosed Scudder's career, and the Okavango was his most significant achievement. Of the three projects, the two completed ones were disastrous and disappointing, respectively, and the unbuilt one was fortunately thwarted. If Scudder was a man in the middle, their examples all pointed toward the same conclusion. As I prepared to visit Scudder one last time, it seemed to me that his career could be read as a struggle to reconcile his belief in dams with the accumulating evidence of their failures. As long as the idea of the good dam existed, his edifice of ideas might teeter but it wouldn't collapse: the good dam reassured him that his career had value.

I had a list of questions for Scudder, and he'd invited me to pose them at his house, then stay overnight and hike with him the next morning. The house is nestled halfway up a canyon ridge in the San Gabriel Mountains, in the town of Altadena, overlooking Pasadena. From the road, the house isn't visible: the front fence is overrun with ivy, and trees overrun much else. Scudder greeted me in his work boots, shorts, and a Bokong Nature Reserve T-shirt. A fire inspector was expected imminently, and Scudder was raking leaves; some of them were in his hair.

We ate lunch in the backyard, under a spreading oak tree, overlooking the canyon. The canyon is still undeveloped, and its aridity and sparse vegetation remind Scudder of southern Africa. Still feeling for Scudder's definition of a good dam, I asked him to give a brief assessment of all the dams he'd studied, and we ran through the list one by one.

Three Gorges, the Chinese dam that will be the world's largest when it is finished around 2009: "Hmmph. Bad project. One, there's no way that the Chinese can resettle adequately one-point-three million people. Two, they have not looked at the impact of it on downstream ecosystems or millions of people who presumably are dependent on the natural flood regime. Three, leading Chinese scientists believe that the navigation component won't work because of environmental degradation and the siltation of the upper portion of the reservoir. Four, the figure I recall is that it only controls about 15 percent of the monsoon flood, and therefore it may not be that much of a flood-control mechanism, yet it will give security to downstream people, which may not be valid. And five, the amount of money which goes into it is not available for better projects."

Longtan, another huge Chinese dam: "That would have been a great project. The Chinese are going ahead without the World Bank and therefore without a panel of experts—oh, it's too bad. It would have been even better than Nam Theun 2 if it had been designed and implemented the way we were talking about when I went there in 1994. The reservoir looks like an African spider, with a central body and legs going out all over the place. That means you have a tremendous area for drawdown cultivation and grazing. By Chinese standards, the people to be resettled are few—only seventy-five thousand. It is a high dam on a tributary, which means that it can be used for controlled releases to a series of downstream dams, so it has the potential for increasing the effectiveness of all the downstream dams. It is a great project, a wonderful project. God, I'd love to be involved in that!"

Grande-Baleine, a component of the vast James Bay Project in Canada: "Bad project, shouldn't be built! And we said so. Hydro-Québec was screwing up the whole entire Cree environment, so that was a bad project!" The project was eventually scrapped.

Hidrovia, a waterway on the Paraná River in Paraguay: "A terrible project, just terrible. This is going to be a huge waterway like the Mississippi, which is going to serve Uruguay, Argentina, Brazil, Paraguay, and Bolivia. And why is it being done? To export soybeans at the expense of small farmers. It will put tens of thousands of small farmers out of business because large farmers will come in, and it will ruin a magnificent river system by turning it into a channelized waterway. All the studies of Hydrovia emphasized either the environmental impact of the project on the river system or the ecology of the indigenous people along the banks. Nobody looked at the impact on the [nonindigenous] poor, who are small farmers. We researched it, and came to the conclusion that it would be a disaster for the small farmers who are the majority of the rural people. And that was the first time this issue had come up, even though people have been criticizing that project on the environmental side for over five years."

Kainji Dam, Nigeria: "Bad project, and it has contributed in a terrible way to flooding downstream. Since it was built for hydropower, the Niger Dam Authority doesn't like to lower the level of the reservoir before the rainy season. So a big flood comes from upstream and they release it and kill people downstream."

And so on—we must have discussed a couple of dozen dams. Inevitably, the path of our conversation led to Nam Theun 2, the dam in Laos whose panel of experts Scudder continues to serve on. If a compelling case can be made for any hydroelectric dam project, then Nam Theun 2 may be the best bet to make it. Indeed, when dam builders complained that no dam project could reasonably carry out all twenty-six of the World Commission on Dams's recommendations on dam building, commission officials cited Nam Theun 2 as a project that has complied with all of them. (The International Rivers Network disagrees, saying that Nam Theun 2 "violates six out of seven" of the commission's priorities.) Like Lesotho, Laos is a poor, mountainous, landlocked country with a potential bounty in high-elevation water: Laos contains twenty tributaries that flow into the Mekong River and together contribute two-fifths of the Mekong's total flow. Because of Laos's need for foreign exchange, the likely alternative to the development of its hydroelectric potential is the dismantling of its forests for export timber. For this reason, Scudder hopes hydropower will become Laos's predominant source of foreign exchange. The Laotian government, a nominally Communist regime with dwindling ties to Marxist ideology, apparently agrees. It has signed about twenty memoranda of understanding with assorted governments and companies that, if implemented, will transform it into Southeast Asia's hydroelectric "battery." Scudder worries that Laos has jumped into the business too hastily, for the agreements call for hydroelectric dams on most Laotian tributaries, when "a more sustainable strategy would be to build a cascade of dams on a smaller number of tributaries," thereby impeding the natural flow of fewer of them and allowing more sediment to reach the Mekong.

The most significant of the agreements Laos has signed requires it to supply 3,300 megawatts of power to Thailand by 2008. Nam Theun 2, the largest planned development project in the history of Laos, will alone deliver 920 megawatts to Thailand and another 75 megawatts to the Laotian grid. (Thailand's megalopolis, Bangkok, alone consumed about 8,000 megawatts in 2003.) The dam will inundate about two-fifths of the Nakai Plateau, a remote region in central Laos. The plateau is already highly degraded: because the Ho Chi Minh Trail traversed a portion of the plateau, Americans bombed it intensively during the Vietnam War, and Vietnamese companies indiscriminately logged it for three years in the early '90s. The watershed above the plateau contains four hundred species of birds, yet the plateau is so devastated that on its riverbanks, where birds ought to teem, they're rare—Lee Talbot, Scudder's colleague on Nam Theun 2's panel of experts, spotted only seven birds during a two-day river inspection. In Talbot's and Scudder's view, it would be hard to find another place whose sacrifice to inundation would destroy so little—it has already been destroyed.

On the other hand, the watershed overlooking the reservoir is as lush as the plateau is sterile. It comprises, in fact, the biggest remaining intact forest in Laos and is one of the biggest in Southeast Asia. The watershed packs dense, mist-enshrouded forest into a terrain the size of Rhode Island, rising to a seventy-five-hundred-foot peak at the crest of the Annamite Mountains, where it meets the Vietnamese border. Its assortment of wildlife is astounding. In the last ten years, naturalists have discovered five new mammal species there, including two kinds of barking deer, a pig, a striped rabbit, and a reclusive two-hundred-pound nocturnal ox. The ox, called a *saola* (technically, a *Pseudoryx nghetinhensis*) is not just a new species but a new genus, which makes it a rare find. The area's trees include pines, tropical mahogany, and a rare and endangered cypress, *Fokienia hodginsii*, which is used to make coffins, furniture, and perfume. According to Talbot, loggers sell the wood for at least $13,000 per cubic yard.

As it happens, the preservation of the watershed is crucial to the functioning of the dam. Because the reservoir floor is a plateau, not a riverbed, the reservoir's average depth is unusually shallow—less than twenty-five feet. The reservoir's shallowness makes it vulnerable to siltation, and the highly erosive nature of the surrounding soil makes it even more so. If the reservoir fills with silt, it will lose hydropower capacity; dam planners therefore have financial as well as ecological motivation to promote the watershed's health. Of course, this cuts both ways: if the watershed is not maintained, the expensive investment in the dam will be wasted.

According to Scudder, World Bank officials understand all this. Before the Bank became involved in the project in the mid-1990s, what Scudder calls the Laotian "military fiefdom," which had political control of the region, began building logging roads into the watershed and extracting trees by helicopter. The Bank made the immediate cessation of logging and road building in the watershed a condition of its participation. The region was already designated a national conservation area, so the Laotian government strengthened protections and expanded the borders of the conservation zone and formed a watershed protection authority to enforce the rules. For the time being, the encroachments have stopped. The irony rankles: we have reached a point in human affairs when a watershed's best chance for survival may depend on the construction of a dam beneath it.

Although the Nam Theun is a tributary of the Mekong, the dam won't even cause a reduction in sediment flow to that magnificent, threatened artery—Theun Himboun, a dam downstream on the Nam Theun, already prevents sediment from reaching the Mekong. (In other ways, the Theun Himboun Dam has complicated the case for Nam Theun 2, most notably by dramatically reducing the downstream flow of the Nam Theun, causing a severe decline in fisheries along the tributary's lower reaches. Scudder calls Theun Himboun's impact on downstream communities "inexcusably detrimental.")

For all these reasons, Talbot, an environmental adviser to presidents Nixon, Ford, and Carter and a former director-general of the World Conservation Union, said of Nam Theun 2, "I know of no other dam project ever, anywhere, where the environmental benefit is so overwhelming." Even so, Scudder wouldn't have supported the project unless he considered it likely to reduce poverty among dam-affected people. The forty-five hundred people who live in the plateau's twenty-plus villages are all indigenous, and are poor even by Laotian standards: in his telling, many have never recovered from the killing of their water buffalo during Vietnam War bombing. As he writes in a working paper for *The Future of Large Dams*, "The people's current poverty is so great that implementation of even a mediocre plan could reduce their impoverishment." And the plan, Scudder maintains, is the best resettlement plan he's seen, with "the potential of helping the large majority of the population to significantly raise their living standards." An anthropologist invited the resettlers to describe their "dream villages," and the plan takes their answers into account. No resettlers are required to move into the territory of an alien ethnic group, eliminating the usual host-resettler conflict. Fishing in the reservoir will be reserved for resettlers only, and the flatness of the plateau will yield extensive recession agriculture and grazing areas during the annual drawdown.

According to the plan, the six thousand people who live in the remote areas above the planned reservoir—a few of whom the Nam Theun 2 experts discovered in possession of bark clothing when they first explored the area—will also benefit from an expansion of schools and introduction of health care. Designation of the area as a conservation zone preserves the villagers' biological resource base, and efforts to integrate village economic activities with watershed conservation could also help them.

The impact on about fifty thousand people downstream is more problematic. Water that generates electricity will not flow through the dam, but instead will drain more than a thousand feet through the bottom of the reservoir to an underground power plant, then will be released into another Mekong tributary, the Xé Bangfai River. The additional water will double the dry-season flow of the Xé Bangfai, probably damaging fisheries on the tributary, and flooding gardens. The developer, a consortium of French, Thai, Italian, and Lao private and government investors, has set aside funds to compensate people for losses as a result of the increased flow, and they may also receive a ration of fish from the reservoir. Plans also call for diverting some of the increased flow for irrigation, which should enable some downstream farmers to plant a dry-season rice crop in addition to their customary rainy-season one. In that case, the dam would become a multipurpose project, generating not just hydropower but also development. "See," Scudder said. "It's the best planned project I have ever seen!"

Finally, the entire $30 million a year that the dam is expected to earn will be placed in a fund for national poverty alleviation. To prevent dam revenue from being diverted to other uses, including corrupt ones, the World Bank will supervise the fund.

Though the project has not yet won Bank approval, the dam is likely to get built: either the Bank will support it, with environmental and social programs intact, or private Chinese investors will build it without them. The Bank is considering loaning $100 million toward the project's $1.1 billion cost, but the reason its participation is considered crucial is that it would guarantee the other investors' stakes. Scudder has worried that the Bank might accede to pressure against the dam from the International Rivers Network or American anti-Communists who oppose any sort of aid to Laos, but as of mid-2004, the Bank was moving toward acceptance.

By now, the debate over Nam Theun 2 sounds familiar. For all of its seeming benefits, the International Rivers Network remains unimpressed: one of its publications calls Nam Theun 2 "another World Bank disaster in the making." Among its many criticisms of the project are that planners never explored alternatives to the dam; that they underestimated the number of people near the Xé Bangfai River who will be harmed by the dam by a factor of at least three; that, as in Lesotho, the Bank will lose its leverage over the authorities as soon as the dam is constructed; and that in its previous hydroelectric projects, the Laotian government has fallen far short of providing environmental protection and social justice. Scudder concedes that the IRN might be right, that the Laotian government might not be capable of administering the programs assigned to it, but Nam Theun 2 represents the best chance he's seen to do a dam right, so he thinks it deserves a chance. He speaks with a wariness he surely didn't possess at Kariba—before the first of his many disappointments, before the good dam had experienced shocks—and his legendary enthusiasm at last sounds faintly labored.

"I haven't been associated with many success stories," he said, "and the few successes have been more about stopping something than creating something. That's why Nam Theun 2 is so important. It would show that dams can benefit both the environment and affected people. I think the International Rivers Network will try to stop it, and I think it will be a disaster if they do. My feeling is, if you don't try to show that something can be done, it won't be."

For dinner, Molly served tilapia, an African fish that is now farmed and marketed in the United States. Afterward, Scudder washed the dishes in a couple of buckets in the kitchen sink, conserving water as if he'd fetched it from a river. At the Scudders' insistence, I slept in their upstairs sleeping porch, separated from the elements by only a mesh screen on three of its four sides—Scudder modeled it after a Rhodesian district commissioner's sleeping quarters he saw in the '50s.

The next morning, Scudder, Molly, and I got up at five and were marching up a canyon trail by five-thirty. Of the three of us, Scudder showed the fewest signs of fatigue. He was carrying binoculars, with which he conducted an anthropological survey of the various backyards we could see into, and the effort seemed to invigorate him. It had been a couple of months since he left his job in Lesotho, severing his last official tie to Africa, but if he felt remorse, it didn't show. We passed a tiny check dam, filled entirely with sediment and wedged into the canyon as if it were a retaining wall, of no use now except as historic relic. "This would explain why the beaches are eroding," he said dryly, for without the sediment that dams prevent from reaching the sea, beaches, including Southern California ones, are eroding and disappearing.

Within a few minutes, he was extolling Nam Theun 2.

PART III: AUSTRALIA

A HEALTHY, WORKING RIVER

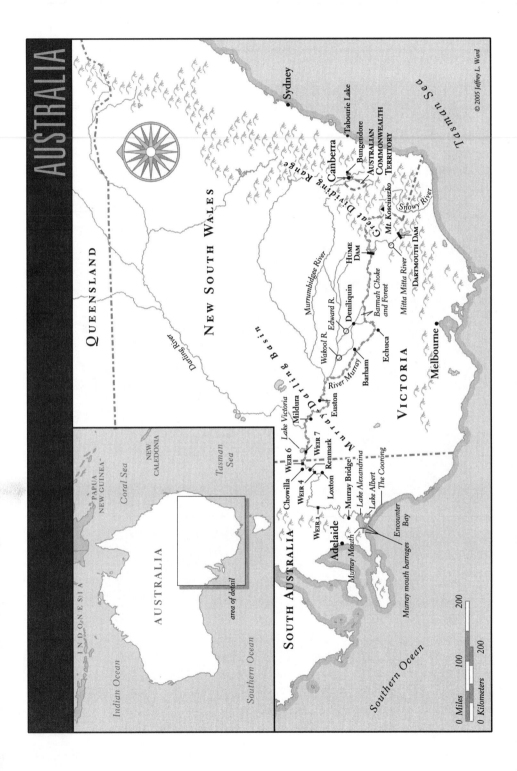

AUSTRALIA

© 2005 Jeffrey L. Ward

QUEENSLAND

NEW SOUTH WALES

Darling River

Murrumbidgee River

Sydney

Tabourie Lake

Bungendore

Canberra

AUSTRALIAN
COMMONWEALTH
TERRITORY

Tasman Sea

Mt. Kosciuszko

Snowy River

HUME
DAM

Denliquin

Wakool R.

Edward R.

Barmah Choke
and Forest

Mitta Mitta River

DARTMOUTH DAM

Euston

Barham

Echuca

River Murray

VICTORIA

Melbourne

Lake Victoria

Mildura

Murray River

Chowilla

WEIR 6

WEIR 7

WEIR 4

Renmark

WEIR 1

Loxton

Adelaide

Murray Bridge

Lake Alexandrina

Lake Albert

The Coorong

Encounter
Bay

SOUTH AUSTRALIA

Murray Mouth

Murray mouth barrages

Great Dividing Range

Darling Basin

Murray

Southern Ocean

INDONESIA

PAPUA
NEW GUINEA

Coral Sea

NEW
CALEDONIA

Tasman
Sea

AUSTRALIA

area of detail

Indian Ocean

Southern Ocean

Miles

0   100   200

0   100   200
Kilometers

11 CHOWILLA

On a long driving trip from one end of Australia's River Murray to the other, I turned off the highway near the town of Loxton to stretch my legs and found myself at a riverside campsite, staring at a singular river red gum tree. It was surrounded by irrigated lawn and distinguished from other red gums closer to the river by a square log railing around it. Red gums are the continent's emblematic tree, the most prevalent of more than six hundred eucalyptus species that have thrived by evolving strategies to cope with Australia's sere surroundings. The red gum's advantage is its intricate root system, both shallow and deep, that enables it to take water from floods in the minority of years when they occur and to absorb groundwater the rest of the time. They're hardy trees that live hundreds, possibly thousands, of years; the sweep of Australian history, from Aboriginal to European domination, is etched on their trunks. Though still living, some are called "scar trees" because they bear the vacant outlines of canoes, shields, and plates that Aboriginals carved from their trunks, and others display the survey markings of the basin's nineteenth-century European explorers.

The tree I beheld was a majestic specimen, eighty or ninety feet tall; its branches were wide enough to create an amphitheater's worth of shade, and its thick trunk, in the midst of a peel, resembled an urban wall layered with weathered posters. The storied Murray some twenty-five yards beyond the tree looked disappointingly tame. The pale green water that moved through it at a stately pace would never be confused with a torrent, and the red gums along its shallow banks were dwarfed by the tree in front of me. Murray Bail, the author of a novel called *Eucalyptus*, tells of red gums that "worm their way greenly into the mind, giving some hope against the collective crow-croaking dryness." This one looked grandfatherly, as if protectively extending its limbs over the campsite's inhabitants. TREE OF KNOWLEDGE, a wooden sign said, and beneath that, in smaller letters, FLOOD LEVELS.

When I moved closer, I saw that the trunk's lower third was arrayed with small metal placards, each labeled with a different year. The fifteen placards represent three-quarters of a century of significant River Murray floods, and each placard's location on the trunk denotes the height of that year's flood. "1970" was posted just a couple of inches off the ground, barely visible beneath the wooden sign. A cluster of placards appeared up to a yard above the sign, connoting the small and medium once-a-year to once-every-ten-years floods, the engines of biological health that replenish the Murray Basin's creeks, lakes, and groundwater. Farther up, six or eight feet above the ground, loomed the big flood years of 1974, 1975, and 1931, and, towering over them, at the astonishing height of twenty-five feet or so, the placard for the once-in-a-century 1956 flood, which in some places lasted two years.

It's indicative of Australia's breath-swallowing aridity that the Murray constitutes the largest channel of the only major river system on the continent, yet its yearly flow is less than a day of the Amazon's. Nevertheless, the Murray Basin is Australia's heartland, its Narmada, its Mississippi Valley, where Aboriginals lived for fifty or sixty thousand years before colliding disastrously with European settlers, and where the settlers' farms and ranches overtook the landscape to the extent that many Australians still think of their country as agricultural. I'd already noticed that quite a few riverfront towns possess some version of the Tree of Knowledge, usually in the form of flood years and heights posted on building exteriors— they reflect an understanding that floods and droughts write the region's history. Staring up at the 1956 sign on the Tree of Knowledge, the most casual onlooker might have tried to imagine how water could rise so high over the broad floodplain's flat terrain. A more attentive observer might have noticed that the last flood memorialized on the tree occurred in 1993, more than a decade earlier, but without an understanding of the Murray's recent history, he could not have appreciated the ominous significance of the gap.

The next day, I drove on another forty-five miles upstream to the Chowilla floodplain, where I discovered that among red gums in the region, the Tree of Knowledge is fortunate to be so robust. Along a six-hundred-mile reach of the Murray, trees are dying. The floodplain is turning into a graveyard of red gums, whose brittle, gray, hollow, and leafless trunks seem to serve as their own tombstones. At Chowilla, erosion has exposed some trees' roots; others are surrounded by their own snapped limbs. The ossified roots and limbs resembled bleached bones, as if they'd passed a point beyond death where plant and animal matter reconverge; in some places, the roots and limbs were intertwined, as in a charnel house.

In its dryness, Australia suggests the planet's future, as the vast human population and the demands of its industries intensify competition for an unchanging quantity of freshwater; in water terms, Australia is a warning, and Chowilla is its immediate expression. The Chowilla red gums are part of a vast death event, encompassing hundreds of thousands, perhaps millions of trees, starting near Euston in southwest New South Wales and extending six hundred miles down the meandering river to the town of Murray Bridge, deep inside South Australia. The immediate cause of the deaths is one of the worst droughts of the last hundred years, which struck the region early in 2002 and has been relieved only sporadically since. But drought is killing the trees only in the way that pneumonia kills a patient with metastasized cancer. Eucalyptuses evolved, after all, on the most arid continent, where drought conditions prevail more frequently than floods: mature eucalypts can handle even once-in-a-century droughts without difficulty. Drought is only the last of many insults inflicted on the Chowilla floodplain in the last two centuries, and is distinguished by being the only one not caused by humans.

The first indignity was sheep grazing. The settlers who brought sheep to the region in the mid-nineteenth century delighted in the Murray Basin's vast grasslands and river water supply but were oblivious to the destructiveness of their ovine import. The sheep stripped the floodplain of its vegetation, including all saplings (thereby eliminating a generation of trees), and their hooves compacted its already fragile soil, complicating groundwater replenishment after rain and floods. Before Chowilla was made a wetlands reserve in 1994, as many as three hundred thousand sheep moved through the area, and reserve or no, a few thousand still graze there.

Tree clearing was the next menace. Throughout the Murray Basin, Europeans cut down fifteen billion trees to make room for agricultural fields, and unwittingly created a saline catastrophe. Australia's trees and other perennial vegetation specialize in the efficient use of water. In the presettler era, of all the water that fell as rain, only 1 percent escaped beneath the root zone of the native vegetation to the groundwater below; once the trees were removed, that quantity leaped to 15 percent, and the water that the trees would have retained soaked through the meager soil and lifted the water table by as much as seventy-five feet. In the process, it has mobilized a vast underground store of natural salt—the Murray Basin alone contains an estimated one hundred billion tons. It has brought water from underground aquifers where it accumulated over millions of years to the surface; there, decades after the trees' removal, the salt has gathered in sufficient quantity to poison huge tracts of land. Irrigation accelerated the process, but even without it, the problem is certain to intensify. The Murray's groundwater is moving toward the river's mouth to the southwest, but at a profoundly slow pace, measured in feet per year; the result is an inexorable tide of salt water that will accelerate over the next century, poisoning lowland soil and water on the way. Because Chowilla was deemed more suited to grazing than to crops, it was largely spared devastating tree clearing itself, but it reaps the salty result of all the cutting upstream. Even under a natural regime, Chowilla collected salt, but now, in the dam era, the proportions have turned lethal. Groundwater and the river bring Chowilla 140 tons of salt a day, while only 20 tons leave it, and the Chowilla reach of the Murray is as saline as ocean water. Red gums are reasonably tolerant of salt, but the intense concentrations in the lower Murray Valley are severely weakening them.

The most serious insults of all were delivered by dams and weirs. Alfred Deakin, Australia's second prime minister, rose to prominence as an irrigation evangelist, and was the architect of the state of Victoria's Irrigation Act of 1886, the seminal Australian water legislation; he saw water as a commodity, divorced from the natural systems that rely on it, and said no price was too high to pay for harnessing Australia's rivers. Over the next century, the Murray most perfectly embodied Deakin's vision. In the Murray Basin, as in most of the Australian interior, water is crucial. After casting out Aboriginals, basin residents kept Aboriginal names for their towns: Yass, for example, is derived from an Aboriginal word meaning "running water"; Jerilderie (jer-RIL-der-ree) means "reedy place." In the 1920s and '30s, the governments presiding over the Murray Basin agreed to promote commerce by building fourteen small dams, called weirs, across the lower Murray. Most river water spills over the weirs, and the amount of water they store is minimal; their purpose is to raise the river's level to foster navigation and water diversions. All but the farthest upstream of the Murray's fourteen weirs include locks, enabling navigation of nearly half the river. By lifting the river's water level eight to ten feet per weir, they've eliminated the river's shallow reaches and smoothed out its highly variable flows—they've turned the river into a staircase of narrow lakes. The Murray briefly supported paddle steamers (fueled by timber hewn from Chowilla and other forests), but the nearly simultaneous advent of railroads and highways caused the embryonic navigation business to collapse. Nevertheless, the locks and weirs have remained in place, even though the only vessels that move through the locks now are recreational houseboats and paddle steamers marketing nostalgia to tourists. For this, the lower Murray channel and its immediate surroundings have suffered a permanent lifting of their water table, in the same eight-to-ten-foot increments by which the locks lift boats. Where the river previously alternated unpredictably between floods and drying phases, now it provides a continual supply of water in all seasons. Fish and plants that depended on the natural cycle for spawning and regeneration suffered severe losses. The elevated groundwater killed trees just upstream from weirs by drowning their roots or poisoning them with salt.

More portentously still, near the top end of the Murray, a thousand river miles to the west, the governments built the massive Hume Dam in 1936, creating what was then the largest reservoir in the Southern Hemisphere. In chess terms, Hume is still regarded as the queen of the Murray's system: it's the system workhorse, relied upon to send water down the river to meet the system's intricate downstream obligations. If the regal title has any applicability to Hume's appearance, it's in suggesting a Botoxed and face-lifted dowager, for in the last four decades Hume has been both enlarged and radically repaired. Hume is a blunt instrument, an earth-rock-and-concrete barrier dam, whose downstream face, topped by a row of gates and supporting superstructure when the dam's height was raised by twenty-seven feet in 1961, seems to smile luridly, and whose spillway is streaked with mascaralike water stains. It is old enough to embody decline, yet too young and ungainly to have a claim on antiquity. Official photographs show the dam at night, bathed in brightly colored lights, but the images suggest a heavily powdered woman, no longer sure of herself in high heels. When I saw the dam, its water level was low enough to reveal the smooth concentric ridges of sediment that had gathered on the reservoir's flanks—testament to nearly seven decades of reservoir operation.

At least Chowilla was spared being turned into a reservoir itself, as the South Australian government proposed in the late 1960s. The Chowilla Dam would have stretched for three miles across the faintly concave floodplain, creating an immense but exceedingly shallow lake—four hundred square miles with an average depth of fifteen feet. The authorities actually carried out some floodplain tree cutting to prepare the reservoir bottom and were held back from construction only by the belated realization that the dam would cause an ecological and financial catastrophe. Researchers pointed out that the proposed reservoir would push up enough saline groundwater outside the reservoir's perimeter to poison the downstream river, and a third of the reservoir's water would evaporate annually—a crippling inefficiency.

Instead of Chowilla, the authorities returned to the top end of the basin and built Dartmouth Dam, ninety miles above Hume, in 1978. If Hume is the workhorse, Dartmouth is the dam of last resort, whose water is held in reserve for dry years like 2002; it's the king to Hume's queen. Dartmouth's role is largely dictated by geography: its watershed is only a fourth as big as Hume's, which means that its reservoir takes much longer—up to ten years—to fill. The water in Dartmouth is therefore more precious, so it's released less often. Dartmouth Dam is even larger than Hume, and remains the tallest of Australia's five hundred large dams. Where Hume is dowdy, Dartmouth looks elemental and precise to the point of abstraction. Even its staircase-shaped spillway, carved out of a mountainside beside the dam's huge rock face, looks razor cut. (A little *too* sharply chiseled, perhaps: the first time the dam spilled, in 1990, the falling water eroded the spillway steps. The designers had anticipated the erosion but put off investing in protective measures. Retaining walls are now being built into the steps to prevent erosion.) Dartmouth is a sleek earth-and-rock-filled slab that points like an arrow into the ravished bed of the Mitta Mitta River, a Murray tributary near the top of the basin—it looks like an aboriginal spear turned on itself, skewering the earth. The dam's clay core is invisible; all onlookers see is a seeming pile of neatly stacked rocks, nearly six hundred feet high, that holds the core in place. The dam's downstream face is so steeply pitched that the rocks seem on the verge of tumbling down, but never do. From above, the reservoir looks like a deer's antler as it follows the Dart and Mitta Mitta rivers up separate canyons.

Together, Hume, Dartmouth, and the weirs on the lower Murray engendered in a few decades the sort of radical change that occurred in pretechnological times over a hundred thousand years. They altered the river's hydrology, plant and animal composition, even its seasons. A flow regime that once qualified as "semiarid"—with water levels mostly in a midrange, except for the occasional flood or drought—has become simply "arid." What the two regimes have in common is that in both, large floods occasionally occur: nothing man-made can stop them. At all other times, the flow is now much lower, but never nonexistent, as it sometimes was before the dam era. Flows that qualified as "droughtlike" formerly occurred on the lower Murray in one year out of twenty; now they happen in three years out of five. And while the big floods keep happening, once every twenty years or so, the more frequent small and medium floods—the lifeblood of the river's wetlands—have been virtually eliminated. Without small and medium floods, the river is separated from its floodplain: the three-directional movement of water—down the river, sideways to its floodplain during floods, and back to the channel through the soil and tributaries—is interrupted. It was therefore inevitable that Chowilla would rapidly decline, as soon as the first long hiatus in big floods occurred. As it happens, Chowilla's last flood was in 1993. As a result of their absence since, Chowilla red gums were suffering from thirst and salt poisoning even before the 2002 drought struck.

Mike Harper, who gave me a tour of Chowilla in his pickup, is one of a few people who noticed in early 2003 that an unusually high number of red gums were faltering. Nearly fifty, graying but still fit, dressed in shorts and athletic shoes on a warm, almost cloudless day, he spoke of Chowilla in clipped sentences, as if the loss of a thing so vast was too painful to ramble on about. He'd grown attached to Chowilla over hundreds of visits, first as a child in the company of his father, a commercial fisherman, and later as a government natural resources manager assigned to look after the area. Already accustomed to seeing signs of floodplain deterioration, he noticed that the pace of decline had quickened. Many red gums were surrounded by mats of leaves, recently fallen from the trees' suddenly barren crowns. Thinning crowns are the first symptoms in red gums' progression toward death. Unless the trees are revived with water, they eventually lose all their leaves, their bark turns a whitish-gray, and their limbs break off one by one. A preliminary study in March 2003 concluded that 80 percent of the red gums in South Australia's River Murray floodplain were "stressed to some degree," and 20 or 30 percent were "severely stressed" and therefore "dry and bereft of vegetation." The difference between a "severely stressed" red gum and a dead one is so fine that it's hard to tell one from the other; what seems unquestionable is that the number of dead trees will grow exponentially until Chowilla experiences another flood. In the meantime, the floodplain is dying. "I've known this country for thirty-five years," Harper said. "It's like losing a big part of you."

Twenty miles outside Renmark, the South Australian riverfront town of eight thousand people where Harper grew up, we drove off the main highway and in a few minutes beheld the floodplain: eight miles wide, it extends from the slightly elevated promontory where we stood, across the broad northern floodplain to the Murray, hidden by trees, and a couple of miles beyond it, the low red cliffs to the river's south. The floodplain is so flat that it looks nearly featureless. That misconception is apparently common, for of the Australians outside the region who know of Chowilla at all, an ample number consider it "stuffed"—without value or inter-est—but they're usually not the Australians with much experience of it. The Murray-Darling Basin Commission, which oversees River Murray operations, has declared Chowilla one of six "significant ecological as-sets," and in 1987 the area was granted protection as a "wetland of inter-national importance" under the Ramsar Convention. At the center of Chowilla's rich floodplain ecosystem sit red gums. Unlike, say, North American conifers, no tree rings indicate red gums' age; in fact, white ants often consume the red gums' cores while the trees are alive. Be-cause the trees' water and nutrients are conveyed upward through a layer just beneath their bark, they can live on comfortably without their cores. Instead, the trees' hollows become homes to parrots, kites, and ernes; opossum, tree frogs, and large iguanalike lizards called goannas; and eight species of bats. Because of the red gums' central role in floodplain biodiversity, some scientists worry that the current dieback could be the prelude to a broader ecosystem collapse.

In Renmark, Chowilla is considered the local recreational area and fills up with holiday campers even now, despite the unfolding holocaust. What campers see, if they pay attention, is that in Chowilla, nuance is everything: the floodplain's flatness means that even minor variations in water flows and land formations have ripples of consequence. Over eons, the Murray meandered from one side of the floodplain to the other, forming and re-forming oxbow lakes, called billabongs, and subsidiary channels, called anabranches, but in the era of dams and weirs, it has stayed still: river regulation discourages variation. Even without a drought, something like two-thirds of Chowilla's one hundred miles or so of anabranches and creeks are dry most of the time, and few of the permanent channels were visible from our vantage point, but their paths were outlined by ribbons of red gums on their banks. Trees populated the landscape according to its subtle contours, curving in faint mimicry of the Murray's meanders. The terrain looked as brown and stubbly as sandpaper. Even from a distance, it was obvious that many of the red gums had lost leaves at their crowns, for we saw more gray than green. Here and there, hardy shrubs called lignum broke through the surface like thumbtacks, and pointed upward a few feet into the harsh, slanting sunlight. Their misfortune is the converse of the red gums': except for the lignums' verdant topmost stems, all of their greenery was missing, having been consumed by sheep.

Harper helped make Chowilla a reserve, but the drought has revealed the inadequacy of the designation. The sign near the park entrance that declares its "floodplain wetland values" is at this point entirely hypothetical. "In Spring," the sign says, "the rising waters of the Murray extend onto the floodplain. The wetlands explode with life as plants and animals take the opportunity to feed and breed on the bounty of the high river." But the river has brought no bounty in a decade. Since red gums customarily live for centuries, their inability to withstand this deprivation suggests that droughts this long were not part of the predam flow regime; in fact, it's likely that between 1993 and the onset of the current drought, tree-sustaining floods would have occurred in at least three separate years if not for the Murray's weirs and dams. As it is, the surroundings certainly give no evidence of recent bounty. As we descended into the floodplain, the health of the red gums plummeted, until the sight of a healthy tree was rare and was usually explained by its proximity to irrigated water. "This was a live tree twelve months ago," Harper said, pointing to a leafless, whitening red gum. As we drove down the middle of an entirely dry creek bed whose shore was lined with hollow, spectral trees, he said, "Back in the '90s, no one thought this creek bed would ever die."

Red gum trunks can remain standing for decades after death; as a result, the trees succumbing to the recent dearth of floods are joining hulks that died over the last seventy-five years of floodplain degradation—"tree ghosts," writer Mary E. White calls them. Harper could date the death of a tree by looking at its location and the color of its trunk; then he could associate it to the corresponding calamity. The trees with white trunks near the river died of waterlogging after Lock Six was constructed in 1930; the trees with darker trunks near a dry creek bed died later of salt poisoning.

Because Chowilla was spared substantial tree clearing, it's one of the few places left in the Murray Basin that provides a glimpse of the terrain that settlers found in the mid-nineteenth century: for both white Australians and Aboriginals, it's a link to the ancestral land. But river regulation is entirely changing the biological landscape, so that it less and less resembles the pre-European floodplain. Along the banks of the Murray and the weir-augmented anabranch called Chowilla Creek, red gums are prospering under the consistent man-made water regime, but as few as fifty feet away, just out of reach of the permanent water, they're dying. Red gums once shunned seasonal lagoons because the trees can't survive even partial water immersion for more than two years, but now, Harper said, since the lagoons have gone dry, red gum saplings are popping up inside them instead of in their customary place on the lagoons' rims. Harper showed me a dry lagoon where floods once occurred every few years—for the first time, red gum saplings were growing in them, a reflection of reduced flooding there. Even that development holds no promise, he said, for the saplings won't survive. Either their need for water will expand as they grow until it collides with the dryness of the soil, or floods will drown them. "After a while," Harper said, "you forget what a healthy tree looks like."

Even the hope for a flood big enough to cause tree rejuvenation in Chowilla seems remote. The Murray receives most of its water from precipitation far upstream, along the Great Dividing Range. When the next deluge finally strikes, basin managers almost certainly will use most of the water to replenish depleted reservoirs at Dartmouth, Hume, and other storage areas along the Murray. That means that a flood delivering water to the anabranches and creeks of Chowilla is probably at least two rainy seasons away. And in the unlikely event of two flood years in a row, the impact will be double-edged. Water will reach plants and animals in desperate need of it, but it will arrive with unprecedented salt loads. In the absence of small and medium floods that would customarily flush salt from the floodplain, the salt has accumulated in increasingly toxic amounts—and a large flood will spread the salt farther up dry creek beds and up over the creeks onto the plain.

Harper showed me an oxbow-shaped lake that had been pumped full of water in a government experiment to find the minimum quantity needed to rejuvenate the trees. The assumption is that once this information is established, river authorities can start mimicking the natural cycle of flooding and drying out that keeps the floodplain healthy, while using a minimum of water. The goal is laudable, but Harper is skeptical of long-term success. The ecosystem will have to become dependent on an artificial regime that must be applied forever, he said. "You might get a good manager for ten years, but the one after him might be a bad one. If we have to manipulate the environment all the time, we're going to fuck it up sometime."

We stopped for a few minutes at Lock Six, the lock-weir complex across the Murray that so emphatically rearranges the Chowilla water table. On the northern side, where we stood, the lock is bounded by a broad, preposterous swath of imported, irrigated lawn. The growth is so luxuriant that trimming it requires a tractor-sized lawn mower with a driver perched atop it; he puttered by as we looked on. The river looked a wan green and was narrow enough to toss a baseball across. The red gums near the banks looked healthy; you had to look past them to a band of trees beyond the shoreline to see evidence of ecological crisis. As we watched, a houseboat from downstream entered the lock and was slowly lifted upward. The campers we'd previously passed will see the dieback, Harper said, because they're surrounded by it, but "the general tourist in a houseboat like this one won't even notice."

Harper's thirty-five years of experience at Chowilla seemed to foster an appreciation of the long view, and that in turn made him melancholic. "Working this game can get pretty depressing," he said, "especially when you see what's happening in your lifetime and you think what it's going to be like in two or three generations."

For lunch, we sat beneath a dying red gum, with its one remaining ashen limb extending over us, as if in vestigial gesture to the shade it once offered. One reason it's impossible to tell when a red gum is dead is that, unlike most North American trees, it doesn't die all at once. A single limb may die without necessarily jeopardizing the others, or it may bear promising green sprouts while all the others are lifeless. Red gums alive now have lived longer than the two centuries of European settlement, so it's impossible to know how long the species lives. Harper guessed that our tree could be a thousand years old. "Oh, the Aboriginals who have sat underneath this tree," he said, sighing, as we began to eat.

12 CANBERRA

In the way that crises focus the mind, Don Blackmore, the chief executive of the Murray-Darling Basin Commission, considered the drought an opportunity. In his office in Canberra, in early December 2002, I watched as he took a phone call from an on-the-air radio interviewer. It was then a year into the drought, and the Australian summer was approaching, which meant that meaningful rainfall was probably a minimum of several months away. For a man delivering bad news, Blackmore's manner was surprisingly jaunty. As usual, he referred to himself and the river management system interchangeably.

"I haven't had one drop leave the Murray and go into the sea since November," he said into the phone, alluding to a period of more than a year. "Everyone from the top of the system to the bottom now faces low reservoir levels."

He was sitting behind his boomerang-shaped desk, adorned in tie, white shirt, and pin-striped suit pants. Only his slight paunch suggested that his regime, like the Murray's, was out of balance. The Murray is one of the most intensely managed rivers on Earth, and the commission is the instrument that manages it.

Blackmore explained that Hume, the workhorse reservoir, was lower than it had been in any previous December. It was down to 16 percent of its capacity, and would run out of water in two or three months. Dartmouth, the reservoir of last resort, was still at 55 percent, but the number was falling 7 or 8 percent a month and would tumble more quickly once Hume was emptied. Blackmore could have added that in many localities across the basin, rainfall totals for the previous nine months were the lowest ever recorded. If not for the dams, the Murray's lower reaches would have been certain to stop flowing during the coming summer; now, because of them, the river was sending water downstream to agricultural and municipal consumers at the rate of 20,000 megaliters — the equivalent of 20,000 Olympic swimming pools — per day.

The last drought of this magnitude occurred in 1982, Blackmore said into the phone, but this one is worse because of climate change. "The temperature over the basin has been one degree higher than in '82, so there is much more evaporation now," he explained. "Rainfall isn't as low as '82, but the demand for water is 20 or 30 percent higher."

The result was unprecedented stress on the water-delivery system. "This year we're going to rewrite the rule book on how we operate the River Murray," he said. His proper Australian accent, derived from the cockney dialect of the continent's first white inhabitants, rendered "rain" as "rine," "water" as "WAH-tuh," and "river" as "RIV-uh." The comment sounded like rueful acknowledgment, acceptance of a challenge, and boast all at once.

To the dairy farmers who were selling off haggard milk cows for the meager price they'd raise in meat—so-called chopper cows—and the rice farmers who'd been forced to skip an entire planting season, the drought was bad, bad news, whereas Blackmore was not entirely displeased, for past droughts led to valuable reforms. A century earlier, a drought forced the bickering states of New South Wales, Victoria, and South Australia to come to terms on sharing the Murray's erratic supply of water: the result was the creation of the River Murray Commission in 1917. A drought in 1967 and a resulting precipitous increase in river salinity forced the commission to change its focus from facilitating more water diversions to maintaining the quality of water still in the system, and another drought in 1982 led the commission to expand to encompass land management. The current drought reminded Australians that they depend on a tenuous water supply, something many would have preferred to forget. Rural meetings on water issues that in wet years drew only a couple of farmers now were packed. Blackmore assumed that once the drought broke, the memory of it wouldn't linger. In his view, policy makers had a year or two to adopt responsible water and land use policies, or they could wait for the next drought. By then, damage to the basin would be even more severe, and the chance of reversing it correspondingly less.

Since Blackmore had started working for the commission in 1984, he'd campaigned to awaken Australians to the deterioration of the Murray and its twenty major tributaries and the intricate set of implications their distress poses for Australian farming, fisheries, drinking water, biodiversity, tourism, and, far from least, national identity. He'd won a succession of improbable water-policy victories, and in the process probably did more than any other Australian to earn the nation an international reputation for forward-thinking water management. (John Briscoe, the World Bank's senior water adviser, declared in a 1997 memo to staff members, "Australia is putting into place every element of what we know to be economically and environmentally appropriate water management.") Yet the victories had not come close to reversing the Murray's declining health, and the global approbation in which Australian water managers basked underlined the incapacity of other nations to face their water problems at all: Australia is a leader in a weak field.

The Murray-Darling Basin—so-called because it also includes the Darling River, a Murray tributary that is longer than the Murray but under natural conditions carries less than a fourth as much water—is big and small at the same time. It's the same size as the Colorado River Basin, as big as Texas and New Mexico or, Blackmore's usual analogy, France and Spain. A commission publication accurately calls it "one of the world's major river systems": among the world's basins, its rivers rank fifteenth in length, and its area is twenty-first. But its flow is paltry. The basin is so flat that its rivers loll, meander, and repeatedly change course on their way to the Southern Ocean. In this way as in many others, the Darling, the longest of the basin rivers, is extreme: it's a respectable seventeen hundred miles long, but it's only a third that length as the crow flies. Most terrain in the basin is semiarid: although it covers 14 percent of Australia, it receives only 6 percent of the continent's runoff. Many basin rivers are flatly seasonal; even big ones have a habit of drying up. At the town of Menindee, the Darling ceased to flow forty-eight times over a seventy-five-year period, including once for a year. Yet the basin is Australia's agricultural heartland. Half of Australia's agricultural production is generated from 1 percent of its land area, and most of that 1 percent is in the Murray Basin. It produces nearly all of the nation's rice and cotton, three-quarters of its wine, more than half its fruits and vegetables, a third of its wheat. More sheep live in Australia than in any other country in the world—and nearly half of them, sixty-seven million, are pastured in the basin, along with a third of the continent's 2.4 million cattle. Virtually all of the basin is too dry to rely exclusively on rainwater, and some areas, such as Chowilla, get fewer than ten inches of rain a year. For most of the last century, the basin's farmers appropriated copious quantities of water from the rivers with minimal restraint; largely to promote water storage, the basin has sprouted 84 large dams and weirs, 2,900 barriers of various kinds, and 650,000 small farm dams. Along the River Murray, water is diverted at more than a thousand sites. Indeed, of all water turned to human use on the continent, more than half is for irrigation in the Murray-Darling Basin.

As deputy chief executive, Blackmore won his argument that his legally ambiguous, obscure agency, then called the River Murray Commission, couldn't tend to its rivers without also changing the way land is used throughout the basin. As a result, the River Murray Commission became the Murray-Darling Basin Commission and expanded from 8 employees in 1984 to 101 in 2003. After Blackmore became chief executive in 1990, he convinced the governments presiding over the basin to accept a cap on water diversions from the Murray, and having made that case in 1995, began explaining that the cap wasn't enough—users had to *return* water to the river if it was to stand a chance of recovering. Improbably, he sold that idea, too. When I met him, in 2002, by his own accounting, in thirteen years as chief executive he'd achieved 80 percent of what he set out to do—but what remained was the last crucial piece. He had to persuade the six governments to agree on the quantity of the Murray's environmental flow—the amount of water that would be given back to the river. Success would add substance to Australia's reputation for smart water management, and it would lend momentum to the international dam industry's turning, the narrowing of the spigot that Hoover and Kariba and Grand Coulee opened wide. It would be an admission of excess, an acknowledgment that the price of taking too much water from the river is later giving it back.

Rehabilitating the river is an infinitely complicated task, also involving timing flows properly and reducing pollutants, beginning with salt, but for now the crucial issue facing the commission, the one that affected most river users, was the number of megaliters to be returned. For all of Blackmore's efforts, only a little more than a fourth of the Murray's pre–dam era flow now reached the river's mouth: the river stands no chance of recovery unless that proportion is raised substantially. Blackmore's goal was to recover enough water for the Murray to become a "healthy, working river." The phrase is itself a compromise, signaling to environmentalists that they will not get their wish for a wholly restored flow and to farmers that they will have to give up a sizable share of the water they used. A technocrat of considerable political skill, Blackmore spoke constantly of finding a "balance" that jeopardized neither the river nor the farming community. Wearying, like the red gums, of stress, Blackmore was considering retirement when his contract expired in April 2004, when he'd be fifty-five. Even now, he didn't have authorization from his government-dominated board of directors to talk about the specifics of environmental flows, to explain who would give up how much water to achieve what probability and magnitude of environmental gain. He had sixteen months to persuade the six governments to agree on a number.

I'd met Blackmore two days earlier, when he picked me up in front of my hotel at five-thirty in the morning on his way to the Canberra airport—we were flying together to Melbourne. A major element in Blackmore's considerable charm is his seeming easygoingness. Despite the hour, he was chipper, pointing out monuments as he drove. Aside from Canberra's inhabitants, who are generally enthusiastic about their city, Australians don't hold the capital in high esteem, but in the pink-tinged gloaming, before the heat had routed the slight chill, it looked ethereal, Xanadu-like. He pointed out the Australian-American Memorial, a sculpted eagle atop a lasciviously tall and slender column, which a commission employee later informed me was referred to locally as "the chick on a stick"; the submerged Parliament House, the largest building in the Southern Hemisphere yet a kind of grandiose bomb shelter, curiously padded with sod; and the three man-made lakes that Walter Burleigh Griffin—an American, Blackmore wanted me to know—designed for this bland, planned, overly ordered city. Bill Bryson describes Canberra as "an extremely large park with a city hidden in it"; Pico Iyer calls it "a Lonely Place institutionalized." According to Blackmore, it was made Australia's capital in 1927 for strictly utilitarian reasons: it was midway between Sydney and Melbourne, and it was outside the range of naval artillery. Canberra's low repute doesn't bother Blackmore: a practical man, he likes the city both for its climate (mild by the continent's extravagant standards) and its lack of congestion, its manageable population of three hundred thousand. Not all of Blackmore's diverse constituents— farmers, environmentalists, municipal water users, fishermen, the six governments that make commission policies—appreciate the irony: the fate of the basin, seat of Australian character, is in major respects being decided in a city reputed not to have any. Situated beside the western foothills of Australia's Great Dividing Range, the low-slung mountains that separate the basin from the continent's eastern coast, Canberra is (barely) inside the basin but not of it, reflecting little of the farming culture that the basin has fostered over nearly two centuries. Some farmers suspect that the city reflects Blackmore's orientation, that he's more interested in forging a deal among politicians than in heeding the farming community's concerns. The accusation points to the central tension of his job, its requirement that he address the wishes of all the basin's stakeholders, from farmers to his many political bosses, while attending simultaneously to the river's health.

As we drove to the airport, nearly alone on the road, I wondered, *Where is everybody?* The thought would occur to me again and again over two visits to Australia, for the country's population density, five people per square mile, is among the sparsest of any country's in the world—directly because of a dearth of freshwater. The Crocodile Dundee–ish image that Australia has propagated—suggesting that Aussies lead swashbuckling, carefree lives amid abundance—entirely overlooks the precariousness of their existence. The basin epitomizes the problem, for many of its towns are losing population as farming declines. Canberra itself is the small capital of a large country inhabited by a small populace, where only twenty million people are spread across a continent as large as the forty-eight states. To Blackmore, that's an advantage. "It's a very quiet country," he said. "Don't tell anybody we exist." He was smiling, but not widely enough to dispel the sincerity of the sentiment.

As our plane took off, Blackmore announced, "I'll do a bit of homework now," but within a few minutes, he was gesturing to the flat, brown expanse of basin visible out the window—our Canberra-Melbourne route bisected the basin's southeast corner. "That looks dry, doesn't it?" He sounded parentally concerned. "We have the highest variability of rainfall in the world, and trying to reconcile that with river flow is . . ." And he laughed at the difficulty of it. The ratio of maximum to minimum annual flow of, say, the Amazon River is a stable 1.3 to 1; for the River Murray, the ratio is a startling 15 to 1; and for the Darling, it is a staggering 4,700 to 1. In some years, the Darling stops flowing entirely, while in 1990 a flood on it stretched across an area as big as Texas and California combined. To cope with such variable flows, the commission maintains an enormous reservoir capacity of 35,000 gigaliters, about three times the basin's annual water diversions.

By the time we'd settled in our plane seats, I'd gotten a good look at Blackmore. In a book published in Sydney called *Watershed: Deciding Our Water Future*, Ticky Fullerton describes him as "suave almost": both words are apt. The suavity is evident in his steel blue eyes, sandy hair combed straight back in neo–George Hamilton fashion, and cheeks pleasantly puffed as if sated with nuts. The suggestion of self-confident urbanity extends two-thirds of the way down his nose, until it meets the "almost" point, where it gives way to a modest but perceptible asymmetrical bloom, reddened and partitioned, intimating working-class roots and pugnacity. Blackmore acquired his badge in battle, breaking his nose three times while playing field hockey. He's a battler, the son of a vegetable farmer, found dyslexic as a child and shunted off to technical school, who nevertheless became an engineer. He worked in dam and irrigation channel design, and even had a hand in the design of a Dartmouth Dam abutment, yet he could see that the dam construction era in Australia was coming to an end. He shifted into water management, set himself the goal of becoming a chief executive by the age of forty, and at the commission came within fifteen months of meeting it.

As Blackmore's career advanced, he shucked not just the practice of engineering but also much of the engineering mentality, with its single-minded focus on mechanical solutions to multidimensional problems—what he calls a "silo mentality." Even as a young engineer, he possessed what must have struck his superiors as a quirky interest in native species. The fashion ever since Europeans' arrival in Australia in the late 1700s was to pretend that Australia was Europe and to revegetate the continent accordingly. Given a chance to choose trees for the perimeter of a reservoir fifteen miles from metropolitan Melbourne, Blackmore chose natives, while his seniors placed exotic oaks and elms around the spillway; the result, Blackmore said, was "a bit humorous." His first environmental epiphany occurred a few years later, in 1984, when he began working at the commission. Salinity was then emerging as a huge threat to the basin, and he was given the job of assessing its impact and devising a solution. He was startled to discover that of the fourteen factors that contribute to salinity—from irrigation practices to reforestation and native vegetation conservation—the commission controlled just one: river flow (which in sufficient quantities promotes salt dilution). Blackmore saw that unless the commission's charter was widened to include land use as well as water, it would have no hope of controlling salinity. He pressed his case so well that after eighteen months of negotiation, the River Murray Commission was reborn as the Murray-Darling Basin Commission, with the right to supervise the use of "Basin resources." "So we set off on a journey," Blackmore said.

The journey included many foreign destinations. As the commission's chief executive, Blackmore was not shy about accepting overseas appointments, even though he reaped criticism for the weeks he spent outside the basin. In 1995, he was selected a member of the World Bank Independent Advisory Panel for the Aral Sea, and found himself inspecting one of the world's most spectacular dam-induced catastrophes. Decades ago, Soviet planners diverted two major rivers that feed the Aral in order to turn the surrounding desert into a cotton cornucopia. At least for a time, cotton bloomed, but the sea wilted: it now contains a third of its former volume and may disappear. All twenty-four of the Aral's native fish species vanished, and the annual fish catch dropped from forty-four thousand tons to none. Each year windstorms pick up millions of tons of salt and dust from the dried lake bed and scatter them over the basin, inducing regional epidemics of cancer, hepatitis, typhoid fever, and respiratory illness. In the Uzbekistan town of Muynak, which once bordered the lake but was now twenty-five miles away, Blackmore inspected a fish cannery abandoned for lack of fish and had another epiphany. "Somebody made a decision to trade this community's wealth off without having a conversation with its citizens. And they knew they were doing it—that's the thing that shocked me. They were prepared to trade environmental and social well-being without any recourse. That's pretty scary stuff . . . I came back with a view that I would never allow that silo mentality to overwhelm a more complete understanding of how systems operate."

At the same time, Blackmore faced what he considers the quandary of regulation, which he explained as we flew toward Melbourne. Regulation is necessary, he said, to provide limits for the basin community—to cap water diversions, for instance, or restrict fishing of native species—but if the regulations are too narrow, they leave no room for changing conditions, including the advancement of knowledge in a field such as river science in which hard data are scant. The U.S. Environmental Protection Agency can limit air pollution by setting emission limits, Blackmore said, but "how do you regulate a landscape?" How do you build into regulations all the dozens of variables that affect the health of a river basin?

Blackmore's answer is "integrated catchment management," or integrated watershed management as it's known in the United States. It's the most sophisticated approach yet devised to face the consequences of the environmental depredations caused by dams, river diversions, and basin land-use practices. It's an attempt to blend all the human uses of a river—water storage, farming, grazing, electricity generation, fisheries, navigation, recreation, and tourism—together with its environmental requirements, by intensely managing the river. It's not a way of having it all, but of having at least a large fraction of each piece of it all. It is not at all "natural"—that is, it does not remotely function the way an undammed river does—but with the aid of computer modeling, it tries to mimic natural processes. Squeezing every last productive drop of water from the river in a controlled, sustainable way is clearly a more prudent approach than ignoring the river's environmental needs, but nowhere in the world has it been accomplished; whether it can be is still an open question. A governing body operating according to its precepts would impose broad limits on its basin's users while trying to engage them all in the search for solutions: to have a chance of working, its first principle must be that the solution will in some way benefit them all. No country has dipped deeper into the bag of techniques and policies that constitute integrated catchment management than Australia, and no person is more responsible for that fact than Blackmore. His Living Murray Initiative, the name given to his fight for environmental flows, calls for feedback from all of the river's users to help establish an appropriate quantity. The old prodam/antidam argument is beside the point in the basin. Its two million people, plus millions more who are sustained by its crops, depend on its dams for agricultural production, livelihood, and drinking water; not even environmentalists dare to propose removing more than a few unproductive dams and weirs. There is no question of restoring the Murray-Darling Basin to a pristine state—and if there were, would "pristine" mean pre-European or pre-Aboriginal? The question, rather, is how to minimize the dams' damage enough to restore health to the rivers. In the fond hope that such outcomes are possible, integrated catchment management has become a catchphrase, cited confidently in the international offices of aid officials as the solution to the world's intensifying water woes. Yet in Australia, where watershed management is state-of-the-art, the commission was still fighting to establish a meaningful environmental flow. It was the culminating battle of Blackmore's career, whose outcome would constitute either his biggest triumph or his saddest defeat.

Blackmore's successes had already lent him a certain mystique. As the story goes, when a senior Australian politician was allowed a short audience with Nelson Mandela during a trip to South Africa, the politician was nonplussed to hear Mandela's parting wish that he "pass on my regards to Don Blackmore." Mandela barely knew Blackmore, having met him only once, at the London unveiling of the World Commission on Dams's final report, but the point still seemed valid: somehow Blackmore got people's attention. His colleagues routinely refer to him as charismatic, for reasons that at first escaped me and that I later decided are distinctively Australian. He is neither disarmingly handsome nor notably eloquent nor irresistibly funny—though he works hard at inducing others' mirth.* His laugh is loud but slightly harsh, as if inspired by dogged resolve more than merriment. ("What gift do *you* get?" I asked him, after watching him at the end of a long day don a Santa Claus cap and boisterously hand out gag Christmas gifts to members of his staff. The staff's "enthusiasm," he unhesitatingly answered.) Perhaps because of Australia's isolation and convict origins, many of its citizens suffer

---

*The best stories about Blackmore are told on him. Denis Flett, a Victorian commissioner, explained, "Don doesn't tell the funniest jokes, but he loves to joke." He and Blackmore were part of an Australian water delegation that visited Vietnam in 1998. Near the beginning of the trip, when the delegation members didn't know one another well, Blackmore recommended a restaurant. "I said, 'You remember I'm allergic to crustaceans, don't you?' Typical of Don, he said, 'Don't worry, Fletty, I'll look after you.' He just assumes leadership even though he wasn't the most senior person there. He ordered the courses, and what happened? The first seven were bloody crustaceans! Seven courses in a row! The seventh course, I've got to say, was crabs, bloody crabs this big"—and he held his hands far apart. "I said, 'I'm going out for a hamburger.'"

Flett believes Blackmore was trying to look after him but didn't know how to communicate his wishes to the restaurant staff. But by the time the noncrustacean courses arrived, the incident had turned humorous, and Blackmore, who took his ribbing good-naturedly, had succeeded in creating camaraderie among the delegation members.

from a kind of insecurity, a skepticism about their capacity to excel, particularly on an international stage. Where many Australians routinely exhibit deference, Blackmore is not afraid of the spotlight. His charm, I decided, arises from a kind of willed optimism, not grace exactly, but ebullience under pressure.

"Another up, another down," he said as we landed.

13 AUSTRALIA

Of the seven continents, Australia must be the most misunderstood, including by its own inhabitants. The European settlers who arrived beginning in the late eighteenth century looked at Australia and saw Europe, or tried to; whatever didn't resemble Europe was subject to renovation or obliteration. Tim Flannery's illuminating book *The Future Eaters: An Ecological History of the Australasian Lands and People* cites some examples of the Eurocentric blather he was taught as an Australian schoolboy in the 1960s. He learned, for example, that the Aboriginals, who inhabited the continent until the Europeans shoved them aside, were "among the world's most primitive people who were now, in a sad although inevitable process, making way for a superior people." He learned that Australian inferiority extended to its animals, the dull-witted "marsupial kangaroos, wombats, koalas and the like," which "were quaint but, in a remarkable parallel with our Aborigines, unable to cope with competition from introduced sheep, cattle and foxes"—all superior placental mammals. Even Australia's splendid wildflowers were deficient, for none could "attain the grace and beauty of an English rose." Compared to Europe, Australia was callow and new: why, it was once called New Holland, and it was surrounded by New Zealand, New Guinea, and New Caledonia. All this, Flannery argues, belongs to the first phase in the arc that Australian settler history has described: from "unbounded optimism" through "bitter disillusion as resources are exhausted" to "a long and hard period of conciliation, during which the land increasingly shapes its inhabitants."

Only as the evidence of the grievous harm the settlers inflicted on a profoundly un-European environment accumulated, throughout the twentieth century, did Australia start to come into focus. The process gained momentum as discoveries in the natural sciences, archaeology, and anthropology bared many of the errors in the Eurocentric assumptions about Australia. The continent, for instance, is far from "new": Aboriginals have probably inhabited it for sixty thousand years, nearly twice the length of human habitation of the "old" countries of Western Europe. In the new understanding, Australia is the most extreme of the habitable continents, the one least suited to humans. Only 12 percent of Australia's rainfall becomes runoff for rivers, streams, and aquifers; most of the rest evaporates, while plants absorb a lesser sum. As a result, though Australia represents 5 percent of the world's land area, its rivers carry only 1 percent of water in the world's rivers. It has a harsh, unforgiving environment and a terrain so deficient in water and nutrients that its plants and animals developed radical, hyperefficient strategies to survive on the bits of sustenance that came their way. Most of the Australian environment is fragile, degraded, and unfit for human habitation. All but two million–plus of the continent's twenty million people consequently live on its coasts, clinging to its edges as if for dear life.

By geological standards, not just cultural ones, the continent is profoundly old. At the beginning of the Cretaceous Period, 144 million years ago, it was still attached to the southern supercontinent known as Gondwana. Forty million years ago, it broke away from Antarctica, the last remnant of Gondwana, and began drifting northward, as Flannery puts it, at a pace "half the rate at which a human hair grows"—enough, in the fullness of geologic time, to separate the continents now by two thousand miles. Vestiges of the connection remain, in the form of ancient dry river channels in southwestern Australia that line up precisely with equally old channels in Antarctica. Even eucalyptus is linked to Antarctica, for fossil remnants of the pollen of myrtle, the family that includes eucalyptus, have been found in Antarctica.

For all of Australia's environmental severity, it has an amazingly rich biological heritage. Its flora and fauna are striking in their singularity, a consequence of their Gondwanan ancestry and forty million years of isolation from other landmasses. The continent is short on mammals but profuse in reptiles, insects, and plants. Its deserts contain more reptile species than any other environment in the world, and its plants outnumber Europe's by many thousands. As for insects, Flannery says, "There are more species of ants inhabiting the hill called Black Mountain that overlooks Canberra than there are in all of Britain."

The Australian paradox—great diversity in the midst of environmental extremity—has depended on the odds-defying changelessness of the continent's climate over the eons. During Australia's forty-million-year-long drift toward the equator, the world's climate has chilled, causing

Australia to perform what Flannery calls "an almost miraculous balancing act." Had it drifted more slowly, the cold would have overtaken it, and many species would have gone extinct; if it had moved more rapidly, much of the cold-adapted Gondwanan flora and fauna would have succumbed to tropical heat. In fact, it drifted just quickly enough to avoid glaciation, even during ice ages—glaciers inevitably take some species with them, so in this way, too, Australian biota eluded extinction. And while other continents endured collisions (such as the one that created the Himalayas) and the blistering of volcanoes that spew out rich lava plains, Australia has been geologically tame. (The one exception to this rule, the chain of low and eroded extinct volcanoes that form the Great Dividing Range, yields most of Australia's fertile soil. Most of it is on the range's gently inclined western and northern slopes, near the headwaters of the Murray-Darling Basin.) Without volcanoes and glaciers, Australian soil has had no means of renewal. As a result, the soil is thin and deficient in nutrients, so sandy and porous that it holds little water.

On top of this, Australia is the only continent whose climate is driven by nonannual cyclical climatic change, in the form of the El Niño–Southern Oscillation. While the El Niño cycle affects most of the globe, no other continent experiences its effects as intensely as Australia. There, it has more impact on rainfall than the changing seasons: its variations produce floods, droughts, high winds, dust storms, and colossal fires. No one knows how long the El Niño phenomenon has existed—Flannery deduces from the adaptations of continental flora and fauna that it has been around for millions of years. The El Niño–Southern Oscillation caroms from drought to vital rain or ruinous flood on an unreliable two- to eight-year cycle, adding considerably to the uncertainty of precipitation on the continent.

The species that have survived, Flannery suggests, all share a few attributes: low reproduction rates, a capacity to "exploit brief windows of opportunity as they open erratically over the land," and extraordinary efficiency in the use of nutrients. While European species commonly wage war against hardy competitors, many Australian species survive by cooperating, by participating in complex systems involving many organisms that together efficiently recycle nutrients. Flannery cites the Western Australian heathlands, where one species of a shrub called banksia "may be able to survive in runoff areas where more nutrients are available than elsewhere. Another may survive in sand at the foot of dunes where water may accumulate. Yet another may survive in barren areas because it can exist on very few nutrients. Because of the complex interplay of soils, nutrient levels and water availability, many combinations of resource availability are possible in a nutrient-poor landscape, leading to the evolution of many specialist species." My favorite example, related to me by Kevin Goss, the commission's deputy chief executive, is an orchid species whose flower resembles a female wasp, thereby attracting male wasps which, in their vain attempts to mate, accumulate and then distribute the orchid's pollen. While the systems are efficient, their interdependence makes them fragile—take away one essential species or merely alter the balance of nutrients, and the entire structure collapses. The river red gum is an obvious example: its diebacks threaten animals up and down the food chain, from insects called lerps through honeyeaters, birds that happily consume the sugary secretions that the lerps deposit on red gum leaves, to opossum, inhabitants of red gum hollows.

Given the lack of nutrients, Australian animals specialize in conserving energy. The koala, for instance, is what Flannery calls "one of the greatest energy misers of all mammals": its females bear only one baby every other year, it moves slowly, and its brain (a huge energy consumer in other animals) is shrunken, occupying only a fraction of its cranial vault. "The koala is clearly an extreme," Flannery writes, "but marsupials in general are not known for their large brains, or intelligence. The fact that they won out over early placental mammals in Australia suggests that it may indeed pay to be dumb in Australia. You save a lot of energy that way."

The Australian environment is so impoverished that few carnivorous mammals could survive in it. At the top of the food chain, they depended too heavily on large quantities of herbivores (their prey), and the grasses that their prey grazed on. Of the sixty or so twenty-pound-plus mammal species that lived in pre-Aboriginal Australia, only two or three were carnivorous, and they eventually went extinct. Carnivorous reptiles, on the other hand, flourished: the pre-Aboriginal Australian bestiary is dramatically scaly. The reptiles' cold-bloodedness was a huge advantage, for they needed less energy to survive and could therefore wait out long periods without food. Land crocodiles bestrode the continent with hooflike feet; across it slithered giant snakes, twenty feet long and a foot in diameter, with heads the size of shovels. The terrain was ruled by the giant goanna, according to Flannery "one of the most formidable carnivores of all time, an echo of the time when dinosaurs ruled the Earth." The giant goanna was a drastically enlarged iguana, twenty feet long and weighing several tons.

Pre-Aboriginal Australia also possessed a large array of herbivorous mammals, including at least twenty species of kangaroo, a four-hundred-pound wombat, and some fifty other marsupial species. Together with the predators, the reptiles, and the plants they depended on, they formed a well-functioning ecosystem honed over millions of years of evolution. Then, some sixty thousand years ago, many of Australia's large animals went extinct. Flannery argues that the Aboriginals hunted them into extinction; other scholars suspect that environmental factors played a predominant role. In any case, the Aboriginals' impact was tempered by their judicious use of fire: they were fire-stick farmer-hunter-gatherers, in effect, who lit small patches of land both to hunt and to release nutrients into the continent's woodland and grassland ecosystems for the middle-sized mammals they now depended on. The Aboriginals became stewards of a degraded land, whose techniques at least enabled them to maintain a new regime that supported the existence of more than twenty middle-sized mammal species over thirty-five thousand years or more. The new system, however, was vulnerable, since it depended on prudent human interventions. As the Aboriginals were supplanted by Europeans, fire-stick farming ceased, and a new era of sporadic vast and devastating fires began. Now it was the turn of the middle-sized mammals to go extinct. Flannery writes, "The arrival of a new people, with no understanding of the ecology of their new homeland, was to prove the undoing of 35,000 years of conservation effort."

The Europeans cut down eucalyptus forests with unimaginably broad swaths. *Murray-Darling Basin Resources*, a commission publication, reports that throughout Australia, despite its aridity and poor soil, a fifth of the continent's native vegetation was removed for agriculture; in the basin, more than half of pre-European vegetation was stripped away. The settlers cleared timber with extravagant wastefulness, never realizing that the supply was finite and that the soil in which the trees had grown would now disastrously erode. Dams and irrigation immediately increased productivity, while the problems they unleashed, such as the ruinous spread of salinity, took decades to surface. Contemptuous of the native species, the Europeans introduced their own, and caused calamitous infestations of foxes, rabbits, feral cats, and carp. Part of the reason these animals spread so quickly was that the ecological niches they filled were vacant.

Even now, it's easy to spot the optimists, the disillusioned, and the reconciled in the Australian population—only the proportions have changed. For example, Alan Jones, a Rush Limbaugh–like Sydney radio jock hyperbolically labeled the most influential political figure in Australia, is a flagrant optimist. In the interests of "droughtproofing" Australia, Jones called for resuscitation of the Bradford Scheme, a 1938 plan to reverse the flow of Queensland's coastal rivers from the Coral Sea, their current destination, to the continental interior. The scheme could serve as Exhibit A in the history of heedless Australian optimism, and it was not implemented for many good engineering and environmental reasons. As for droughtproofing, as Blackmore put it, 99 percent of Australia isn't irrigated and never will be. When an ad hoc group of eminent environmental scientists issued a statement explaining why droughtproofing was folly, Jones dropped the subject. Known as the Wentworth Group because its members decided to speak out while having drinks together at the Wentworth Hotel in Sydney, the ten scientists issued a conservation manifesto that Blackmore could have endorsed:

> Our land management practices over the past 200 years have left a landscape in which freshwater rivers are choking with sand, where topsoil is being blown into the Tasman Sea, where salt is destroying rivers and land like a cancer, and where many of our native plants and animals are heading for extinction.

> We are taking more resources from our continent than its natural systems can replenish. That, by any definition, is unsustainable.

> The current crisis is an opportunity to design a new way of doing business to build resilience into rural and regional communities, enabling them to cope with the variability of our climate . . .

> Despite water being our most scarce natural resource, we treat rivers as drains. If we keep doing this, neither our rivers nor the rural communities who depend on them will have viable futures. Everything we do in the landscape impacts in some way on water quality—even in the driest parts of the continent.

Among the members of the Wentworth Group were Peter Cullen, a freshwater ecologist whose work provided the scientific underpinnings of Blackmore's campaign, and Leith Boully, the only nonscientist among the signators, who was chairwoman of the commission's community advisory committee. It's safe to say that the commission was in the vanguard of conciliation forces.

14 BUNGENDORE

In contrast to the basins' mercurial rivers, the intensity of Blackmore's professional life ebbed and flowed with nearly tidal precision, according to the scheduling of meetings. The biannual sessions of the Murray-Darling Basin Commission's Ministerial Council, the body of federal and state government ministers that lays down commission policy, produced the highest tides, while the quarterly meetings of the "commissioners"—chief executives of federal and state water, land, and environmental agencies who comprise the equivalent of the commission's board of directors—ranked second. Blackmore's attention was focused on a commissioners' meeting only a week away, in the town of Renmark, where he hoped to get authorization to talk about the specifics of environmental flows. The task was highly sensitive, since it involved explaining in broad strokes how much water farmers would lose so that the River Murray could gain some. The Ministerial Council had already set an October 2003 deadline for deciding how much water would be devoted to environmental flows, and Blackmore meant to ensure that the council met its goal. Leith Boully said at the time, "Don would feel personally cheated if he weren't able to reach some conclusion before he retires." Blackmore himself called the emergence of environmental flows "a sign of the maturity of Australia as a nation."

The legitimacy of Blackmore's campaign for environmental flows rested heavily on a single scientific study, the ponderously titled "Independent Report of the Expert Reference Panel on Environmental Flows and Water Quality Requirements for the River Murray System." In the 1980s, as Blackmore's experience at the commission deepened, he realized that it lacked a sophisticated scientific understanding of the impacts of regulating the Murray's flow. The commission needed what Blackmore calls a "knowledge base"; to get it, the commission became one of the biggest investors in a newly formed research unit at the University of Canberra called the Cooperative Research Centre for Freshwater Ecology. The Expert Reference Panel report is the culminating scientific document of Blackmore's leadership of the commission, and it reflected his moderate approach. Completed in February 2002, the report concluded that the River Murray became unhealthy by the 1960s or '70s and was continuing to decline — despite the imposition of the cap on water diversions in 1995. A more radical, less pragmatic study might have explored how to return the river to self-sustaining healthfulness, but since that would have required a drastic reduction in irrigated agriculture, Australian politicians would have rejected the idea summarily as a political and economic disaster. Instead, the report considers how to produce "a healthy, working River Murray system" — the phrase is key, and is cited over and over in discussions about River Murray environmental flows. The report defines "a healthy working river" as a river requiring management "to provide a sustainable compromise, agreed to by the community, between the condition of the river and level of human use." The report acknowledges that by itself, increasing the quantity of water flowing through the Murray would not return the river to health; such issues as the flows' seasonal timing, their hour-to-hour variability, the separation of the river from its floodplain by construction of levees and roads, and water quality are also vital. Nevertheless, the report focuses on the volume of environmental flows. As Gary Jones, the report leader, explained, "The toughest decision you have to make is the volumetric question, because that's the one that really affects everybody." With current flows at the barrages just above the Murray's mouth averaging about 3,000 gigaliters a year, a little more than a quarter of the average pre–dam era flow of 11,300 gigaliters, the report found that the river's flow would have to increase by about two and a half times to at least two-thirds its former level to reach "a high probability" of a healthy river; if the flow level was maintained below half its natural rate, the probability of a healthy river was "low." The report examined in detail the likelihood of four basinwide environmental flow options: returned flows of 350 gigaliters ("low" probability), 900 gigaliters ("low-moderate"), 1,950 gigaliters ("moderate"), and 4,000 gigaliters ("high"). Of the two lowest flow regimes, 350 and 900 gigaliters, the report was even more explicit: it said they "provide little confidence that a healthy river would be achieved in the future."

On the basis of the study, the Ministerial Council, the group of politicians that decides the commission's policies, issued its landmark

"Corowa Communiqué" on April 12, 2002, a hundred years after the first interstate meeting on cooperative management of the River Murray. The communiqué was simultaneously exciting and cautious. It announced approval "in principle" of environmental flows, but it discarded as politically impossible the 4,000-gigaliter option, the only one with a high probability of success. Instead, it directed the commission to investigate further the three lower options, which translate to 350, 750, and 1,500 gigaliters, respectively, on the Murray alone. At least at first, the other basin rivers would not be part of the environmental flow plan. The communiqué also indicated the council's desire to make its final decision on environmental flow eighteen months later, at its October 2003 meeting: that looming deadline gave environmental flows its current momentum.

To make the deadline, Blackmore was counting on getting approval from his commissioners to talk about the numbers—in effect, to begin negotiating them—with the basin's assorted constituents, from farmers to environmentalists. Now, five days before the commissioners' meeting, Blackmore was showing some of the numbers to his staff, essentially thinking out loud. It was on the first day of a two-day commission executives' retreat in the one-intersection town of Bungendore, a half hour's drive outside Canberra. The faint smell of fire was in the air, for the day marked the kickoff of what was becoming the annual Australian fire season. All around Sydney, 160 miles northeast of Canberra, some fifty brush fires were consuming homes, disabling high-voltage transmission lines, and sending giant plumes of smoke for miles over the Tasman Sea. As electricity faltered, traffic lights failed, and Sydney's rush hour turned chaotic; wholesale electricity prices experienced a 10 percent surge. "Firestorms paralyse Sydney," summed up the *Sydney Morning Herald*. And that was not all: directly east of Bungendore, along the coast, other fires consumed thousands of acres. The fires were another demonstration of the failure of European environmental practices. Once the Aboriginals' tool, fire had not been used judiciously for a century and a half, and now it had become an omnivorous destroyer.

Standing at the apex of a U-shaped table, surrounded by a dozen or more of the commission's managers (all but one of them male), Blackmore showed a neatly typed chart on a slide projector. It was titled "The Opportunity in Water Recovery," and it contained a list of the possible sources of water for environmental flows and estimates of their potential yield in gigaliters of water:

	How Much
voluntary contributions	10–20 GL
system efficiency improvements	100–200 GL
farm upgrades	80–150 GL
retiring unsustainable irrigation	150–200 GL

OR

general reduction	500–1,000 GL

Blackmore considered the first two categories painless. By "voluntary contributions," he had in mind a program of government inducements to appeal to farmers' altruism by paying them below-market rates for water they weren't currently using. "System efficiency improvements" referred to steps the commission could take to improve the efficiency of water deliveries, such as converting open channels to pipelines to prevent evaporation. The problem was that these two steps brought the commission nowhere close to even the lowest environmental flow scenario, 350 gigaliters, with its "low" probability of returning the river to health. "Farm upgrades"—subsidizing farmers at least half the cost of converting to water-efficient systems such as drip irrigation and laser-guided field leveling in exchange for recouping some of the saved water—at least had the virtue of enlisting farmers voluntarily, but its water yield was limited. That left only what Blackmore called the "directed" intervention of "retiring unsustainable irrigation"—that is, banishing from agricultural use fields that are bound to become unsustainable. For example, a field contaminated with salt might remain productive for another twenty or thirty years, but if it lacks drainage, it will eventually become so saline that it won't be worth farming. The commission had already identified tens of thousands of acres of unsustainable fields, an unpopular act in itself. Requiring that such land be retired, Blackmore told his staff, was in social and political terms "the hand grenade with the pin pulled out." To get to a meaningful level of environmental flow, he said, the only alternative to all four of these steps was a fifth one—an across-the-board reduction in water allocations to all farmers, regardless of their operations' water efficiency. It would produce a bounty of water for environmental flows, but it was "dumb," Blackmore said, and it would produce vast resentment among farmers.

Though the chart's implications were distressing, it gave new specificity to what had become a stale conversation about the merits of environmental flows. Kevin Goss, Blackmore's deputy, later explained the chart's significance. "We were all caught up in what are the right principles, how would you acquire the water, and so on. Don jumped in and said, 'Here are the ways you can do it—boom, boom, boom.' "

15 THE RIVER MURRAY

If rivers are "the souls of the landscape," as a poet has put it, then the Murray confirms the notion. It is, for one thing, ancient. It's sixty-five million years old, which makes it three times as old as, say, the Columbia River. Compared to brash tropical rivers in their teeming neon greens, the Murray, with its soft, murky palette of Rembrandt golds and eucalyptus, looks grandfatherly. It is probably the site of the world's longest continuous human habitation, from the Aboriginal arrival in Australia (following the world's earliest known sea voyage) to the present. During the sixty thousand or so years that Aboriginals roamed the Murray Basin, they were nomadic but hardly aimless. They moved across the land in pulsing patterns, as if riding a tide, gathering for a ceremony or a seasonally abundant food supply, dispersing as the supply dwindled or the floodplain filled. Because of the Murray corridor's relative abundance (arising from the year-round presence of water), it was one of the few areas in Australia to house semipermanent campsites that could accommodate gatherings of several hundred people; what attracted them were wild fowl and freshwater crayfish in the springtime and fish in the summer. From reach to reach, the Murray still glistens with middens, the massive agglomerations of shells discarded by generations of riparian diners: you find yourself envisioning sated Aboriginals tossing shells over their shoulders as they eat and imagine the action repeated over eons. (The biggest of the middens, near Weipa on Australia's northern coast, weighs two hundred thousand tons.)

The most astonishing of the creatures the Aboriginals dined on was the Murray cod, the river's emblematic fish. The largest specimen ever caught weighed in at 249 pounds, which makes the species one of the world's largest freshwater fish. The Murray cod possesses sixty-million-year-old ancestors, and, in proper Australian faunal fashion, is markedly docile. John Koehn, a freshwater ecologist at the Arthur Rylah Institute for Environmental Research outside Melbourne, told me he'd hooked Murray cod, then gently eased them to the side of his boat, where they allowed him to reach into the water and pat their heads. After stunning the fish with an electric stimulus so that they could be tagged, he'd returned them to the river and held them in place until water began flowing through their gills. "They just lie there and look at you, and they don't worry about things too much," he said. "They drift around, and then eventually off they go." On the other hand, if he tugged at a Murray cod's line instead of coaxing it, it would go "absolutely berserk," making tagging impossible. In Aboriginal dreamtime stories, the thrashing of a giant Murray cod establishes the river's meandering course. Before European settlement, the river teemed with cod: the men who traveled with explorer Charles Sturt to the river mouth in 1830 caught so many that they tired of eating them. Now Murray cod are seen infrequently. For two years in the mid-1990s, scientists sampled twenty randomly chosen sites along the river and failed to catch one. In 2003, the Australian government declared that the Murray cod's survival was "a matter of national environmental significance" and listed the fish as threatened.

Sturt, the intrepid explorer who "discovered" both the Darling and the Murray, embodies a major strand of Australian self-image. As a young Briton, he crisscrossed the ocean for the empire. He fought for Wellington in the Peninsular War against the French and for the Canadians in the War of 1812 against the Americans; he served time in Ireland and sailed to New South Wales as a convict ship guard. Then he became an explorer. Tales of Australia's European explorers commonly combine bravery and buffoonery, as their blunders taught them that Australia was radically different from Europe. As in most Australian explorer tales, water is at the heart of Sturt's story, for his quest was to find the prodigious river or lake that European coastal settlers persuaded themselves must exist in Australia's interior. His exploration was complicated by the confounding complexity of basin rivers: sometimes they were flooded, at other times they were entirely dry, and between these extremes, a maze of branches masked their main channels. Sturt, of course, found no inland lake, but in 1829 he became the first European to lay eyes on the Darling, which he named after the governor who selected him to lead the expedition; a year later, he named the Murray for George Murray, the British colonial secretary whose favor he courted in a vain attempt at knighthood. As the Australian writer Bill Peach concludes, "Our greatest river is named . . . for a dim British politician who showed not the slightest interest in Australia."

There's a Panglossian quality to Sturt, as he lurches enthusiastically from crisis to crisis. Since his first expedition took place in the third year of a drought, drought naturally became his first lesson. He soon discovered that river levels were too low for his boat, which meant that it had to be carried overland, and drinking water became sparse. By the time he came upon the Darling, his party was desperate for water. He wrote, "Our difficulties seemed to be at an end. Our men eagerly descended to quench their thirst. Shall I ever forget the looks of terror and disappointment with which they called out to inform me that the water was so salt[y] as to be unfit to drink!"

Yet a year later, Sturt was on a second expedition, this time intending to find the Darling's outlet, Australia's grail, the presumed inland sea. In expectation of water, Sturt and his party of eight—two soldiers, two convicts, a carpenter, the son of New South Wales's colonial secretary, Sturt's personal servant, plus Sturt—carried sections of a prefabricated twenty-five-foot whale boat. When the party's wagon wheels bogged down to their axles in soft, impassable soil, the group assembled the boat and rowed down the Murrumbidgee (mur-rum-BE-djee), a Murray tributary, to the Murray. The men floated by magnificent 250-foot-high cliffs and diplomatically negotiated their way past hostile Aboriginals. After identifying the confluence of the Darling and Murray, they sailed on another five hundred miles to the Murray mouth, achieving yet another European first. But like so many other stories of Australian exploration, this one ended in disappointment and suffering, as the environment registered its contrariness. Before it reaches the Southern Ocean, the Murray disperses into a shallow lake of shifting sandbanks, named by Sturt Lake Alexandrina, and a narrow, ninety-mile-long wildlife-rich lagoon, the Coorong (COO-rong). Sturt realized that oceangoing vessels could penetrate neither body of water, which meant that the Murray would never connect inland Australia to the sea. Worse, the men had to row themselves back upstream for nine hundred miles, while living much of the time on flour and tea. The men endured rapids, Aboriginal threats, a puncture in the boat's keel, and scurvy. One of the convicts went insane, and Sturt was partially blinded.

Nevertheless, Sturt returned to the expeditionary trade one more time, in 1844, still in search of a vast inland sea—the water fantasists had shifted the presumed location to the northwest. This time he went up the Darling River and beyond it, toward central Australia. Early in the trip he wrote excitedly, "We seem on the high road to success . . . We have strange birds of beautiful plumage and new plants . . . It will be a joyous day for us to launch on an unknown sea and run away towards the tropics." Sturt's party included sixteen men, eleven horses, thirty bullocks, two hundred edible sheep, a year's worth of supplies, and, as usual, a boat. That boat must have looked odd in the blistering desert that Sturt wandered into, where the temperature burst his thermometer when it hit 127 degrees. Sturt wrote, "I cannot find language to convey to the reader's mind an idea of the intense and oppressive nature of the heat that prevailed." The party hunkered down around a shriveling water hole and waited out the six-month-long summer. One man died. By now obsessed, Sturt and an aide left the party to continue the search. He managed to reach the Simpson Desert, where he confronted "stupendous and almost insurmountable" sand dunes as long as a hundred miles. Here he gave up. Historian John H. Chambers reports that Sturt grew so emaciated that when he "finally stumbled into his home in Adelaide at midnight in mid-January 1846, his wife fainted away."

The transition from exploration to exploitation was short. Throughout the nineteenth century, the three European settler colonies watered by the Murray—South Australia, Victoria, and New South Wales—maintained fractious relations, and the Murray was a principal source of dispute. South Australia, the colony containing the Murray's last four hundred miles and mouth, had two interests in the river. As Australia's driest state, it coveted the river for its drinking water and also saw it as the state's commercial link to the inland. Its interest therefore was in promoting navigation by supplementing the river's erratic flow. After successfully negotiating the Lake Alexandrina sand bars, the river's first two shallow-draught paddle steamers began plying the Murray in 1853 and instantly inspired visions of the river as a navigation hub: a 1906 book even calls the Murray "Nature's Gateway to the Interior." Hundreds more riverboats were built in the Murray's inland ports, but they negotiated the river with difficulty. Chambers writes that "the system had a massive navigational problem: it was often too shallow for vessels drawing even a few feet; often too narrow, where overhanging trees could damage parts of the ships or cargoes; the Murray was navigable only from June to December even in normal years, the other rivers for even shorter periods; steamers sometimes stuck in the Darling mud for a year, waiting, as one historian puts it, 'for the river to arrive'; sandbanks shifted dangerously; driftwood accumulated on bends. And the rivers so meandered that, to progress a straight-line mile, a steamer traveled three miles of river." Out of this frustration grew South Australia's enthusiasm for weirs.

The two upstream colonies, Victoria and New South Wales, had little interest in navigation: they wanted to use the river's water for irrigation. To them, the flow of water to the Murray mouth was typical of nature's profligate wastefulness, which humans were morally compelled to redress with irrigation. As Paul Sinclair explains in *The Murray: A River and Its People,* "The idea of making the deserts bloom was an expression of faith in the power of European society to control and master the natural world." Victoria consequently established farmers' irrigation trusts and railroads to ship agricultural products to Melbourne. For its part, New South Wales rejected any Victorian claim to Murray water and declared that even along the twelve hundred miles of river that formed the border between the two colonies, it alone possessed the right to divert Murray water. To this day, the border incorporates the New South Wales position. Instead of running along the center of the river channel, as most river boundaries do, it follows the Victorian bank—the entire river channel therefore belongs to New South Wales.

Over the last half of the nineteenth century, the three states quarreled over their rights to charge tariffs on one another's river traffic, until drought brought the colonies together. To be sure, it was a spectacular seven-year drought, ending in 1902, and the colonies, promoted to states in the new Australian federation, brimmed with mutual mistrust. The three states sent representatives to a conference on basin management in 1902, then took thirteen years to reach a settlement. With some justification, current commission publications call the River Murray Agreement, the pact that emerged in 1915, "pioneering." It created the River Murray Commission, the forerunner of the contemporary commission, and charged it with running the main stem of the Murray—all its tributaries, including the Darling, were excluded. The agreement counted on satisfying all three states' water demands by using technology to expand the water supply. The River Murray Commission was directed to oversee construction and management of a large dam near the Murray's headwaters, a storage reservoir at the other end for the exclusive use of South Australia, and up to twenty-six locks and weirs on the Murray that would make the river navigable for more than half its length.

According to the pact, South Australia was guaranteed an annual allocation of water at least big enough to refill Lake Victoria, the natural lake that was turned into South Australia's water storage. Lake Victoria lies across South Australia's border, in New South Wales, but South Australia does not trust New South Wales in any matter involving water and won the right to manage Lake Victoria anyway. The two other states, New South Wales and Victoria, agreed to share equally the remaining portion of Murray flow. Since the commission's decision-making depended on consensus among the three states and the federal government, its power was constrained by the political realities in each state. Over the last two decades, the commission has changed its name, enlarged its staff and responsibilities, and expanded its board of directors to include Queensland and the Australian Capital Territory, encompassing Canberra, but its task remains delicate. With few policy-making powers of its own, it relies largely on the authority of its science to build consensus among its six disparate governments, pressing them to value basin-wide benefits over statewide political concerns.

The Murray originates in the distinctly un-Alpine Australian Alps, whose highest peak, Mt. Kosciuszko (KOZZ-ee-AHS-koe), is a mile and a half shorter than Mt. Blanc. The range is, nevertheless, a significant repository of snow, most of which melts and descends into the basin. Only a seventh of the Murray-Darling Basin's watershed contributes water to it on a regular basis, and most of that seventh is in the Great Dividing Range, of which the Australian Alps comprise the most elevated terrain. The first two-hundred-mile stretch of the Murray, from its source near Mt. Kosciuszko to the town of Albury, is by far the steepest, embodying three-fifths of the river's vertical descent though comprising only a seventh of the river's length; the headwaters collect runoff from a little more than 1 percent of the Murray-Darling watershed, but they provide nearly two-fifths of the river's flow. As the Murray slows and flattens, it slides westward for some eleven hundred miles through a floodplain of intensifying heat and aridity. In this zone, the Murray's water becomes precious. Much of it is lost to evaporation: indeed, throughout the lower floodplain, evaporation claims at least two and a half times as much water as rainfall provides. Other water is lost to plant transpiration, soil infiltration, and, increasingly over the last century and a half, human use.

Through much of the floodplain, the Murray's path is ornate, forming and re-forming elaborate meanders that from the air look like unraveling scrolls. The meanders endlessly reshape themselves, carving a curve into one bank, depositing silt on the other, constantly abandoning old beds for new ones. A scroll segment that gets left behind becomes a mere pond, a watery squiggle, separated from the river, and therefore dependent on floods for water replenishment. The name for this kind of pond is billabong, a resonant and evocative Aboriginal-derived word whose mention suggests all that is singularly Australian. With their brackish, multihued water, billabongs don't look inviting, but they are crucial to the riverine ecosystem. Extrapolating from the findings of a wetlands census of the River Murray, scientists estimate that the entire Murray-Darling Basin contains more than fifty thousand billabongs. They are natural biochemical test tubes, in which nutrients brought by the occasional flood mix with a profusion of life-forms, fostering a microscopic abundance that is partly returned to the river during ensuing floods. Billabongs are far more productive than the river, and their animal population density is a hundred times greater than the river's reservoirs. Unlike the river's water, theirs is calm, which makes them superior breeding habitats. They teem with microorganisms consumed by insects that in turn are eaten by aquatic and terrestrial birds, fish, and freshwater crustaceans.

The Murray also throbbed. Just as the Aboriginals moved in pulses, so did the river: to a degree, the two movements were linked. Sinclair advises, "The pre-regulation Murray was more like a long continuous ocean shoreline than a conduit carrying water from the mountains to the sea." Between floods, the Murray often appeared to wither, occupying a dwindling portion of its riverbed until, during extreme droughts, it was reduced to a chain of saline water holes. Inevitably, however, a flood occurred, and the river's banks no longer sufficed to contain it. The increased flows renewed life not just in the river but also in the swamps, the billabongs, the riverine lakes, and not just in reservoirs of water but also across the floodplain, as far as it reached, as much as several miles from the main channel in both directions. Basin floodplains typically possess between a hundred and a thousand times as many species as their rivers. In *Running Down: Water in a Changing Land*, Mary E. White explains, "Within hours of wetting, floodplains teem with tiny aquatic species, mostly microscopic, starting the food web—and opportunistic species come from afar. New waterborne species are introduced in the floodwaters, isolated pockets of water are connected with each other and with the parent river, and breeding cycles of plants and animals are switched on." Increased water levels triggered spawning in many Australian fish species, and the floodplain inundation meant that their larvae would enjoy an abundance of food. Sustained by the short-lived profusion, water birds and frogs bore offspring. Whether the flood was small (once every year or two), medium (once every five or ten years), or large (once every twenty to a hundred years), it played a vital role. The small and medium floods met the billabongs' water needs, and the big floods did the same for the floodplain.

Drying, too, served a purpose. During the periods between floods, bacteria slowly decomposed floodplain and wetland vegetation, which relinquished its carbon and nutrients to the soil. When the next flood occurred, the carbon and nutrients were swept up in the water, available for consumption by plants and animals. If the drying process was interrupted before completion, its benefits were erased, as soil bacteria ate the nutrients before they could enter the water column. In contrast, extreme drying killed the bacteria, enabling the release of more nutrients during the next flood.

After a left turn at the town of Morgan, about seven-eighths of the way down the river, the Murray slithers south through limestone cliffs on its way to the mouth and the blue Coorong. There, as many as 250,000 birds once fed during the summer. The basin is saucer shaped and inward draining: the river's flow is so sluggish that mere escape to the ocean is problematic.* The Lower Murray's gradient is so slight that until barrages were installed in the early twentieth century, tides influenced water levels thirty-five miles up the river.

The river system that remains is a Rube Goldberg contraption of dams, weirs, locks, barrages, flow regulators, canals, pipelines, pumps, and more than a thousand water diversions superimposed on tracts of swamp, marsh, meander, creek, anabranch, billabong, and gorge. It is an overtaxed apparatus bristling with valves that tries simultaneously to support farmers, city dwellers, and the basin's environment, even when their interests profoundly clash. Upstream, the machines deliver water to

---

*Despite this, Fodor's 2002 Australia blithely alludes to the "mighty Murray," which tells you all you need to know about guidebook prose and the temptation of alliteration.

farmers, whose use lowers flow and intensifies salinity; downstream, where people drink the water, other machines *reduce* salinity. The farmers worry about salinity's impact on their soil, but they're less interested in other aspects of water quality, including the effects of the chemical and biological runoff from their fields and feedlots, while the municipal consumers in South Australia—at "the bottom of the drain"—deplore the horrendous quality of their drinking water. Both the farmers' and city dwellers' concerns have overridden the environment's.

Even the Murray's source has been overridden. In 1974, after twenty-five years of construction, the Commonwealth, Victoria, and New South Wales governments completed the massive Snowy Mountains Scheme, which straddles the basin's perimeter at the crest of the Australian Alps—it entails sixteen large dams, seven power stations (supplying peaking power to five of Australia's eight largest cities), eighty miles of tunnels, and fifty miles of aqueducts. The Snowy scheme effectively enlarged the Murray Basin by diverting water that would have flowed southward, down the three-hundred-mile Snowy River to the sea, and instead sent it westward, into the Murray and the Murrumbidgee. The scheme diverts 572 gigaliters per year to the Murray and 550 gigaliters to the Murrumbidgee—comprising a tenth of the Murray-Darling Basin's water supply. The result is that from its source, regardless of season, the Murray teems with water; even during a drought, the river carries more water than it would have in a pre–dam era average rainfall year. In the predam regime, snowmelt poured down the river in the winter and spring, and dwindled steeply until the river reached its annual autumn low; now the flow peaks in the summer, when farmers want it, nullifying the seasons. In droughts (including the current one), when the Murray should have been nearly dry, it continued to brim with water. The riverbed never dried out; the floodplain was rarely flooded.

At least the Murray remains a conduit. On the other side of the Australian Alps, the upper reach of the Snowy River was reduced to a trickle, with all but 1 percent of its flow from the project area diverted to the Murray-Darling Basin. Tributaries farther downstream still contributed to the Snowy's flow, so that at its mouth it possessed more than half its predam volume. Nevertheless, the upper reach, once known for its world-class trout and bass fishing, suffered an environmental collapse. Part of the river turned into mud pools; in some stretches, it disappeared or turned saline. The water table dropped, killing livestock farmers' fodder plants. "No one thought we were going to lose the river," said a resident cited by the Snowy River Alliance, a community group. In response to the controversy, the New South Wales, Victorian, and Commonwealth governments agreed on an environmental flow plan, which went into effect in August 2002. Its long-term aim is to restore 28 percent of the predam flow downstream from the Snowy Hydro dam, but that number may be optimistic: for now, the restored water has doubled the flow at one upstream reach from 3 percent to 6 percent of its predam flow. Yet the Snowy's environmental flow plan put the Snowy in direct competition with the Murray-Darling Basin, for the water used to augment the Snowy's flow was rediverted away from the basin. Blackmore had no objection to the Snowy plan per se, but he was not pleased that it was implemented before the Murray-Darling Basin's.

All the way down the river, the changed Murray regime registers ripples of consequences. Consider fish. The water that would have flowed from the Australian Alps in spring is held in the Hume and Dartmouth reservoirs until farmers want it, through the summer. And the water that flows in the summer is typically released from chilly reservoir bottoms and is therefore as much as eighteen degrees (Fahrenheit) colder than natural summer water. That's too cold for many native fish to spawn. The river's man-made barriers prevent fish from swimming up and down the river to spawn and mature, while navigational needs forced the removal of snags—fallen tree limbs and other obstructions that provide fish habitats. The commission long ago installed fish ladders—water-filled ramps meant to ease fish over man-made obstructions—at two weirs, but the design relied on data about North American salmon and was thus pitched too steeply for the Murray's much less energetic fish. As a result, the ladders were a total failure. The lack of substantial flows and the construction of barrages near the Murray mouth have depleted the once-magnificent Coorong estuary; there, the silver perch and catfish that once amounted to 40 percent of the estuary's biomass are now rarely seen. Of the river's twenty-nine species of native fish, seven are considered threatened; at the current rate, half the Murray-Darling Basin's native fish will be endangered within fifty years. Meanwhile, exotic species have flourished. Thirty years after their introduction to the Murray in the early 1960s, carp alone comprise a large majority of the basin's fish population. For the river, that is exceedingly bad news. Carp are bottom-feeders whose eating methods undermine basin riverbanks, kill aquatic plants and habitats, and stir up settled sediments, releasing nutrients that help cause toxic blue-green algae attacks. In 1991, the Darling River experienced the world's longest blue-green algae bloom ever recorded—more than six hundred miles.

For all of the interventions meant to facilitate human use of the Murray, the path to South Australia's guaranteed summer allocation of 7,000 megaliters per day—seven thousand Olympic swimming pools—is still filled with obstacles: South Australia drinks from the Snowy bathtub through a slender Murray straw. The 150-mile reach downstream from Hume Dam to the highest weir on the Murray, Yarrawonga, is the first constraint, limiting the flow to 25,000 megaliters a day. A third of the way down the river, the flow encounters the Barmah Choke, at the heart of the world's largest and most splendid remaining river red gum forest. The word "choke" is apt: it's a dramatic constriction that confines the riverbed to a fifth of its upstream capacity. When the summer flow exceeds the choke's 10,000-megaliter-a-day flood level, the Murray produces man-made summer floods, drowning the surrounding forest when it is used to drying, causing another jumble of adverse environmental effects. The commission is constantly forced to weigh the well-being of the forest against farmers' requirements and the imperatives of the South Australian water entitlement. The commission sends another 2,000 megaliters a day down the Edward River, a 340-mile-long anabranch that bypasses the choke, and has even resorted to using spare capacity in irrigation channels, but these diversions still are not enough, particularly since a canal downstream from the choke diverts 4,400 megaliters a day to farmers in Victoria. The last constraint is South Australia's water storage, Lake Victoria itself. Lake Victoria, once a beguiling natural lake, was buttressed with levies and turned into a storage in 1928, nearly seven decades before anthropologists realized it was also the site of a huge Aboriginal burial ground. Now strict rules designed to protect the burials require water managers to mimic natural flows: lake operating levels now are more often full in winter and low in summer, as in the predam regime. All these constraints put a premium on the timing of water releases from Dartmouth and Hume. In years with substantial rainfall, they may discover that they've wasted water by releasing it too early, while in dry years they must be sure that enough water is flowing down the Murray to meet their delivery obligations. All this makes managing the river an intensely fretful occupation.

16 THE BARMAH CHOKE

In midautumn, when I visited the Barmah Choke, the river's flow should have been near its annual low, but of course it wasn't: the water lapped at the choke's diminutive brim as it hurried down the channel to meet all its farming and municipal drinking requirements. "Barmah" is an Aboriginal word meaning "meeting place," and the reason for the name is obvious: it's a forest of red gums straddling the river just upstream from a major fork, a natural gathering place. For such an old river, the choke is strikingly young; its formation is recent enough to be described in Aboriginal lore. About twenty-five thousand years ago, between the contemporary towns of Deniliquin and Echuca, a forty-five-mile-long slab of basin floor rose up as much as forty feet, blocking the river with a natural dam. In the thousands of years since, one part of the river has turned south, forming the Murray's modern channel; another part went north, creating what is now the Edward River anabranch; and a third part backed up, producing at first a lake, now a wetlands and a forest. Designated by the commission as one of the Murray's six iconic sites and awarded protection under the Ramsar Convention, the choke evokes what all the Murray wetlands once looked like at the apex of spring: refulgent, green, teeming with wildlife. Even now, the forest supports 350 indigenous plant species, 200 bird species, and 50 mammals. For migratory birds and native fish, it's a crucial habitat, a pocket of biological richness surrounded by degraded, treeless land. Like the other Murray wetlands, Barmah functions as buffer zone, as shock absorber. During wet periods, the wetlands mitigate floods' effects by flattening their peaks and storing their water; in droughts, they provide refuge for wildlife. Wetlands afford breeding and feeding habitats for animals on every rung of the food chain; they trap sediment and recycle nutrients. The trouble is that in the regulated Barmah, springtime is perpetual. During the current drought, when land a few miles away was sere, the choke looked exaggerated in its greenness, as if caked with makeup, a Hollywood version of itself, headed toward becoming an eco-Disneyland.

For all the esteem it reaps in environmental and water-management circles, Barmah is not choked with tourists. I happened to visit the choke on the day before Easter, the biggest Australian holiday, when Echuca, the nearby river town/tourist attraction/paddle steamer nostalgia factory, was clogged with cars bumper-to-bumper. Even so, a half hour's drive away, at the downstream end of the choke, I found mostly empty seats on a late-afternoon voyage offered by the choke's only ecocruiser, the thirty-seat *Kingfisher*. The choke, so narrow at points that people can stand on opposite banks and talk across it, feels at first like an oasis. For the full length of the two-hour cruise, both shores were lined with river red gums: they sprouted from the soil like sprigs of elongated broccoli. Unlike in Chowilla, we saw few unhealthy trees. There were healthy red gums in three dimensions, extending outward from the banks in both directions, as far into the forest as we could see; long green grasses filled the gaps between trees. We saw black swans form a perfect V as they flew overhead, and spotted, high up in a tree, a sleeping koala (though all I could make out was a brown blob). Some red gums leaned far out over the river, while others had fallen all the way in: the blackening hulks, no longer removed to foster navigation, provide fish habitats. Near the end of the cruise, high in a red gum that arched over the river, we passed under a sea eagle's nest and, beside it, saw the elegant white-headed bird itself—a nice moment.

The choke's allure veils the evidence of its degradation, but the cruise offered numerous hints. I don't include here the high-decibel motorcycles that periodically buzzed through the campsite behind the boat landing, nor the couple on jet skis who roared through at high speed, oblivious to the bank erosion prompted by their waves—those offenses were obvious enough. I also don't include the river's recreational fishermen, revealed at nearly every bend aboard their boats or onshore, shirtless and pink skinned and flabby, as startling as rhinos in Antarctica. I don't even include their intended catch—Murray cod, golden perch, the endangered trout cod—presumably what we saw cooking on their portable barbies. It's carp, however, that dominate the Murray: by one rough estimate, they comprise 70 percent of the river's fish biomass. Unlike native fish, which depend on river variability to provide spawning cues, carp benefit from the managed river's stability. While native fish fare better in the choke than elsewhere in the Murray, the snake population here has crashed, probably because settlers did their best to exterminate them. We did come across a dead snake, apparently killed by a fisherman and put on display, coiled around a twig, for our delectation. Large lakes on both sides of the choke once supported a commercial fishery and a medicinal leech operation, but both died for lack of product. Hundreds of thousands of birds formerly populated the choke; now bird sightings are events. We spied some of the twelve hundred cows that graze the Victorian side of the forest, but the damage cows have perpetrated on it—consuming vast expanses of reeds to the point of local extinction and trampling what vegetation they don't eat; eroding riverbanks, compacting soil, releasing sediment into the river, and polluting waterways with their waste—went unmentioned. As if to score a point in the cows' favor, a passenger asked whether it was true that the cows' grazing fees are applied to their upkeep, and the tour guide concurred. She didn't explain, however, that the fee—less than twenty American dollars per animal per year—is six to ten times less than grazing rates on private land and covers only about a tenth of the cost to the Victorian government of managing the animals: it's likely that grazing wouldn't be profitable around the choke if not for the government subsidy. And if the cows were on private land instead of Barmah state land, the Victorian government would have paid the landholders to fence off grazing land from wetland riverfronts; public land, meanwhile, goes unfenced. From the *Kingfisher*, we saw no evidence of logging, but that's only because loggers don't operate next to the river. Two-thirds of the logging on the Victorian side of the choke produces low-value firewood, most of which ends up in the fireplaces of gas-heated Melbourne homes. The loggers take mature trees, depriving many animals of habitats.

Even so, the forest's fundamental problem is the river's flow regime. Over centuries, sediment has raised the choke's riverbed above the surrounding plain; this elevation combined with the choke's narrowness is what makes flooding so frequent here. But flooding no longer happens as it used to. All but the biggest winter and spring floods have disappeared in the dam era; floods lasting four months, which are big enough to cover half the forest, occurred three out of five years before dams, but one year out of five since. At the same time, small localized floods covering less than a tenth of the forest happen eight times as often as before regulation. These floods reflect no natural phenomenon: they occur unseasonally, and arise from the way the river is managed. In the summer and autumn, when farmers want irrigated water, it is funneled down the choke at its maximum capacity. Farmers' water orders are taken into account to maintain the maximum level, but if it rains, farmers can cancel their orders. The problem is that it takes water four days to flow from Hume Dam to the Barmah Choke, and once it leaves the dam, it can't be retrieved. The farmers reject the water, but it continues to flow down the Murray, and since the choke is already at full capacity, it floods. These so-called rain rejection floods are usually small, and reach only the lowlands near the river. For many years, water managers thought they were environmentally beneficial, since they bring additional water to the forest, but in the last two decades that assumption has been proved wrong. The drop in large spring floods causes the forest's high ground to dry out, while the increase in small summer and autumn floods waterlogs lowland close to the river. On the edges of the forest, far from our vantage point on the *Kingfisher*, red gums are suffering from floods' absence; close to the river, the red gums are supplanting grasslands because of floods' overabundance. Formerly a mosaic of habitats, the floodplain has grown more homogenous and less vital. Hydrologists have pointed out that river rejection floods can be eliminated either by lowering the maximum flow allowed through the choke, giving the channel room to conduct the rejected water, or by temporarily stowing the excess water in an upstream storage lake. It may be indicative of how little flexibility exists in the system that the commission has so far embraced neither option.

17 THE MITTA MITTA VALLEY

The road from Hume to Dartmouth, from the queen to the king of dams, leaves behind the desolate Hume reservoir, from whose exposed floor hollow remains of long-drowned trees still rise, wraithlike. In the midst of drought, the water level was so low and the reservoir floor so flat that its waterless portion went on for fifteen or twenty miles; as I drove by, I could identify the reservoir's edge only as the point at which the dead trees stopped and the live ones began. It was a relief when the road finally abandoned the reservoir and began its climb through the narrow Mitta Mitta Valley, the seductive, pastoral exception to the basin's rule. Most of the basin is flat, parched, treeless; the Mitta Mitta Valley is sloped, amply watered, abundantly treed, and luxuriantly grassy. Downstream, as the Murray accepts the multihued flows of its tributaries, it turns increasingly muddy, until it arrives in South Australia as pea soup; here, high in the basin, rocks a few feet beneath the surface of the Mitta Mitta River are blurred but visible. Perched on the western face of the Australian Alps, the valley catches thirty-five inches of rain a year, triple the amount downstream. Many reaches of the 135-mile-long river are narrow enough to throw a beach ball across, but each spring, as snow on the Great Dividing Range melted, the tributary could be counted on to flood; in fact, "Mitta Mitta" is derived from an Aboriginal word meaning "thunder," apparently a reference to the sound of the river near its source, where it tumbles over large boulders. The combination of rainfall and flood made the valley a lush grassland, which European settlers turned into cattle country. Unlike other farmers in the Murray Basin, those in the valley had no use for irrigation—they said it wasn't a part of their culture.

All that changed in 1977, when the Dartmouth reservoir began to fill, and the river's flow was cut to 120 megaliters a day, a relative trickle. From one end of the valley to the other, the Mitta Mitta River was isolated from its floodplain, and the perennial grasses that had covered the floodplain died. Over a seven-month period in 1982 and 1983, a substantial amount of water—up to 12,000 megaliters a day—was released from Dartmouth to provide drought relief downstream, but for Mitta Mitta farmers, it brought only a reprieve. Within a few years, a majority of the valley's seventy or so full-time farmers went out of business or drastically cut their operations, while the remainder installed irrigation systems. For all of them, irrigation was a huge expense, which they incurred even as their incomes fell. The irony offered no consolation: thanks to Dartmouth, the only farmers in the basin with a naturally abundant water regime now had to irrigate to survive. Earnest irrigation authorities encouraged the farmers to install flood irrigation, the simple but wasteful technique of delivering water across graded fields in thin, broad sheets. Alas, the farmers had to intensify fertilizer use to complement irrigation, only to discover that flood irrigation promoted leaching of the fertilizer into the groundwater and river. By the time this was understood, the farmers had gotten used to flood irrigation and had little interest in spending money on more efficient spray or drip techniques. The onset of irrigation forced the farmers into the murky, frustrating, bureaucratic world of water rights, for which they'd previously had no need. All three Murray states allocate water rights to their farmers, permitting them to withdraw a specified amount of water from Murray storage that correlates to their farms' size and crops. Up to the 1970s and sometimes beyond, farmers could secure water rights with few restrictions; as the realization of scarcity set in, obtaining them became increasingly thorny. To secure allocations, Mitta Mitta farmers had to show where they would install irrigation. That set up a vicious circle. If they lacked the money to install irrigation, they couldn't get a water allocation, and without the allocation, they'd never make enough money to start irrigating. On the other hand, the farmers who got the allocations discovered that water was cheap. In a bit of holdover logic from predam days, water's price in the Mitta Mitta Valley reflects little more than the cost of extracting it from the river—and since valley farmers pump water from the river themselves, the price is minimal. With water prices low and flood irrigation prevalent, water efficiency in the valley remained a low priority.

Halfway up the valley, I stopped at Mac Paton's farm. The handsome brick farmhouse sits upon a knoll overlooking green pasture bisected by an S-shaped strand of river, and struck me as the most successful approximation of a bygone Britain I'd seen in the basin. Paton's great-great-grandfather, a Scot, started the farm in the 1880s, and the farmhouse was built in 1906. The spacious kitchen, through which I was led, evinced a Dickensian eventfulness, filled as it was with jovial women in the act of cooking. "I'm the proverbial mother-in-law!" an older woman declared, before she and her daughter, Paton's wife, burst into laughter. The picture windows in the living room looked out over rolling hills to weeping willows planted at the river's edge. Behind the house, a grand copper beech tree, another import, shaded a lawn tennis court that looked invitingly playable, if brown and slightly neglected.

Paton, who is in his late fifties, is fair-haired, ruddy-faced, and strapping, with Northern European looks that are complemented by Australian congeniality. "We were the victims of a dam here," he said, without seeming to bear a grudge. With six hundred milk cows on 640 acres, he runs the biggest dairy farm in the Mitta Mitta Valley, and he's the chairman of a committee representing the farmers to Goulburn-Murray Water, the government-owned irrigation provider in Victoria. When the farmers realized that the dam had upended their water regime, they launched a campaign for compensation. Compared to the thousands of farmers on the flatlands downstream, they had scant political influence; their only asset of much significance was a study that Paton heard about "through the back door." As it happened, Dartmouth was one of the first Australian dams whose construction was preceded by a study forecasting its environmental impact. According to Paton, the document's predictions lent legitimacy to the farmers' complaints and led to the funding of more studies confirming the forecast's findings. Armed with all that, the farmers complained to Goulburn-Murray Water, which tried unsuccessfully to resolve the issue by offering to increase the farmers' water entitlements. The farmers turned to the commission, and won Blackmore's support; he acknowledged the injustice dealt them and addressed it. Fifteen years after Dartmouth was completed, the Ministerial Council gave the commission authorization to start negotiations with the farmers. Since many dams throughout Australia have caused downstream damage that was never acknowledged, let alone compensated, dam officials outside the basin feared that the commission would reach a precedent-setting agreement with the farmers. Instead, the agreement called for "ex gratia" payments totaling $2 million to about ninety farmers. The commission avoided admitting liability while winning the Mitta Mitta farmers' goodwill.

Over time, irony piled on top of irony: Mitta Mitta farmers got used to irrigation and started to fear floods. This was understandable, since the character of floods changed under the man-made flow regime. The floods that struck the valley in the pre-Dartmouth era crept across the gently sloped pastures in a layer at most a couple of inches thick, and dissipated in a day or two. But during the last flood, in 1996, Dartmouth Dam held back water to reduce the flood's peak, then gradually released it, extending the flood's duration: in the valley, the flood lasted three weeks, and the pane of water was three or four inches thick. By the time the flood subsided, the pasture grass had died, and the riverbanks were badly eroded. As if that was not reason enough to dislike floods, their increasing rarity left young farmers without experience in dealing with them. As at the Barmah Choke, sedimentation over centuries has lifted the Mitta Mitta River's bed above the surrounding grassland. When the 1996 flood began, some cows took refuge on the higher ground of the riverbanks, where they became stranded. Releases from Dartmouth take at least a couple of days to reach the valley, so farmers had time to move their cattle to high ground away from the river, if only they'd thought of it. But with floods becoming less and less common, Paton fears that by the time the next one arrives, none of the farmers will have flood experience, "and they'll have no idea what they're in for."

In the spring of 2002, with dam levels low and a flood unimaginable, the Mitta Mitta Valley was again a zone of contention because of the drought. The dispute that broke out was a variation on the endless succession of interstate water squabbles that began in the nineteenth century. All that had changed was that the drought rendered water particularly precious, while removing the last megaliter of flexibility from the water-delivery system. When Hume Dam's core wall suddenly moved a quarter of an inch in 1996, the commission made extensive repairs, then overhauled Hume operations. Under new rules, the Mitta Mitta River's regulated flow was never to exceed its carrying capacity of 10,000 megaliters—ten thousand Olympic swimming pools—a day. The provision aimed to reduce the possibility of more man-made floods in the valley and cemented good relations between Mitta Mitta farmers and the commission, but it also turned the Mitta Mitta River into the Murray system's first bottleneck. Because of the drought, farmers downstream were growing desperate for water. In Victoria, a dairy industry estimate pegged the loss to the average dairy farmer throughout 2002 at $85,000. Farmers without water loaned their cows (and the milk the cows produced) to farmers who had water. Others sold their cows, even though that jeopardized their capacity to recover when the drought ended. So many dairy farmers sold their cows that the going price dropped sixfold. The abattoir in the Victorian town of Tongala, which specializes in the production of ground beef, doubled its daily kill. In New South Wales, only ten or twelve of the one thousand rice farmers in the state's largest irrigation district were even trying to grow rice. *The Australian*, the country's national newspaper, said the country's clergy were telling their congregations to pray for drought-affected people.

Yet the commission could not get water from the Dartmouth down the river as fast as Blackmore would have liked, for a set of snarled, interlocking reasons arising from the scarcity of Murray water and the clogged path downstream. Since South Australia is guaranteed a specified monthly allotment of water from the Murray, New South Wales and Victoria are pitted against each other in a never-ending competition for the remainder. Of the two states, Victoria claims the higher moral ground: it consumes less water than New South Wales in return for a greater reliability of water delivery—it annually diverts on average about 15 percent less Murray water than New South Wales so that it will be more likely to have water in dry years. By comparison, New South Wales is a go-for-broke guzzler, which takes all the water it can get, even if that means receiving much less in dry years. In bountiful years, New South Wales farmers cash in; during droughts, many curtail operations. Victorians call New South Wales farmers "immature" because they seem unable to restrain their water appetite; New South Wales farmers answer that their state policy at least allows them to decide for themselves when they'll use their water, while the Victorian government makes that decision for its farmers. Both positions reflect the size of farms and the composition of crops grown. In Victoria, the small farms that the state government bequeathed to World War I and II veterans were best suited to horticulture and dairy, while the larger farms of New South Wales were more appropriate for cereal production. Thus, in Victoria, farming is perennial. The predominance of pastures given over to fruit orchards, grapevines, vegetables, and grazing cattle means that farmers need water every year, without exception, or their plants and animals die. In New South Wales, annual crops reign, so the reliability of water deliveries from year to year is less crucial. Despite being a prodigious water consumer, rice is New South Wales's predominant crop, and wheat is a distant second. The great advantage of New South Wales's cereal crops over horticulture and viticulture is that in drought years farmers can simply skip planting them. The farmers don't make money, but they save the cost of seeds and planting.

A system of entitlements determines the amount of water that the states receive annually; the states then divide their proportions among their farmers. In three out of four years, New South Wales farmers receive their entire water entitlements or more; in a few years of plentiful rainfall, their allocations have approached 200 percent. But allocations in 2002 reflected the risks of New South Wales's policy, for its farmers were receiving only 10 percent of their entitlements, while Victorian farmers were getting 129 percent of theirs. Snowy Hydro Limited, owner of the sixteen hydroelectric dams atop the Australian Alps, might once have come to New South Wales's rescue, but in the previous year it was "corporatized." Though it was still government owned, it was newly charged with maximizing profit, and therefore usually timed its water releases to coincide with peaks in electricity demand. Since demand is highest in the summer, when air-conditioning drives up electricity use, Snowy Hydro was presumed to be hoarding its water for release then.

Thus it was that in October 2002, Murray Irrigation Limited, the farmer-owned New South Wales irrigation company, approached the Mitta Mitta farmers and the commission with a proposal to flood the tributary. It would pay the farmers nearly $200,000 to allow a 10 percent increase in the Mitta Mitta's flow, an extra 1,000 megaliters a day, for ninety days. Since the New South Wales farmers had already used up their annual water allocation, Murray Irrigation asked the commission to make an exception, allowing the new water to be subtracted from next year's allocation. Blackmore hated the idea. He didn't want to jeopardize the commission's hard-won amicable relationship with the Mitta Mitta farmers, and he didn't want to reward the New South Wales farmers for losing a gamble on their highly unstable water supply. The Mitta Mitta farmers sided with the commission and rejected Murray Irrigation's offer; Mac Paton called the offer a bribe.

The tale might have ended there, except that now, two months later, Blackmore himself was proposing to flood the Mitta Mitta. The combination of the drought and the shift in Snowy Hydro's release pattern left the commission without water to meet its obligations in all three states. Predicting flows on the Murray rests on a set of highly imprecise mathematical variables that try to account for the river's myriad diversion and frequent surprises: this is a river, after all, that can flood on one end while the other end experiences a drought. ("Water systems at either end of the hydrological cycle bring up new things all the time," Denis Flett, the Victorian irrigation company chief executive and a Murray-Darling Basin commissioner, told me. "You think you know the bloody beast, and suddenly you've got a circumstance you've never encountered before.") Now the commission's computer model was predicting a 30 to 40 percent chance of a shortfall on the Murray of 40 gigaliters for February, only two months away—that is, the commission might fall 40 gigaliters short of the allocations it was obligated to deliver to the three states. If the shortfall occurred, Blackmore would face three unpalatable options. He could impose restrictions on water supply to all three states, even though doing so would intensify farmers' water shortages. He could reduce the flow to South Australia to a level below its entitlement—something that had never happened before—and then try to make up the amount in future months. Or he could pay a high premium for more water from Snowy Hydro.

To avoid having to choose among those options, Blackmore, Flett, and David Dole, a senior commission manager, met beneath the beech tree in Mac Paton's backyard with fifteen or twenty valley farmers and hammered out another deal. To feed the downstream demand, Blackmore proposed a "trial," which would raise the Mitta Mitta River's flow by 300 megaliters a day, enough to lift the river's level a mere two inches, which probably would keep it inside the river's banks. If it did overflow, the trial would immediately stop, Blackmore promised. In fact, he said, any Mitta Mitta farmer could object at any time, and the effort would end; otherwise, the commission would raise the level a second two inches, and no more. The four inches would add 600 megaliters a day, three-fifths of the total the New South Wales irrigators proposed, to the existing 10,000-megaliter-a-day flow, and would make up nearly half the possible 40-gigaliter shortfall downstream. The commission promised to repair all damage caused by the flooding, but this time it ruled out even ex gratia payments. New South Wales had offered to pay the farmers, whereas Blackmore did not, but Blackmore's proposal would not flood valley fields, whereas the New South Wales proposal would have. Eager to maintain their good relations with the commission, the Mitta Mitta farmers accepted Blackmore's offer, not New South Wales's. Blackmore contended that the first offer asked the commission to advance next year's water to New South Wales to make the state good after it followed a reckless policy that failed, while the second offer responded to an emergency and served farmers in all three states.

Not surprisingly, the commission's agreement with the Mitta Mitta farmers did not go over well in New South Wales. George Warne, general manager of Murray Irrigation Limited, argued that the only difference between the two deals was that one drew from next year's water supply, and the other from this year's. Warne wrote in a sarcastic e-mail to David Harriss, a New South Wales deputy commissioner, that he hoped the water Blackmore secured was "for exclusive use on 'non-shitty' (their words, not ours) irrigation practices performed in World-best-practice-Utopia (ie Victoria and South Australia)."

At the same time, Blackmore's proposed solution showed the complexity of the demands placed on the commission. If his long-term goal was environmental flows, he was acting here at cross-purposes to it, for the emergency he was responding to was agricultural, not environmental, and the increased supply of water would not help the Murray ecosystem. An unregulated Murray would have reflected the drought with extremely low flows, but now, despite the drought, the Murray flowed unnaturally well.

It is a measure of the destructive power of a collapsed dam that the first item on the agenda at the December 4, 2002, board meeting of River Murray Water, the commission's business arm, involved neither the Mitta Mitta Valley nor environmental flows, but terrorism. To the extent that the September 11, 2001, World Trade Center and Pentagon assaults unnerved Americans, the Bali bombing of October 12, 2002, did the same for Australians. Eighty-eight of the 202 people killed were Australians, and the Australian government assumed that its citizens were the intended victims. It's easy to envision dams as targets of terrorism, for the torrents they could unleash would be devastatingly lethal—but not during a drought. Both the Dartmouth and Hume reservoirs were sufficiently low that even if the dams tumbled down, no flood would ensue. Blackmore pointed this out to the nine other men who sat around a sleek elliptical table in the commission's fourth-floor conference room, and the board's president Roy Green mildly proposed to write a letter requesting police to make their dam patrols more frequent and less predictable. Then the discussion moved on. (Left unspoken was the unsettling idea that if Dartmouth or Hume were severely damaged even during the drought, the months or years that their repair would require would be sufficient to send Australia's agricultural economy into a depression, not to mention delaying environmental flows.)

Of all the rivers in the Murray-Darling Basin, the Murray remains the only one that the commission actually manages; state authorities run all the others. Even on the Murray, state employees carry out many of the commission's decisions. Nevertheless, because of the river's unique status, what happens to it has broad implications. If the commission succeeds in making the Murray "a healthy, working river," then the commission's authority is more likely to spread to other rivers in the basin, and its ideas to other basins. River Murray Water runs the river's dams, stores and delivers its water, manages its salinity, navigation, recreation, and tourism, and gets paid by state governments that in turn recoup their costs from client irrigation companies. Unlike the commission office's regulatory and resource-management policy division, River Murray Water is meant to function as a business. Its eight board members include water officials from the three Murray states and the Commonwealth government, an outside business management expert, and three officers of the Murray-Darling Basin Commission.

The next item on the meeting agenda was the drought. "We are facing very difficult circumstances," Blackmore warned the group. "Assuming no change in rainfall in the next few months, this will be as difficult as any period we've ever experienced . . . It's very easy to think, 'Are there things we could have done to avoid it?' but it's not very productive."

Blackmore got more specific. "Should we have started releases from Dartmouth earlier? But it didn't make sense at the time. We're all trying to work out how we got into this situation. It's bloody dry. In the southern basin it's been three degrees [Celsius, or 5.4 degrees Fahrenheit] hotter than it has been historically. We're always going to run on minimums, we're always right on the edge, and these other things tip us over. We're definitely in uncharted territory, and our models don't pick this up all that well . . . Our problem will be a sequence of very hot days in January and February, [when we'll be] regulating cautiously and starting to lose flexibility [in water supply and delivery]. How much do we release out of Hume to buffer that?"

The board was headed toward a charged discussion of Blackmore's Mitta Mitta proposal, but first it wandered into speculation on the likely water release behavior of Snowy Hydro. At a ten o'clock coffee break, Blackmore introduced me to David Harriss, the New South Wales deputy commissioner, Blackmore's potential antagonist in the Mitta Mitta discussion. Blackmore had warned me that to the extent I was perceived as his amanuensis, he would face ribbing from his mates, so I paid attention when he told Harriss that I was writing a book partly about him.

"A short book," Harriss said.

"Postage-stamp-sized," Blackmore replied.

After the break, the group took up Mitta Mitta. Harriss was still angry that the commission hadn't approved the New South Wales plan two months earlier and read an e-mail from a farmer whose drought woes caused him to threaten suicide. (Later, Blackmore called Harriss's use of the message "crap.") Since the commission didn't approve New South Wales's deal, Harriss said, New South Wales wasn't likely to approve the commission's. He said he'd check with his superiors and would announce the state's final decision at the Renmark Commission meeting six days away.

Blackmore was on his third or fourth cup of coffee by this time. As he later told me, he considered New South Wales's position "head-in-the-sand." He started to defend the trial, but Roy Green, the feckless commission president, interrupted. "Let's stop talking about it," he said. "If New South Wales doesn't want this to proceed, then it won't proceed."

18 TABOURIE LAKE

Blackmore and his wife, Sharron, spent an occasional weekend outside the basin, at a humble beachhouse in a community called Tabourie Lake, a five-minute walk from the Tasman Sea—Blackmore's only hope of not continually thinking about the basin lay in leaving it. But on the weekend I accompanied the couple there, he continually checked the radio and television to make sure fires didn't threaten the road to Canberra—he had to get back on Sunday afternoon. Blackmore had promised that he'd devote Saturday to a fishing expedition, and indeed fishing is a Blackmore family pastime. His father often took his brother and him on three- and four-day trips on the Murray and other rivers. Fishing stuck with Blackmore, and so did the rivers—only camping didn't take. On Saturday morning, we got off to a late start and found ourselves standing on a secluded crescent of sandy beach, looking for worms. It was already midmorning, past ideal worming time, but some worms still lingered just beneath the sand's surface, waiting to catch the scent of a morsel carried over them by the waves. Blackmore wore a fleece jacket, trunks, and baseball cap. He hadn't shaved. He was staring at the sand, looking for the tiny air holes that gave away the worms' presence; when he spotted them, he ran to them and bent over. With one hand, he held a pair of pliers attached to his wrist by a pink Velcro strap—he said he "borrowed" the strap from his granddaughter's toy purse "for a higher purpose." The other hand held a knotted nylon stocking containing a few fragments of redolent mullet. He waved the stink bag over the worm's suspected location, until, once in a while, a deceived worm momentarily poked its head up; then he tried to catch the worm with the pliers. Most times he failed, but he caught five or six decent-sized worms. Worming, he said, "is a displacement activity for your brain." Maybe so, but the act, groping in water and sand for something solid, something living and elusive, struck me as a variation on his occupation. For nearly an hour, we tried casting into the oceans with the worms as bait, and caught nothing.

Later that day, I disrupted Blackmore's vacation respite by conducting an interview with him about the basin, but he seemed not to mind too much. In fact, I soon discovered his predilection for twenty-minute answers to each question. We sat at a table on his outdoor deck, as countless spectacularly colored birds—Australian king parrots, rainbow lorikeets, and buffoonlike pink-breasted gallahs—ate birdseed from the deck's railing, where Sharron had strewn it. The birds' enthusiastic chirps may have helped soften Blackmore's mood. Yet for all his words, he seemed to hold back slightly, in tacit acknowledgment of the sensitivities of his constituents, from farmers to environmentalists. He has a politician's knack for delivering thoroughly considered assessments in a nimble and self-deprecatory manner, as if to neutralize the sting in his opinions. Halfway through, he said, "I don't think I've answered one of your questions yet, so I've done well!"

Since the World Commission on Dams had drawn Blackmore from its "prodam" group of commissioner nominees even though he maintains that he is neither pro- nor antidam, I wanted to probe his views on the subject. I told him about Thayer Scudder's assertion that 70 percent of large dams shouldn't have been built and asked him if he agreed. His view is "slightly" different, he declared, with what I quickly understood as bureaucrats' understatement. Of the world's forty-five thousand large dams, he said, some forty thousand are relatively small—they're located on tributaries and cause little environmental damage. The five thousand largest dams, on the other hand, have had a "significant" impact. Bad small dams exist, but they're not the rule. "In Canberra, to use a local example, we have two dams in our major nature reserve which clearly have changed it, but we have three valleys that aren't dammed and we have a huge area around them that is protected for conservation—it's pretty hard to describe those things as a disaster. If there was fish migration, it was relatively small, there are no ephemeral wetlands, and there are certainly no indigenous people living there. The dams provide the community with water that doesn't have to be treated, so on balance, I'd say their impact is more than positive. There are a lot of dams like those. It's very glib to say that the balance sheet is negative for those sorts of dams, because I wouldn't be able to flush the toilet or switch the tap on without them.

"But if you've got large dams that block off whole river systems—the Zambezi, the Nile, the Columbia, and it goes on and on—and they're being put in place without due regard for siltation, fish migration, the rights of people downstream, recession agriculture, things provided as free agricultural services, then you've got to say the jury's still out."

I noticed that his criticism of large dams was cautiously couched in hypothetical terms, and considering the offenses he cited for the worst dams, his conclusion that "the jury's still out" seemed tame. Nevertheless, his observation about the relative performance of different-sized dams is intriguing. The World Commission on Dams's "cross-check survey" of 125 large dams acknowledges that the smallest of large dams, those between fifty and one hundred feet high, comprise two-thirds of the world's 45,000 large dams, but its conclusions do not distinguish among sizes. With the few resources they have, dam opponents focus their energy on projects they consider unusually destructive or strategically significant, which means that most smaller dams go unchallenged. The argument that even proportionally speaking, smaller large dams cause less damage than large large dams is plausible, perhaps even probable, but untested.

"I'm not pro- or antidam," Blackmore continued. "I'm pro–sustainable development for the benefit of the planet and humans. And I didn't say 'humans and the planet'—at the end of the day, you have to put them in balance."

Blackmore cited India's failure to curb its population growth after compulsory vasectomy policies in the mid-1970s collapsed. The result of that failure was a greater demand for water and the destruction of more natural areas. As Blackmore put it, "The inevitable result of that piece of public policy is dams."

He explained. "It's pretty hard then to go in and say dams are bad. What was probably not the best decision India has ever taken"—more understatement—"was not to balance the country's population against its carrying capacity."

I asked Blackmore what an Australia without dams would look like. He answered that its population would have been in the single-digit millions. "What's the first thing the people who came here in the 1800s did? They started to secure the water supply, because dams are the only stabilization we've got against the Australian climate. The question is whether dams' impacts are unacceptable, and clearly, in some cases, they are. We had a debate about building a dam on the Franklin River in Tasmania. The government wanted to build a dam to produce green energy, and the green movement didn't want a hydroelectric dam at all."

He laughed.

"They wanted the wild river to continue to exist, and they won. I think the decision was the right one, because we have relatively few wild rivers left in the country, but as a consequence of that choice, we put more coal or more natural gas through a turbine, and the balance sheet is that we generate some more greenhouse gas. There is always a trade-off."

19 LAKE VICTORIA

From prop-jet altitude, the lower Murray looks life-giving: it resembles a slender ribbon slackly tied around a vast brown parcel of parched land. The green belt of farmland on either side of the Murray does not fade gradually from riverbank to interior the way a natural channel does; its edges are geometric, jagged, precisely and abruptly indicating the extent of irrigated fields. Where verdancy ends, degraded scrubland begins. As we descended over Mildura, a Victorian farming hub of fifty thousand people, the farmland seemed to expand, until it stretched to the horizon. Mildura is the center of the Sunraysia farming region, the area where Alfred Deakin helped establish irrigation. (The name "Sunraysia" derives from "Sun-Raised," the winning entry in a 1919 national publicity stunt that offered a £50 prize for an appropriate trade name for the district's dried fruit.) Water is dear in Sunraysia: the summers are hot, with hundred-degree temperatures common, and the evaporation rate is six times higher than the average annual rainfall. In many Australian irrigation districts, gravity is sufficient to deliver water to farmers, but in Sunraysia it must be pumped as much as a hundred feet above the river. Only high-value crops justify the cost, so Sunraysia produces four-fifths of Australia's dried vine fruit, a third of its table grapes, a quarter of its citrus, an eighth of its wine. But economic and water efficiency don't necessarily coincide, as the predominance of overhead sprinklers on the fields I saw around Mildura suggested. Drip irrigation would cut water use by as much as half, but farmers eschew it because installation is costly.

I was traveling with Kevin Goss, the gangly commission deputy chief executive. Goss is the thoughtful, low-key counterpoint to Blackmore, the politician and practiced extrovert. In 1998, Blackmore was looking for a new deputy when he heard of Goss, then running a sustainable rural development program in West Australia. Blackmore called Goss and proposed that they meet that night for dinner, then flew across the country to keep the date. Goss got the job a few weeks later. Now Goss was giving a speech at a former cattle ranch where we were stopping on our way to Lake Victoria; the plan was to linger at the lake for an hour or so, then hurry back to make the afternoon flight out of Mildura.

The ranch, called Ned's Corner, was once an iconic part of the vast farm holdings of Sir Sidney Kidman, the "Cattle King" (and distant relative of actress Nicole Kidman). In the 1930s, his properties extended across four Australian states, and were as big as the British Isles. Ned's Corner itself is the largest privately held landholding fronting the River Murray and one of the largest in Victoria. Yet after 150 years of sheep and cattle grazing, it's thoroughly degraded: its 115 square miles (half the size of Chicago) are so bereft of topsoil and poisoned by salt that even hardy saltbush is sparse, and cars raise clouds of dust as they come and go. The occasion of Goss's speech was a celebration of the purchase of Ned's Corner by an environmental group, the Trust for Nature, which hopes to rehabilitate the land and restore native habitats along a twenty-mile stretch of the Murray. Whether it can succeed is uncertain, for one purpose of the ceremony was to raise money simply to cover the $1.4 million purchase price. The trust planned to remove the last of the livestock from the ranch so that native vegetation could start to re-generate.

The property had the feel of a ghost town. I looked inside a sprawling sheep-shearing shed where the holding pens and shearing tables were rusted and dusty but still in place. The structure was dark and shadowy, except where shafts of light penetrated holes in the corrugated tin roof— the place felt archaeological. A few hundred yards away, I wandered into the gutted residences where the cowboys—stockmen, in Australian parlance—once lived. Now the building was shorn of doors. An actual stockman with leathery skin and a broad cowboy hat appeared, and pointed to a room he said he lived in ten years before. "It's just fallen to pieces," he said. "Bloody disgrace."

From the makeshift dais where Goss spoke, the Murray was not quite visible. Behind the speakers, a few minutes' walk away, I could see the tree line that rimmed the river's shore. The speakers took turns standing in the shade of one tree, while we listeners sat a slight distance away in the shade of another. Behind the speakers, a banner carried the Trust for Nature slogan, which does not translate well into American English: "Protecting Bush Forever."

After the speeches, we ate white-bread sandwiches, and located Peter

Mills, a commission employee who works part-time at Lake Victoria. As soon as we could, the three of us left for the lake.

Lake Victoria is a fundamentally important water storage, without which Adelaide and South Australian agriculture would wither. It is also sacred ground. The combination left the commission mired in controversy, and frustrated Blackmore for nearly a decade. Until April 1994, when Lake Victoria's water level was lowered eleven feet so that a flow regulator could be repaired, the knowledge that it was the site of Aboriginal burial grounds did not extend much beyond archaeologists, local landowners, and local Aboriginals. Yet, in the words of a commission document, until Europeans arrived, "Lake Victoria was an important point in the network of Aboriginal settlements and relationships from the Murray mouth to the upper Darling." Though archaeological evidence of habitation goes back only eighteen thousand years, it is plausible that Aboriginals lived at Lake Victoria for forty thousand years or longer, which would make it one of the longest continuously inhabited hunter-gatherer sites known to scholars anywhere in the world. In 1830, Charles Sturt needed only to climb one modest crest to catch sight of the lake, but he didn't know it was there and missed it. Even so, he encountered at least two hundred Aboriginals a day as he passed by the lake. He wrote, "They sent ambassadors forward regularly from one tribe to another, in order to prepare for our approach, a custom that not only saved us an infinity of time, but also great personal risk."

Overlanders, pioneering cattle and sheep ranchers who drove their stock long distances to open new grazing land, followed the explorers. Joseph Hawdon became the first European to see Lake Victoria while driving a few hundred cattle some five hundred miles across the Murray-Darling Basin. He wrote, "The scenery is very pleasing, and the air is perfumed by a sweet odour from the herbs and flowers growing on the margin of the Lake. To this splendid Lake, of which I had the pleasure of being the first European discoverer, I gave the name of 'Lake Victoria' in honour of our most gracious Queen."

As Sturt's experience showed, the route that Hawdon pioneered passed through densely populated Aboriginal areas, but it still became the standard inland link between the continent's eastern colonies and South Australia. The overlanders did not just usurp the Aboriginals' territory and trample on their burial ground; their livestock damaged the Aboriginals' food supply and began the corruption of Australian soil. Not surprisingly, the area became the site of one of the bloodiest encounters between Aboriginals and Europeans. Both sides carried out atrocities and unprovoked attacks. When the Aboriginals killed four overlanders trying to cross a tributary near Lake Victoria in 1841, South Australian police launched a retaliatory expedition. After further skirmishes resulting in casualties on both sides, on August 27, the police closed in from opposite sides on a group of spear-bearing Aboriginal fighters and gunned down at least thirty. The Rufus River Massacre, as this event became known, marked the end of Aboriginal resistance to European settlers in southeastern Australia.

When the River Murray Commission turned the lake into a water storage nearly a century later, it acted without regard for the area's Aboriginal history. The commission added thirty-two miles of levees to the lake's shoreline, thereby lifting its water level from a varying sixty-nine to eighty feet above mean sea level to a more constant eighty-nine feet. The addition doubled the lake's water capacity while inundating much of the land that Aboriginals once occupied. The obliviousness continued during World War II, when the Royal Australian Air Force used the lake for gunnery practice. One plane would fly over the lake, casting a fleeting shadow across it, and another plane's student pilots would shoot at the shadow. Pilots had trouble judging their height above the water, and crashes into the lake were common. Six airmen died.

As Europeans spread across the Australian continent, they killed Aboriginals, expelled them from desirable land, even abducted their children and forced them into servitude; the devastation was so thorough that Aboriginal cultural transmission broke down. Thus, only the tiniest shards of dreamtime stories about Lake Victoria survive. In 1994, most Aboriginals outside the area had never heard of Lake Victoria, and the news of its cultural significance certainly hadn't spread to Blackmore. "We knew that in all of that country there were some Aboriginal artifacts, but we hadn't envisaged burials in quite the way we found them," he said.

Lake Victoria's obscurity ended as a result of fieldwork led by a Canadian-born archaeologist named Colin Pardoe. Pardoe has lived in Australia since 1980, and became an authority on Aboriginal burials; before 1994, he excavated burials all along the Murray, including at Lake

Victoria. What was happening in many areas was that overgrazing over decades robbed the land of its vegetation, leaving it vulnerable; when droughts struck, the topsoil blew away. Pardoe estimates that the soil layer in many areas he has studied is now a foot lower than it once was. As a result, dozens of Aboriginal archaeological sites have been exposed.

Soon after Lake Victoria was lowered in 1994, a group of Aboriginal elders asked Pardoe to move and re-inter some twenty burials that had been exposed around the lake. Pardoe rounded up fieldwork money, assembled a team of archaeologists, students, local Aboriginals, even two cooks, and set up camp around the lake. It didn't take long before the researchers realized that the burial ground was extensive. At one spot, across a creek from where they'd been excavating, they found more than eighty burials. And when they excavated one burial, they found another beneath it—in fact, according to Pardoe, they found four layers of burials. All together, the team's cursory investigation located 312 burials.

"We were sitting around the campfire one night, and I idly said, 'Gee, I wonder how many burials there are in this cemetery,'" Pardoe said. "We made some rough calculations and came up with a number in the tens of thousands. Everybody was saying, 'That can't be right.' I remember laughing, and I said, 'We're going to have to be serious about this now. Let's get some actual measurements of the landforms instead of making rough estimates.' So we got the data up. I realized that we needed to have a defensible number because this was going to be widely discussed, and it was going to have dramatic implications."

Pardoe arrived at an estimate of between six thousand and sixteen thousand burials, and settled on ten thousand as a defensible figure—enough, he says, to make Lake Victoria the largest hunter-gatherer burial site found anywhere in the world. A report for the commission that Pardoe coauthored states, "Lake Victoria is unique because here for the first time, there is a clear indication of the size and scope of Aboriginal burials that reflects the intensity of Aboriginal land use in the past." Pardoe's findings, combined with his previous work on burials, led him to an assertion gaining support elsewhere in Australia—that the conventional notion of Aboriginals as entirely nomadic is mistaken. At least in the Murray-Darling Basin, Aboriginals weren't buried where they died, as nomads usually are, and they had notions of consecrated ground, ownership, and territoriality.

Another archaeologist who worked extensively at Lake Victoria after

Pardoe, Jeannette Hope, disputes Pardoe's numbers. Six thousand is a more reasonable estimate than ten thousand, she says, and even in Australia, other cemeteries may contain as many burials—but even if true, she vouches for Lake Victoria's international significance. According to conventional archaeological wisdom, the accelerating growth of human population that began several thousand years ago resulted from the rise of agriculture, but Hope says the abundant evidence at Lake Victoria suggests that despite being hunter-gatherers, not farmers, the Aboriginals there experienced the same sort of population increase. "The basic question becomes," she said, "What is it that led to both agricultural and hunter-gatherer populations increasing over the last few thousand years? If hunter-gatherers do it, too, then it can't be just the result of the invention of agriculture!"

As soon as Pardoe put out a press release announcing his numbers, the campsite was deluged with hundreds of visitors, notably including journalists. Rumors spread that Lake Victoria contained 250,000 burials, ten times Pardoe's initial overlarge ballpark estimate. Pardoe said one visitor took a look at the graves, started "weeping like a baby," and immediately produced his checkbook, asking, "How much money do you need?" And when Pardoe uncovered a burial that showed signs of being associated with sorcery, an Aboriginal colleague grew fearful and asked him to close it up without carrying out an intended analysis. Pardoe acceded. In fact, the team uncovered so many burials that the investigation was overwhelmed. The team lacked the resources to study all the burials at Lake Victoria, and in any case found no justification for unearthing them all. But the project's impact was substantial. "Everybody started to see," Pardoe said, "that it was a very big thing, because one could quite clearly make the argument that this was a piece of consecrated ground that was being affected by water-management practices."

That left Blackmore in the middle of the very big thing. "We knew we had to protect the burials," he said, "but we didn't know the extent of them, and, given the dynamic nature of the lake, we didn't know how to protect them."

After that, the atmosphere at Lake Victoria turned, in Pardoe's word, "squirrelly"—so unpleasant, in fact, that Pardoe ceased work on Lake Victoria and never returned. Aboriginal elders assembled and broke into factions. The "radicals" considered the lake's use for water storage "desecration" and wanted the lake immediately decommissioned. The "conservatives" recognized Lake Victoria's importance in water storage and accepted the argument that inundation at least prevented looting: they argued that negotiation with government authorities would produce better burial protection than decommissioning. Blackmore hired outside engineers to build erosion-protection works around the biggest burial ground, but, according to Hope, who represented the commission at Lake Victoria from 1995 to 1999, while the works "protected the burials, they didn't cool the politics. Over the next few years, that just got worse and worse and worse."

At least nine government agencies got involved. Investigations in 1996 uncovered what a commission report calls "extensive cultural artifacts such as shell middens, hearths, and stone tools"—in fact, Lake Victoria contains tens of thousands of such artifacts. The National Parks and Wildlife Service—which despite its name is a New South Wales state agency, not a national one—accepted the commission's contention that inundation would protect the burials but declared that it would also disturb the artifacts, thereby "knowingly" damaging them. The service therefore cited 1974 state legislation requiring the commission to obtain a legal "consent to destroy, deface or damage" the artifacts. Blackmore hated the language and considered the act that created such a consent "flawed as a result of it"; besides, he believed that the commission's effort would protect the sites, not destroy them. The environmental impact statement that the consent required wasn't completed until 1998; then the National Parks and Wildlife Service rejected it. Instead, it issued a consent accompanied by seventy-seven detailed instructions about how the commission must manage not just the lake but also long stretches of the River Murray. The commission appealed.

Meanwhile, the commission held frequent consultations with the Aboriginal community, but the sessions were complicated by increasing enmity between the factions. At one point, meetings with the two groups had to be held separately because the factions made physical threats or refused to talk to each other. Some conservatives defected to the radicals, and then, after a while, switched back again. The radicals got help from white activist lawyers, who, Hope said, ranged from "nice, sensible people to nasty young upstarts who you had to feel were getting onto the native title bandwagon as a career move." Despite their sophisticated legal representation, the radicals miscalculated and walked out on the negotiating process in 1996; from then on, the conservatives dominated it. The consultations often wandered down tangents, as when one Aboriginal woman argued that the lake ought to be decommissioned because that would force downstream basin vineyards out of business, and that was good because alcohol ruined Aboriginals. The Aboriginals' disarray at Lake Victoria, Hope said, was "hugely complicated, terribly tragic, and highly symptomatic of the divisions and frustrations in these poverty-stricken minority groups. A high degree of alcoholism, a high degree of family fragmentation, tight families but children taken away. The whole catastrophe, basically."

Once more, Blackmore was involved in establishing a balance between cultural and environmental needs on one side and water supply demands on the other. He had few options. The alternatives to maintaining Lake Victoria as South Australia's principal water storage appeared nonexistent. In fact, over the previous fifty years, the commission and its predecessor, the River Murray Commission, considered proposals ranging from a Chowilla dam to using a nuclear explosion to create a new storage basin, and rejected them all on environmental, social, or logistical grounds. The commission was already forcing reduced water demand in the basin through introduction of such programs as the cap, and other demand-side solutions big enough to obviate Lake Victoria were inconceivable. Besides, even if alternatives to Lake Victoria storage existed, decommissioning the lake and returning it to its preregulation water level would have exposed new shoreline areas to erosion, hampered the growth of stabilizing vegetation, left exposed many more of the lake's burials and artifacts to looting, and increased River Murray salinity. The commission therefore supported continued operation of the lake at its full supply level. Most of the erosion that had left burials exposed occurred because of the constant variation in the lake's level as it met water-supply needs downstream or was replenished from upstream. To reduce erosion, the commission proposed limiting the period of drawdown and keeping the lake full as much as possible. Though some Aboriginals considered inundation of the burials a form of desecration, the commission argued that inundation with properly managed drawdowns would protect the burials more effectively than permanently lowering the lake's level.

It wasn't until May 2002, eight years after the burial exposures were recognized, that the New South Wales National Parks and Wildlife Service accepted the fifth draft of a commission management plan for the lake, resolving the conflict over the state's requirement for a "consent to destroy." The final draft incorporated about twenty of the original consent's seventy-seven conditions and eliminated those that did not pertain specifically to Lake Victoria. By then, the commission had spent $5 million in research and works. The commission removed nonnative livestock from the lakeshore, either by piping water to nearby properties so that animals would not need to drink at the lake, or by buying grazing land outright. The commission also changed its operating regime to allow the lake's shoreline to dry out in autumns when the system was flush with water, and to delay refilling for as many months into the following spring as possible. The new regime has promoted growth of shoreline vegetation, but it has added yet another layer of complexity to the water-delivery system: the burials will probably end up being protected, but the regulations will hamper the administration of environmental flows.

Near Lake Victoria, we drove off the main road onto a dirt one, and got lost. Not terribly lost—the road appeared to have been recently used, probably by fishermen. Then we found the bridge we were looking for. It was a rickety span that dipped downward from the bank and precariously hovered no more than a couple of feet over a small anabranch. It was barely wider than our four-wheel drive, and it wobbled and groaned as we drove across it. I marveled that it didn't collapse. The anabranch was no more than ten feet wide and a foot or two deep, and it was studded with snags, in the form of fallen river red gums that provided habitats for countless marine creatures. On the banks and extending fifty feet behind them, river red gums flourished; farther away from the anabranch, the terrain was all scrub. The creek's nearly still water looked layered, colorless at the surface but a clouded lima-bean green underneath.

We found our intermediate objective, River Murray Lock Seven. It possessed the usual expanse of healthy green lawn; in fact, at this moment—one o'clock in the afternoon, when the sun is high and evaporation is peaking—the lawn was being extravagantly watered with rotating sprinklers. At the lock, we met up with Sid, an Aboriginal employee of South Australia Water, the agency that carries out the commission's decisions at Lake Victoria. Sid drove us the last few minutes to the lake. We passed a dead kangaroo on the side of the road.

The lake startled me. It's a shallow, nearly perfect ellipse, eight miles wide and fewer than ten yards deep. It looked real and unreal, beguiling and repulsive, prehistoric and contemporary, dead and alive. Looking across it from its southern bank, all I could make out of the opposite shore was a thin, pinkish-brown line, seemingly all that kept the blue of the sky and the blue of the lake from merging. Over it, a wispy cloud reiterated the horizon. Trees that once lined the lake's shore, having drowned in the 1940s and '50s after a decade or two of inundation, were now reduced to wizened, leafless trunks and limbs, which rose out of the water like Neptunian tridents. Though dead half a century, those trees still provided hollows where small birds nested and raptors conducted raids. Closer to us, in a neat row, we saw the top portions of vertically arrayed logs that engineers stationed in the lake as a screen, keeping broken tree limbs from floating into the regulator and jamming it. We saw pelicans, glossy ibises, and water hens. We heard the screech of a spur-winged plover, the honk of black swans, a frog's bellow. As we walked along the shore, we saw half-buried plastic tarps placed over exposed burials. Without even trying, we found on the ground an Aboriginal grindstone as big as a fist.

Sid rarely spoke. He walked with his arms held stiffly at his sides, as if his jeans, short-sleeved work shirt, and thick buckled belt did not quite contain him. Middle-aged, he'd lived around Lake Victoria all his life, so I asked him if he knew any dreamtime stories about the lake. He said no.

During his tenure as chief executive, Blackmore has made a point of installing plaques at commission installations, and now Mills steered me to Lake Victoria's. One said:

IN REMEMBRANCE OF

THE RUFUS RIVER

MASSACRE

ON 27TH AUGUST 1841

AND THE 2 TRIBES

BARKINDJI/MARUARA

WHO OCCUPIED THE

LAKE VICTORIA AREA

THIS STONE UNVEILED BY

REPRESENTATIVES OF THE

BARKINDJI/MARUARA COMMUNITIES

ON THE 26TH OCTOBER 2002

Next to it was a second plaque, dated a year earlier. It was dedicated to the "memory of those brave airmen who lost their lives during bombing and gunnery practice at Lake Victoria," and it listed the six names. Even the plaques struck a balance.

As we left the lake, we spotted a scar tree and stopped the car. The gray eucalypt was, if not thriving, quite alive, despite the presence of a broad gash on one flank, revealing the horizontal grain of its core—as shocking, in its way, as images of a pulsing human heart. The scar's outline disclosed a shield.

In David and Robin May's comfortable farmhouse near the Wakool River in New South Wales, it is possible to forget that the history of Australian agriculture is a story about water. On the night I spent there in April 2004, fear of renewed drought unsettled the basin, and the Mays' twenty-two square miles of farmland looked gray and barren. But the bathroom shower flowed without restraint, and a vase of roses graced the kitchen table, around which we downed wine with our lamb dinner. In the next room, Campbell, the Mays' younger boy, played a TV video game: the house felt suburban, though the surroundings were anything but. David May's lips were chapped, his hair was tousled, and the top three eyelets of his work shirt went unbuttoned, but he looked at ease in his rumpledness. Farming here is a seriously straitened profession with diminishing prospects, but May remains enthusiastic about it. "I'm the sort of person," he explains, "who likes growing things."

Back in 1947, when May's father, Greig, started the farm, it consisted of nine thousand acres of sheep and cattle pasture. When irrigation became available, many of the Mays' neighbors resisted it—out of fear of change, May believes—but Greig embraced it, and began growing rice in 1951. The commonwealth government had already shown the way. In 1943, it assigned two hundred Italian prisoners of war, all captured in North Africa, to contribute to the Allied war effort by growing rice in the Wakool area, and found itself bestowed with seventy-seven hundred tons a year from five thousand acres. Rice helped perpetuate irrigation in New South Wales after World War II, when irrigation districts were plagued with financial problems. As a 2002 water-policy history written for the commission explains, rice "played an important role in maintaining the political commitment to irrigation development in the face of the accumulating evidence of its marginal economic viability." Success now depends on producing the world's highest rice yields, something Australian farmers have been achieving for the last decade or two. They don't compete with Asia's vast bulk rice producers; the niche they've carved is for premium rice, for which Asians pay twice the usual price. Their greatest agricultural asset is the region's unrelenting sunlight—clouds are much rarer than in stormy Asia. But rice is a notorious water guzzler: though Australia's rice farmers use half as much water as the international average, they still require more than a thousand pounds of water for each pound of rice produced. For equivalent profits, rice in Australia requires nearly four times as much water as dairy products and nine times as much as vegetables.

The Mays eventually put the farm on a three- or four-year rotating crop schedule. Each spring, they planted rice, and in the autumn, they harvested it; then, taking advantage of lingering moisture in the soil, they planted a cereal crop, usually wheat or barley in winter, undersown with clover; after the cereal was harvested in early summer, the clover emerged, and the field was turned into livestock pasture for two or three years. At first, the Mays planted rice in their richest soil, but as it was porous, most of the copious quantity of water applied to the fields leached into the groundwater, further raising the already elevated water table and spreading pollution from fertilizer and pesticides. Agricultural restrictions eventually forced the Mays to move their rice crops to fields of heavier, clayey soil, thereby eliminating leaching and reducing water use. (Even so, rice production continues to account for something like half of the groundwater leaching from irrigated areas in New South Wales.) Along the way, the Mays learned some environmental lessons. Greig, acting on an aesthetic of tidiness that then prevailed, habitually used his tractors and bulldozers to remove vegetation from beneath the trees lining the region's waterways, while David learned to leave the vegetation alone to promote biodiversity.

As the years passed, the farming community declined. Mechanization reduced the number of farm laborers, and scarce water pushed land out of production. Local sports teams, the heart of rural social life, closed down for lack of young players. By the time nine out of ten farmers on an irrigation channel had gone out of business, irrigation companies found it uneconomical to continue feeding water down the channel to the tenth. Departing farmers often faced an agonizing choice: should they sell their land to the highest bidder, whose sole interest was the land's water entitlement, or should they sell to a much lower bidder who intended to keep working the land? In New South Wales, many farmers realized that the only way to succeed was to get bigger. May bought out nine nearby farms and the water rights that go with them, and expanded from nine thousand acres to fourteen thousand. At the same time, the farm's workforce dropped from eighteen people to four or five. As May's experience reflects, farm consolidation has accelerated over the last two or three decades, but most farms remain family-owned.

As in the Mitta Mitta Valley, irrigation turned the water regime upside down. The farm suffered through bouts of sterile rice and years when livestock commodity prices dropped out of sight, but thanks to irrigation, droughts were, of all things, a blessing. "The dams had water in them, and we had access to it," May said. "Dry times were the only times you could make some money, because you're the only ones who can grow food." But the 2002 drought was an exception. With New South Wales's water allocation cut to 10 percent, May didn't even try to plant rice; he saved his meager ration for wheat and barley, which use nearly half as much water but are still less profitable. After considering laying off a couple of his workers, he decided instead to put them to work on improving the water efficiency of his fallow fields. Whereas May's plantings and irrigation once followed his farmland's contours, he has gradually "laser-leveled" his farm, using Global Positioning System technology to flatten the fields and shape them into precise, easily cultivated rectangles. By now, nearly half his fields are laser-leveled. The land is so flat that no puddles form, and the clayey subsurface retains the irrigated water; in some fields, May has cut water use by nearly half. With a satellite guidance system attached to the steering wheel of his truck, he can spray agricultural chemicals at night, when they're more effective, enabling him to cut chemical use by a third. May borrowed $140,000 to pay for the latest installment of laser-leveling. The loan amounts to a bet that the drought won't last. "You've got to be pretty optimistic to be a farmer," he said.

The next morning, May took me on a drive through his farm. The farmhouse is surrounded by trim green lawn, and the willow-lined irrigation canal behind it brimmed with water the color of milky green tea, but the rest of the farm looked as featureless as the moon, as if agriculture, instead of complementing nature, erased it. No rain had fallen in four months. The grassless land looked bald and exposed. The farm was a relentlessly geometric universe, as lifeless as parking lots; even the farm reservoir is an oval indentation enclosed in a rectangle of land. The reservoir customarily contains both farm drainage and purchased water, but it must be used quickly, before it evaporates. Now it was empty; I could see faint bathtublike rings of sodium deposits encircling its seven-foot walls. We drove by some of the farm's three thousand ewes; they looked perplexed and alien atop their crusty, geometric plot. They had just been separated from their newest lambs, which would be slaughtered over the next few months.

The 1995 cap was meant to stop expanding the amount of water taken from the river, but it fell unevenly across the basin. (In fact, irrigation administration is such a thorny, intricate task that no plan could have been devised with comparable impacts across the basin.) In New South Wales, some farmers in possession of water entitlements had never bothered to use them, or used only a share. On the grounds that these unused "sleeper" and "dozer" rights amounted to an untapped source of water, the New South Wales government had encouraged other farmers to expand their operations beyond a level justified by their water entitlements. But with the advent of a market in water rights at the same time as the cap, the sleeper and dozer rights abruptly had value, and their owners happily sold them. Since the water represented by the unused rights was already being distributed to other farmers, the sleeper and dozer rights were effectively counted twice. The New South Wales government had no choice but to end its overallocation. Farmers like May, who'd grown accustomed to receiving more water than their entitlements justified, were forced to cut back to their entitlement level — May lost a fifth of his water. To make up for it, he bought water on the open market at premium prices, nearly doubling his water costs. While ambitious farmers like May were penalized, less enterprising farmers weren't. Farmers with sleeper and dozer rights received a windfall, and farmers who'd never expanded production beyond what their entitlements justified felt no immediate impact: all the cap did was limit them to their existing level of water use. To May, this made no sense.

Blackmore's relationship to farmers is layered, and the cap further complicated it. After all, he grew up with sixteen members of his extended family on a farm outside Melbourne, in an atmosphere that his sister Julie describes as "idyllic," and he clearly feels affection for farmers. When he had to select an engineering specialty, he chose water partly to stay connected to farmers. Yet his profession often cast him as the farmers' antagonist. When the 1982 drought struck, Blackmore was managing a complex irrigation system at Swan Hill, Victoria. Desperate for water, some farmers were caught stealing it, and Blackmore had to prosecute them. "That caused a bit of stress and strain," he said. A decade later, when the commission's Ministerial Council decided on the cap, it didn't consult farmers beforehand; instead, it seemed to accept the farmers' anger as the price of initiating decisive action to save the river. Blackmore stressed that future policy would involve extensive consultation with farmers, but most of his attention seemed to be focused on forging an agreement among his disputatious bosses. May's attitude is probably typical: he regards Blackmore with both respect and suspicion.

Of course, the cap was only the beginning. May assumed that once environmental flows started, he'd lose more water. As quickly as he moved toward water efficiency, the government seemed to nullify his gains. He was buffeted between his farm's interests and his hard-won environmental understanding, between the needs of his irrigation district and the needs of the entire basin. For now, he called himself three-fourths environmental skeptic, one-fourth environmentalist. He'd been pointed in that direction by Murray Irrigation Limited, his water supplier, which trumpeted the skeptical results of a couple of reviews of the scientific documents supporting environmental flows. One of the reviews was commissioned by Murray Irrigation itself, and the other was written by the environmental director of a right-wing think tank called the Institute of Public Affairs. Lo and behold, both reports found the science of environmental flows deficient, and called for delay in implementing the project. But the reports had little impact. Blackmore dismissed the Murray Irrigation study as "a put-up job" and argued that holding off on environmental flows was comparable to "fiddling while Rome burns." Even Prime Minister John Howard, usually an opponent of government regulation, agreed. When members of a parliamentary committee cited the Murray Irrigation study in a vain attempt to delay environmental flows, Howard answered, "Ignore them." Now May was saying that the review "discredited" environmental flows, that the river's condition was improving, that scientists even overestimated the river's carp and underestimated its Murray cod because they didn't know how to fish. The environmental flows policy, he said, was dictated by urban dwellers in pursuit of environmentalism's "warm, fuzzy feeling," who would experience none of the damage it would inflict on farmers. Environmental flows weren't sophisticated, he said; they amounted to "throwing water at rivers."

I told him that the environmental destruction I'd seen along the river at places like Chowilla looked quite real, and asked what he thought should be done about it. There would have to be "trade-offs," he said. Trying to protect all of the Murray's environmentally significant reaches would leave all of them half dead, so it would be better to focus on just a few—Chowilla, in other words, might have to be sacrificed.

"We've just been through the worst drought in history and we're talking about areas of the river that are suffering," he said. "Communities are suffering as well—everyone is suffering. The hardest thing to get through to governments and bureaucrats is that unless farmers are making some money, you won't get the desired environmental outcomes—actually, it goes completely the other way." Indeed, farmers in need chop down trees to sell as timber, and delay installing water-efficient technologies. On this point at least, May finds agreement among some environmentalists, who envision farmers playing the role that Aboriginals once did, tending to the land's needs.

Yet there is a limit even to May's enthusiasm for farming, and it is generational: he doubts that his two boys ought to take over the farm. "It's getting tougher and tougher, because other people have so much say in your business," he said. "I'd like them to come home, but the opportunity to look at some other career will be more beneficial to them."

After the tour, we met the rest of May's family and assorted friends for a picnic at a bend in the Wakool River, the Murray anabranch and irrigation conduit that forms the May farm's southern boundary. The bend marks a family gathering spot, where May's father has fished for sixty years. Here, no agricultural fields were in sight, and the country resumed its vestigial stately character. Thanks to the Wakool's artificial abundance, the trees looked healthy, and the landscape was daubed in a spectrum of soft greens, golds, and browns. The river barely flowed. Its banks were steep, and red gums with gnarled, exposed roots leaned halfway across the channel. From one of them, a rope dangled; Campbell and half a dozen other boys used it to swing over the river, then let go. Add a straw hat, and there was Huck Finn: it looked more like a Mississippi swimming hole than any place on the Mississippi now does. At the apex of the bend, we unfolded a table and some chairs and ate white-bread sandwiches cut into squares. While we chatted, I found myself looking down the bend as far as I could see, envisioning the river's erstwhile beauty, imagining our present exquisite prospect all the way down it.

21 RENMARK

Most of Australia's interior is too sparsely populated to support regular commercial flights, so people in a hurry routinely charter small planes. Blackmore, however, loathes small planes, after enduring one too many close calls in them. As a result, his December 2002 journey to the commissioners' meeting in Renmark, where he hoped to get approval to discuss the specifics of environmental flows, was a day-long choreographed procession. Instead of flying from Canberra to Renmark in a chartered plane, an hour and a half and done, we flew in commercial jets first to Sydney, then nearly back over Canberra to Adelaide, South Australia's capital. Then the six of us—Blackmore, assorted commission personnel, and me—drove north. Adelaide is reputedly handsome, but what we saw of it were its low-slung outskirts, populated with Subways and Pizza Huts and a chain liquor store called Sip 'N Save. The evidence of the region's aridity registers even in its electricity poles, made of concrete instead of precious timber. At the same time, we passed parks whose greenery seemed to defy the drought. "They're using their water for gardens, by the looks of it," offered David Dole, who was driving. Dole, a gracious, sad-faced man of sixty-four, had once been Blackmore's boss and now worked for him as general manager of River Murray Water; Blackmore called him "my conscience." Adelaide is the only basin capital whose drinking water comes from the Murray, and its 1.1 million citizens consequently care passionately about the quality (preferably high) and turbidity (with luck, low) of the river's water. Turbidity—muddiness, more or less—is a serious issue in South Australia, for the lowest tributary in the Murray watershed is the dramatically murky Darling River, which lends the Murray a brownish hue and makes the Lower Murray three times more turbid than its middle reaches. Upstream, Dole said, conditions are reversed. There, salinity is the issue, while turbidity hardly matters. The problem this year wasn't turbidity, but its near opposite: algae blooms, which flourish in clear water. The combination of low flow, high water temperature, and elevated levels of phosphorus and nitrogen in the soil produced a "high likelihood of intermittent algal blooms this summer," Dole said. The low flow was a consequence of the drought, while the high nutrient levels resulted partly from the upstream use of fertilizer on eroding land. Blue-green algae exist naturally in the basin, but in conditions like this year's, their population can explode. Blue-green algae can be lethal to both humans and animals, and they can contaminate irrigated wine grapes. Algal alerts could go on for several weeks, Dole said, and if that happened, people would have to avoid contact with contaminated water. Fishing, boating, and swimming in the Murray would have to stop; the recreation industry would suffer. Algae blooms' sole virtue may be that they make the commission's point: if the Murray dies, the basin dies with it.

We drove through the town of Elizabeth, named for the queen. As the only Australian state that did not arise from a convict colony, South Australia holds itself apart. Even so, it appeared to face the same drought the other states did. The fields were brown, and the canola and cereal crops planted on them were stunted and dry.

I asked Dole to describe his career in water management.

When he left university, he said, the prevailing ethic in water management was the domination of nature. "We were changing natural conditions for the good of people, without a great deal of understanding of what the environmental impacts would be. When I look back on this, I realize the rivers were the silent victims of the great wealth that has resulted. We have to re-create a balance.

"Then came the realization that we ought to make better use of what we have." The distribution of water from reservoirs to plant roots was "disturbingly inefficient," and there was strong resistance to any suggestion that the users of water should even identify the full costs of its delivery, let alone pay them. Farmers rejected the notion that they bore any responsibility for the environmental damage that irrigation caused — "there was a serious discrepancy between the real cost and what they were paying."

We were passing fields punctuated by rolled bales of hay to be sold for fodder — the drought made the bales lucrative.

"I got my graduate degree while large land clearing was going on with bulldozers and chains and steel balls," Dole said. "I was disturbed by it, but I didn't know what to think about it. More disturbing is that it's still happening in Queensland." The northern tip of the Darling River reaches into Queensland, which has qualified the state for participation in the commission, but as the basin's least developed state, it wants to reap the river's temporary bounty before acceding to all the commission's rules.

It wasn't until after we'd driven for an hour or so that we went over a crest and entered the flat basin. The Murray was a few miles to our east, but invisible. We were moving through a zone of mallee, an Aboriginal word that refers at once to a group of about a hundred short, multistemmed, drought-resistant eucalyptus species, the vegetation those trees dominate, and the landform on which the trees grow. The area once was probably covered with mallee trees, but they were cut down long ago to provide fuel for Murray paddle steamers and irrigation pumps. Now much of the land was barren.

Our first destination was Lock One, at the first of the Murray's fourteen weirs. Dole said South Australia was in such a hurry to promote river navigation that it started building the weir in 1913, two years before the River Murray Agreement was even signed. Adelaide never became the trade center that promoters of a navigable Murray envisioned, but South Australians remain attached to their weirs. They've become community centers, and they provide a measure of control over the river, when necessary countering the effects of upstream water diversions. The weirs' disadvantages are less obvious. One is danger, Dole said. "They're designed with a certain degree of heroism in mind": the seasonal dismantling and replacing of the weirs' steel trestles require fearless divers who work in low visibility and in the midst of considerable flow. Of course, "the weirs are also distinctly unfriendly to fish," Dole said. "The temptation now is to remove weirs for part of the year and let the banks dry out—we're putting our toe in the water about that."

A sign said:

WELCOME TO BLANCHETOWN
HOME OF LOCK ONE

"There's the river in front of you," Dole said, and I beheld a broad sward of carefully mown grass, and beyond it, the lock, the weir, and the flaccid Murray. The opposite bank was lined with willows, which, compared to the softer palette of the native eucalypts behind them, looked garish in their neon green. Clouds, putting in an infrequent appearance, lent the river a slate-blue cast. Upstream from the weir, a single horizontal pane of water glided almost imperceptibly toward the structure, then, sliced into identical arced slats by the weir's twenty or so gates, descended smoothly for a few feet to the bottom step on the Murray staircase: from Blanchetown, the Murray met no more man-made barriers until the barrages near the river's mouth. A yard or two downstream from the weir, bobbing pelicans and cormorants, having adapted to the altered regime, waited patiently for their prey, fish stunned by running upstream into the weir or tumbling downstream over it.

I wasn't yet used to the popularity of lawns in rural Australia, and asked Blackmore if he intended to replace it with something requiring less water.

"No!" Blackmore answered, with a vehemence that surprised me. "In an area as arid as this, people come here to picnic."

The lawn ended at the edge of the lock, a hollow cement rectangle with doors, through which passed the occasional houseboat. Just upstream, yellow buoys extended from the bank across the river to the lock on the weir's western flank. They were meant to catch houseboats and their drunk or otherwise oblivious inhabitants before the boats crashed into the weir.

We walked across the weir, and the crew members demonstrated how they removed trestles to lower the weir's height. Blackmore explained the design of the hundred-yard-long fish ladders that would eventually be installed here.

Blackmore conducted me toward Lock One's plaque, set in a raised stone foundation in the middle of the lawn. The commission's plaques serve dual purposes: they acknowledge the workers who build and run the installations while alluding to the commission's role in lessening their unintended consequences. Lock One's inscription called the creation of the River Murray Commission in 1915 "one of the great engineering initiatives" of Australian federation and briefly described its material achievements, but cut short the triumphal tone by the last sentence: "The challenge remains for current and future generations to ensure that continuing operations sustain the health of the river's ecological community."

"Not bad for a group of rapers and pillagers," Blackmore said.

Blackmore got pointed down the road toward community involvement more than two decades earlier, at Swan Hill. He still calls his three years there the most enjoyable of his career, since he dealt both with "real people"—farmers—and issues of nationwide significance. One of the most pressing was salinity. To combat it, Victorian water planners devised a "salt interception scheme"—a ten-thousand-acre disposal area where highly saline water would be diverted, and salt concentrated: the idea was to sacrifice the disposal acreage so that land and water downstream would be spared the salt. The scheme was part of a plan that would reward Victoria for its salinity-reduction projects by allowing farmers 150 miles upstream to release *more* salt into the river. Everyone benefited, it seemed, except the local farmers, who feared despite the government's assurances that salt would spread from the disposal area, contaminating nearby properties. They "felt no ownership of it—in fact, like any dump, they hated it," Blackmore said. He had to buy land for the disposal area from farmers who didn't want to sell, and if they refused, he had to deliver notices requiring them to vacate their property. "They wouldn't accept the notice," he said, "so you had to touch them with it and have that witnessed. It was pretty bitter." The $6 million project was completed in 1986, yet it has never operated. Though local farmers lost the first class-action suit in Australia's history in their efforts to stop the project, its enduring unpopularity dissuaded planners from turning its valves. Blackmore acknowledges that from a technical point of view, the project was "totally competent," but the omission of affected farmers in its planning was crippling. "That taught me," Blackmore said, "that *that* was a stupid way to do business."

Our next stop was Stockyard Plain, a former off-road-vehicle track that had been transformed into a state-of-the-art salinity-disposal basin: it was Blackmore's answer to the Victorian debacle. To reach it, we drove north, down a straight, flat two-lane highway surrounded by desultory mallee scrub and occasional fields of stubble, the desiccated remnants of an abandoned wheat crop. The wind was blowing, and the day had grown hot, though still well below the blazing seasonal average. We passed kangaroos and ostrichlike emus, shy creatures with ungainly strides driven to unaccustomed human proximity by the drought. At a designated dirt road intersection, we found Peter Forward, the South Australian Water Corporation's salinity-control manager, who greeted Blackmore warmly, then led us to the site. Though Forward wore the blue work shirt of SA Water, his direct employer, he was a Blackmore success story. Blackmore had hired him a few years earlier, when he found himself "in the wilderness" after an SA Water corporate reorganization. Since then, he'd flourished.

By the time we arrived, we were seven or eight miles from the River Murray, seemingly descending into intensifying blight: a salt-disposal basin is, after all, a compromise, a relinquishment of land for a greater good. At the gate of a chain-link fence that Forward opened for us, a sign said, POISON LAID ON THIS PROPERTY. The poison was for foxes and feral cats, while the fence itself was designed to keep out rabbits. Before the advent of the disposal basin, rabbits by the thousands consumed Stockyard Plain's vegetation. The land was stony and overgrazed to boot, and therefore had no agricultural value. Given its degraded state, the commission felt confident that even a salt-disposal project would improve it, and bought the forty-six-hundred-acre property for the modest sum of $85,000, less than $20 per acre. The scheme began operation in 1990, when pumps started diverting highly saline groundwater beneath two reaches of the Murray to the natural depression at the core of Stockyard Plain. The two groundwater-pumping installations, seventy miles of corrosion-resistant pipe, and the seven square miles of Stockyard Plain itself add up to the largest salt-interception scheme ever attempted, anywhere. The two Murray reaches were chosen for their unusually high salt concentrations; the goal was to reduce their salt loads by 75 percent and 65 percent, respectively. This time, Blackmore met with local farmers, convinced irrigators that the salt-interception scheme was in their interests, and paid landowners for their properties. Since the scale of the groundwater systems was unprecedented, by necessity the builders invented a suite of new techniques. Among them was an accurate method for measuring river salt loads and a way to fend off naturally occurring bacteria — so-called iron bacteria — that turn soluble iron into sludge, and blocked nearly half the pumps' flows within fifty days of their installation. As a result of all the innovation, the Institution of Engineers Australia gave the scheme its 2002 "excellence in engineering" award, which usually is given to more visible projects such as dams and bridges. Kevin Goss said later, "It was gratifying to see something that is 98 percent underground win." Blackmore and Forward wore tuxedos when they accepted the award.

Australian salt statistics are mind-boggling. Of the Murray-Darling Basin's one hundred billion–plus tons of salt, only something like 0.0005 percent is mobilized each year. That number may sound minuscule, but it still amounts to five million tons—and the rate will double over the century as the consequences of 150 years of tree clearing expand. Of the current five million tons a year, about two hundred thousand are now diverted to Stockyard Plain, and the understanding reaped from it will be incorporated into a new set of schemes that should divert another 125,000 tons a year. Yet for all that, what astonished me as much as the numbers themselves was the way their meaning disappeared into the landscape: Stockyard Plain is part sleight of hand.

From the gate, we drove toward the basin through concentric zones. The outermost ring is the "buffer zone"—the land intended as a barrier to keep basin salt from escaping into the river system, to my eye no more or less degraded than the terrain outside the gate. The buffer is supposed to keep the basin's salt out of the Murray for fifteen hundred years, according to the plan; after that, the concentrated saline groundwater beneath the basin may begin leaking back into the Murray. The next concentric ring is the "salinized zone"—the land already poisoned by salt. As we approached the salt lake it surrounded, we could see the salt more and more easily: it looked as if the brown ground was flecked with snow. The mallee trees were long dead, with gray, brittle limbs that will eventually snap; only salt-tolerant minishrubs survived. Yet the innermost circle, the basin itself, looked eerily vital. It was lively and lifeless at the same time, awkward in its man-made-ness, not arising from its surroundings but overlaying them like a fried egg atop a grill, and its color was an unlacustrine pale green. Its water is one and a half times as saline as the ocean, but in the heat it looked confoundingly inviting. The presence of some of the tens of thousands of birds that now make Stockyard Plain their retreat attested to this. The South Australia Ornithological Association has counted 156 species here, including eight species that migrate from the Northern Hemisphere. All but two, starlings and Eurasian coots, are native; among the most elegant of them are black swans, which were prominently represented in front of us. Some birds, Forward said, used the basin as a refuge but got their water from plants and rainfall; others could tolerate the lake water despite its intense salinity. A species of seaweed, its spores apparently transported by birds, prospered in the water, and a tiny snail was multiplying along the shore. The commission intends to rehabilitate Stockyard Plain's buffer zone, among other ways by replanting endemic vegetation such as eucalypts that were removed when the land was used for grazing. Conspicuously missing were fishing birds such as the pelicans and cormorants I'd seen at Lock One—the basin has no fish.

Forward showed us where the piped water from the Murray flows into the basin. It looked suspect as it gurgled through a narrow channel into the basin, turning from a faint blue to green. A flock of birds had stationed itself just beyond the mouth, apparently attracted by the incoming water's warmth. We walked up a small hill overlooking the lake. As Blackmore surveyed the fruits of management, he looked, as he himself might have put it, chuffed. "This isn't amateur hour here, mate," he said.

It's tempting to think of our next stop, Banrock Station, the vineyard where commission officials were meeting for dinner, as Stockyard Plain's mirror image: though the two tracts are nearly the same size, one is headed for concentrated salt degradation, while the other aims for environmental restoration. Passing in opposite directions, it's as if they cancel each other out. Banrock Station represents the environmental salient of BRL Hardy, one of the world's ten largest wine producers and the biggest wine company in Australia. Its physical plant consists of a hillside green with grapevines (contrasting dramatically with the surrounding brown bushland), topped by a sleek ecofriendly visitor center, all overlooking a luminous lagoon, and beyond it, a sinuous reach of the Murray. The farmland that BRL Hardy bought in 1994 for Banrock Station was thoroughly corrupted. Sheep and rabbits had grazed it for a century, so devouring its native vegetation that the soil eroded, the land formed gullies, and weeds covered the gullies. Salt had accumulated in the property's wetlands as a result of the construction of a nearby River Murray weir, which raised the water table and helped suppress the small and medium floods that once flushed salt from the floodplain. The new owners removed the sheep and drained the lagoon. Since native species had evolved strategies for coping with drought, they survived, while the exotic species that thrived in the dam-era Murray suffered: the draining killed sixty tons of carp. By a fortuitous circumstance, rabbits also declined, as large numbers succumbed to calicivirus, accidentally spread to the mainland during a trial rabbit eradication program on an island off the coast of South Australia in 1995.

The environment that emerged at Banrock Station is as much manmade as "natural." It's a microcosm compared to Blackmore's domain, but the objective is the same: instead of integrated management of an entire watershed, it's integrated management of a thirty-four-hundred-acre bush and wetland parcel. Banrock Station's managers installed flow regulators and fish barriers at the two spots where water flowed from the Murray into the lagoon and back again. With those tools, they simulated small spring floods that aerated the soil, flushed salt from the floodplain, and increased germination of trees, sedges, and grasses, while preventing more exotic fish from entering the lagoon. Every two years, the flow was turned off, and the lagoon dried out. The managers ran baiting programs to kill foxes. They replanted three layers of vegetation, from low-lying saltbush and blue bush ground cover through six-foot-high shrubs to eucalyptus trees. The managers aim to plant five or ten thousand seedlings a year, encompassing twenty species of vegetation, and in the next five years they'll reintroduce small marsupials such as bilbies, bettongs, and numbats. The restored vegetation has reduced soil erosion, enabling small animals once more to make nests. With the lagoon's carp population reduced, aquatic plants returned. So did native fish, frogs, insects, reptiles, and, most gloriously, 166 species of birds so far, about as many as at Stockyard Plain: grebes, warblers, herons, and ducks; crakes, plovers, and kites; white-bellied sea eagles, sacred kingfishers, laughing kookaburras, hooded robins, and peregrine falcons all came back. In 2002, Banrock Station received a compliment more savory than any of its wines: along with a government agency in India and a Central European consortium of nonprofits, it was given the 2002 Ramsar Wetland Conservation Award for "its innovative approach to supporting the sustainable use of wetland resources." The award, given once every three years, was established in 1996 to recognize "significant contributions to wetland conservation and sustainable use in any part of the world," and came with a $10,000 check. In the same year, Banrock Station joined the list of nearly thirteen hundred international wetlands protected by the Ramsar Convention.

The visitor center is a kind of advertisement, suggesting that the environmental awareness that informs it also characterizes Banrock Station wine: it's an attempt to make environmental concern a selling point. Tony Sharley, Banrock Station's manager, declares, "Our aim is to be the number one environmental wine brand in the world." That strategy makes all the more sense because Banrock Station wines aren't otherwise exceptional: they're intended as affordable exports, selling for less than $10 a bottle in the American market. Opened in 1999, the visitor center is an airy two-story building that manages to look much bigger from within than without; it doesn't quite disappear into its surroundings, but it finds an understated accommodation with them. From its expansive bar, with its floor-to-ceiling windows, drinkers can look out over the lusciously green vineyard and the reverberant wetlands and entertain the notion that life is bountiful. The building uses native eucalyptus, culled from sustainably managed forests. It is designed to minimize energy consumption: its capacious eaves catch cool breezes from the lagoon, and its roof is doubly insulated, shielding rooms from the summer sun but retaining heat in the winter. Solar panels near the building provide 40 percent of Banrock Station's energy requirements; when they generate more power than the facility needs, the excess is sent to the grid, and the electricity company pays for it. Rooftop water harvesters collect rainwater for drinking. Wastewater is recycled and used to irrigate the grapes.

The vineyards have been overhauled. The Banrock Station approach is technological and highly quantitative. Managers divided the vineyard into thirty-six units, then studied the soil characteristics of each one. They sank instruments deep into the earth to measure groundwater changes, and in each unit they installed an electromagnetic soil moisture gauge; then they used the information to determine how much water each unit should get. Instead of the overhead sprinklers that the previous owners used, they installed drip irrigation. Drip reduced evaporation, and it lowered salt transfers to the river to an undetectable level. With all this technology, Banrock Station cut its water use to half the regional industry average.

The sight of the place invigorated Blackmore. It made him, he said, "an absolute optimist" about the future of the basin. The connection Blackmore felt was not exactly surprising, considering that Sharley is the commission's former catchment scientist, whom Blackmore holds in high regard. In a certain way, that makes Banrock Station Blackmore's baby. He wandered breezily through the building, greeting arriving commissioners and staff in ingratiating master-of-ceremonies style, issuing frequent loud but slightly forced laughs. The assembled group was predominantly male and generally cheerful, given to ample bellies perhaps more attributable to beer than to wine, in no way connoting by appearance the membership's high political rank. After a few minutes, Sharley ushered us into a sunlit conference room for a welcoming ceremony.

With his boyish face framed by fleecy, short-cropped brown hair, Sharley looked younger than his forty-three years, so that his evident self-assurance arrived as a surprise. During a short slide presentation, he proclaimed, "On an environmental scorecard, ten years ago Banrock Station reached one of ten environmental goals; now it's five out of ten; and in twenty-five years, it will be nine out of ten."

At Sharley's invitation, we took a walk along the narrow boardwalk that twists through the wetlands—indeed, Banrock Station's two wetlands trails, one and a half and three miles long, attract some twenty thousand tourists a year. According to Sharley, the visitor center draws an annual total of ninety thousand visitors: apparently most people come for the wine. We had only enough time to walk to a handsome wooden blind that seemed to hover over the lagoon, from which vantage point we stared through binoculars at feeding hardhead ducks, egrets, and black swans. Banrock Station had obliterated the drought—its bounty of water in the midst of so much barrenness underlined its human management.

In the bar, we tasted the wines and ate kangaroo prosciutto hors d'oeuvres, as David Harriss, the New South Wales deputy commissioner who'd caused the flap at the River Murray Water board meeting in Canberra, declared that he and his minister had decided not to budge—at the meeting, they'd oppose Blackmore's Mitta Mitta deal.

Blackmore circulated through the room like an olive in a martini. Occasionally he'd emerge from some mirthful huddle to identify the commissioners or offer a bit of wisdom. He thought 80 percent of the commission meeting would not be controversial, and the other 20 percent would involve environmental flows. The commissioners might instruct him to explore all the options he'd listed in his chart at Bungendore, or they might tell him that some weren't worth the political price. "The art form," he said, "is how you make the cake out of the options."

During the cornucopian buffet dinner that followed, I sat next to Geraldine Gentle, the voluble Queensland commissioner. The Murray-Darling Basin Commission, she said, grew out of a long-standing Australian penchant for creating long-lived special authorities: for example, the precursor of Australia's Productivity Commission was established in 1921. The role of commissions was to be a thorn in the side of government, an independent source of advice and analysis. The fiasco that followed the 2000 Bush-Gore election could never have happened in

Australia, she said—the Australian Electoral Commission wouldn't have allowed it. Even though the Murray-Darling Basin Commission is a creature of governments, it reflects the independent commission tradition.

"On the commission," she said, "there is a sense of being in the vanguard of a movement," and that is a strength and a weakness. The commission's success caused state governments to initiate their own sustainable management research, and the powerful Council of Australian Governments had just announced its own salinity and water quality plan for the whole country. "The commission's success pulled the rug from under its feet," Gentle said. "It was left wondering what its role would be. It was no longer the only game in town. Some people in the commission felt their role was being undermined, and then they started working it out. They realized that at the end of the day, it was the commission's framework at the heart of all that was being done."

The meeting that began in Renmark the next morning was off-limits to journalists, so I spent part of the day with Jeff Parish, the chief executive officer of a South Australian irrigation company called the Central Irrigation Trust. As it happened, Parish faced abdominal surgery the next day and seemed grateful for the distraction. White-haired, nearly sixty, and pared to a noticeable thinness by his ailment, he nevertheless cheerfully gave me a tour. He showed me a spot where a carelessly planned road embankment had stopped up a creek, blocking water flow and jeopardizing the movement of small animals (hence its name, Tortoise Crossing); a small salt-disposal basin; and, from a distance, Lock Four. Upstream from the lock, the trees were long dead, waterlogged when construction of the lock lifted the water table beside the river. Now the dead trees extended bare limbs toward the river: another irrigation graveyard. Immediately downstream from the lock, the shoreline looked healthy; the disparity between upstream and downstream, Parish said, "is the difference between a managed river and a natural one." We crossed the River Murray on a bargelike ferry big enough to handle four cars or more, but we were its only passengers. The Murray looked simultaneously full and vacant.

At the apex of a thirty-seven-year water industry career, Parish presides over a trust that delivers water to more than sixteen hundred member farmers and three thousand nonmember households and businesses. Central Irrigation Trust serves nine irrigation districts that line the Murray shoreline like ribbons and cover about two-thirds of the Murray's South Australian length. Partly in anticipation of a 2003 South Australian River Murray catchment plan that requires all irrigators to achieve 85 percent water efficiency by 2005, Parish had modernized his equipment to a degree unsurpassed in the Australian irrigation industry. He replaced open water channels with pipes, thereby eliminating evaporation and reducing leaks. He automated pumping stations so that they could be remotely monitored and controlled, and he installed solar-powered water meters. His farmers could even place their water orders electronically, using the Internet or a telephone keypad, or, if they weren't technologically inclined, they could speak to a receptionist: in fact, the trust's two receptionists recognize most callers' voices. All those measures smoothed water delivery from river to farms, until only the farms themselves remained notably inefficient. The trust addressed that issue by launching a program using aerial photographs and Geographic Information Systems technology to help identify fields where water is being wasted. Armed with that data, trust representatives pay visits to the owners of the water-inefficient plots and suggest methods to improve. Parish revels in the system's imaging capabilities. "Take any complex plan and see it as clear sheets of plastic. One layer shows the geographic features and divisions of land, and over that lay the pipes and outlets, then the farms and plantings and houses—you can add as many layers as you want. We can produce a plan instantly that shows which farmers grow which crops, how much has been replanted in the last five years, which are 50, 60, 70 percent efficient, all in different colors. It provides visual impact. This thing is so powerful that we can start with a photograph of the quarter of South Australia we occupy, and we can zoom in until we can see the shadow of my car in this car park."

As we drove, I asked Parish for his assessment of the commission. "I think it's done good things under a serious handicap," he said. "Governments have been reluctant to bite hard on the things that need to be done to the river system. They're not taking decisions quickly enough."

Parish's chief example was the focus of today's commission meeting: environmental flows. He could have been speaking for the entire state of South Australia when he said, "When you travel through the system as I do, you learn that everybody has a concern for the river. It's just that if the system is badly run, the downstream user will be more of a victim than the upstream user"—and too much water was being extracted upstream.

In response to this, the commission was doing an "okay job," he said. "Don is strategically and politically very conscious and astute. He knows when he can win and when he can't. His eternal challenge is to keep all the bastards happy. He has a hard life, meeting everyone's expectations when they come from such different viewpoints. I think the commission knows what's needed for a healthy river. Their problem is getting people to do it."

On the evidence of the Murray-Darling Basin Commission's organization chart, far-reaching reform looks unlikely. The commission's structure echoes Australia's parliamentary form, in which politicians from the ruling party become ministers who set their ministry's policies. Directly beneath the ministers are chief executives, often bureaucrats who have risen through the civil service, who carry out the minister's policy directives. The interaction of minister and chief executive marks the point where politicians collide with the bureaucracy—the design is intended to join political considerations with practical ones, but it is susceptible to stalemate. As if the commission were a ministry, the 1992 Murray-Darling Basin Agreement burdens it with a double dose of political and bureaucratic control. Its paramount authority is the Ministerial Council, consisting of thirteen ministers representing relevant ministries (land, water, environmental resources) in the six governments belonging to the commission. Not surprisingly, the council, the realm of politicians, makes commission policy. But its decisions require unanimity among the state and commonwealth governments, and that kind of unanimity is hard-won.

Directly beneath the Ministerial Council on the organization chart is its board of directors, confusingly also called the Commission and composed of commissioners. They're the bureaucrats: eleven chief executives drawn from the same set of ministries as the council ministers, plus the appointed (and sometimes semiretired) Murray-Darling Basin Commission president. Just as within each government ministry the chief executive advises the minister, so the commissioners advise the Ministerial Council. Like the Ministerial Council's decisions, theirs must be unanimous. To fulfill his agenda, Blackmore thus faces two sets of daunting obstacles: he must win the unanimous support of both commission bureaucrats and council ministers. His intricate dance is complicated by an immense number of partners, since the composition of ministers and commissioners changes with every state and federal election. In thirteen years as chief executive, Blackmore had worked with seventy-one council ministers and fifty-three commissioners: he averaged ten new bosses a year. The potential for obstruction is enormous, and can be overcome only when ministers and commissioners put aside some of their provincial concerns for the perceived benefit of the entire basin. That sometimes happens, but state politics is a constant presence at the meetings.

The December 10, 2002, commissioners' meeting at Renmark marked one more step in Blackmore's patient march toward environmental flows: he wanted approval to talk about the numbers. Without it, he'd have little chance of arriving at a final decision on quantity by October 2003, as the ministers' 2002 Corowa Communiqué called for. But Blackmore already knew that the Ministerial Council had limited his prospects, by declining so far to put up money to compensate farmers for reduced water rights and retirement of unsustainable land. Unless the Ministerial Council promised some money, the environmental flow quantities would be dependent on voluntary actions and would be minimal.

The meeting went long. I waited outside the conference room for a while, then walked across the street to the grassy park beside the River Murray. Inhabited shoreline up and down the Murray is festooned with neatly mown lawn, the gesture of a settler populace both eager for community and nostalgic for Europe. The river at Renmark is so broad and thoroughly tamed that it looks like a lake; the ripples that disrupted its flat surface were barely visible. I walked along a paved walkway along the riverbank, passing weeping willows and secluded houseboat moorings. I saw a sign that said, "DO NOT FEED THE PELICANS," but no pelicans. Another sign summarized Renmark's flood history: the river and floodplain here are both so wide that the 1956 flood lifted the river's level only eleven feet or so, but that was enough to inundate the town. A plaque commemorated a restored 1911 paddle steamer called the *Industry* that still went on twenty excursions a year "when the general public can experience the atmosphere of a by-gone era"—but the boat was not in sight. In half an hour or so, the only river vehicles I saw were two houseboats motoring in opposite directions.

When the meeting finally adjourned, after seven hours, Blackmore emerged looking energized. He said the session was "rocky" but didn't elaborate. Instead, we quickly loaded into two cars and began the three-hour drive back to Adelaide. Since I rode with Kevin Goss, I asked him to explain what happened.

The meeting exhibited the usual parochial concerns, Goss said. As David Harriss promised, the New South Wales government blocked Blackmore's emergency Mitta Mitta "trial." "This was payback."

And New South Wales and Victoria got into an argument about the baseline to be used for the cap.

"Logic would tell you," Goss said, "that when the governments took the big decision in 1995 to put a cap on diversions, the level of diversions would be frozen at that time—you'd use the 1995 levels as a baseline. The problem is that it implies that at the time of the cap, New South Wales and Victoria had each achieved the same amount of good policy in constraining diversions. Victoria said no, New South Wales gets an advantage by using 1995 as a baseline because New South Wales used water more wastefully then, and in fact New South Wales relies on using some of Victoria's unneeded water from time to time.

"Victoria wants to use 2002"—the current drought year, that is—"as a benchmark, but that would undermine New South Wales," since Victoria in 2002 had no water to spare for New South Wales. "Victoria wants 2002; New South Wales wants 1995."

I asked Goss how the argument was settled.

"It's not resolved," he said. The commissioners chose the easy path, by forming a working group to figure out a solution.

In distant brown fields, we saw emus, motionless and seemingly dazed. The land our road bisected was flecked with salt and populated by dead trees. Aside from the occasional irrigated vineyard, the only living flora were sparse salt-tolerant shrubs. From time to time, we caught a glimpse of the river. Incised in a distant canyon, it looked more like thread than ribbon.

The environmental flows discussion took up most of the meeting. Goss said Blackmore displayed a form of the environmental flowchart he'd shown at Bungendore, only now it had acquired two more options for acquiring water, both voluntary and yielding small amounts. Even with the additions, the chart served its purpose of showing the commissioners that the "easy" environmental solutions wouldn't suffice to achieve a substantial environmental flow. Blackmore got authorization to talk about the numbers, but he still didn't know whether a big sum was attainable. "Two steps forward, one step back," Goss said.

Blackmore emphasized the positive. Midway through the drive, he called for a switch of cars on his cell phone and climbed into ours. Soon he told Goss, "The end result is that except for one thing—"

"The flooding of the Mitta Mitta," Goss filled in.

"—we got everything we wanted."

Part of Blackmore's eternal dilemma as chief executive extended from two contradictory requirements of the job. He was expected to possess a sophisticated scientific understanding of the river, which would guide him in anticipating its problems and devising solutions. At the same time, he had to show healthy respect for the views of the basin's farmers, whose grasp of river science lagged far behind his. He thought he knew at least some of the answers, while the farmers were just getting to the questions, yet he had to lead them to the answers as if they'd discovered them on their own. These dueling objectives made his job highly political, and left him open to the accusation that he spent more time applying the art of persuasion on his official bosses on the Ministerial Council and commission than on farmers. He was called a "deal maker," a "consummate bureaucrat," and more bitterly, "arrogant," a man who "listens but doesn't hear." It was partly to escape his dilemma that a week after the commission meeting, he had breakfast in the Melbourne Airport Hilton with David Buckingham, one of Australia's most illustrious political consultants. Buckingham, though only fifty-five, had been a diplomat, a prime minister's adviser, a government division head, and chief executive officer for, consecutively, the Minerals Council of Australia, the Business Council of Australia, and the continent's largest public relations firm. Now Blackmore was hiring him as a commission consultant.

In theory at least, basin views are represented on a twenty-person panel called the Community Advisory Committee, which includes a generous contingent of farmers and a less generous one of environmentalists. The committee reports directly to the Ministerial Council, but since the council bypassed community consultation when it made its pivotal decision on the cap in 1995, farmers distrusted the council. The council made clear that it would consult the committee before making its environmental flows decisions, but many farmers feared they'd be bypassed again. Blackmore wanted Buckingham to work with the committee, negotiating the principles that would guide environmental flows.

Blackmore sported a new haircut and wore his office regalia, with long-sleeved shirt and tie. Buckingham, the older man by three years, looked comfortable in a dark blue suit.

On the table, Blackmore placed a stack of commission documents eight or nine inches tall (higher, I noticed, than the stack he'd given me), and wore his reading glasses low on his nose. "I'm going to start with the answers and work back to the question," he told Buckingham. "How's that for a novel approach?"

The answer, in fact, took up most of the breakfast. Blackmore held up each document, explained its significance and which parts Buckingham should read, then dropped it on the floor, forming a new stack. Buckingham took notes. Blackmore started with his keystone principle, "All participants should be project beneficiaries," then explained the numbers in the Expert Reference Panel report. Once again, he displayed a form of the chart I'd first seen on an overhead projector in Bungendore, when it was titled, "The Opportunity in Water Recovery." Now it was called simply "The Living Murray." It was far more detailed than the Bungendore chart because it was designed for a more exclusive audience. It broke down the three environmental flows scenarios under consideration—350, 750, and 1,500 gigaliters—into the amounts of water the various recovery options would have to yield. Yet unlike the neatly typed Bungendore slide, the chart looked handwritten, as if to suggest that Blackmore had just thought it up: the handwriting reinforced the tentativeness of the numbers. (When I asked him about it, Blackmore called the handwriting a "deliberate strategy.") Blackmore described both the easy voluntary measures and the hard compulsory ones that the ministers would have to accept to reach the 1,500-gigaliter scenario. There were only two hard measures, but they would provide more than half the 1,500 gigaliters: a blunt across-the-board cut in water allocations (690 gigaliters) and a program to pay farmers for retiring unsustainable irrigated fields (150 gigaliters). "I don't know that the governments will have the guts to do this," Blackmore said. It was only after he'd finished explaining the chart that he declared, "That's the answer, like it, love it, or hate it." The answer, in other words, meant accepting politically unpalatable options. Later on, Blackmore confessed that he himself was shooting for 1,000 gigaliters, enough to give the Murray a "low-to-moderate" chance of surviving. That was Blackmore's answer within the answer.

Just before the end of their breakfast, Buckingham listed the itinerary of his impending European honeymoon trip: London, Rome, and Venice.

Blackmore knew he was only fifteen months from retirement and sounded dreamy. "Rest is good," he said, not quite in reply. "I'm ready for about thirty years of it."

From the hotel dining room, Blackmore and I walked to the elevator. As it rose, he said, "I've been working for ten years to do this. I'm not going to solve a single issue, water—we're going to fix water and communities and the environment" together. The "we" and "I" were, as usual, interchangeable; the ambition remained large. Yet the increment Blackmore was betting on, 1,000 gigaliters of environmental flow, would yield at best a moderate chance of producing a healthy river. Even if Blackmore got his 1,000 gigaliters, future generations would have to build on the commission's work.

Two floors up, Blackmore walked into a conference room where a subgroup of the Community Advisory Committee was about to hold a meeting. He sat at yet another U-shaped table and listened uncomfortably as Leith Boully, the committee chairwoman and a Queensland cotton farmer, led a discussion about the principles that should guide the environmental flows process. The representatives of farmers and bankers sat on one side of the table, while the environmentalists sat on the other. Tony Sharley, the Banrock Station manager who here wore an environmentalist's hat, expressed regret that the Ministerial Council had not gone beyond its three prescribed scenarios—350, 750, and 1,500 gigaliters—of water returned to the Murray. The council should have also studied 3,000 gigaliters, since the Expert Reference Panel report showed that it was the only scenario with a high likelihood of success—"Otherwise," he said, "we're just kidding ourselves." A farmer answered that if 3,000 gigaliters was put on the table, "some people"—presumably including him—"will get very nervous about the signal that's being sent." If the farmers got nervous enough, they'd refuse to participate in the process. To Blackmore, this all sounded like old news, and he struggled not to let his impatience show. As the committee members took turns proposing principles of varying applicability, Blackmore leaned over and whispered, "The only principle that matters is, 'Everyone is a beneficiary.' " The comment was prescient to a degree, but in seeming to dismiss the committee's exercise as arid, it smacked of presumption.

When the conversation turned to a voluminous listing of key basin issues, Blackmore's frustration increased. Clutching the chart he'd shown Buckingham, he jumped up to the front of the room and began writing the numbers on a whiteboard. He declared that "the issues for me are the absolute lack of understanding of the opportunity that environmental flows represent and the lack of vision about water in the Australian environment"—and for a few minutes he was off, once more explaining the environmental flows numbers to skeptical farmers. Just as suddenly, he stopped—he was putting the answer before the question, as at breakfast. Abruptly, he offered an apology for taking the stage: "I just wanted to be a mongrel because that's what I get paid to be." Worried that he'd said too much, he quietly left at the lunch break. Blackmore's behavior had not endeared him to the group.

Over the next nine months, Blackmore hammered out a deal. The surprise in it was the large input of the body that Blackmore slighted: the Community Advisory Committee. Its farmer representatives saw that achieving measurable improvement in the River Murray's health would probably require more than 1,500 gigaliters, the highest level under consideration for environmental flows. But even 1,500 gigaliters would push many farmers out of business, yet it might not be enough even to validate the environmental flows approach. Indeed, in months of work with the flow numbers, the commission's computer modelers had been unable to predict specific benefits: the river system was too complicated for that. To the farmer representatives, the commission seemed to be advocating flow numbers without certainty of ensuing good.

The Community Advisory Committee advocated another approach. In fact, it arose from scientific studies conducted in Boully's home watershed, the Lower Balonne in Queensland. Instead of broadly aiming to protect the whole watershed, the studies identified specific "ecological assets" within the watershed that environmental flows could benefit; the thinking was that if these sites benefited, then the whole watershed would share in the improvements. With the blessing of the Community Advisory Committee, Boully presented an expanded version of the Lower Balonne concept at a commissioners' workshop in October 2002.

She argued that the best way to achieve a healthy, working Murray was to focus at first on the health of such sites as the Barmah Forest, the Chowilla floodplain, the Coorong, and the Murray mouth. If successful, this "first step" would demonstrate the validity of the environmental flows approach; then ensuing steps could address the entire Murray basin. The idea caught on. The commissioners endorsed it, and Community Advisory Committee representatives convinced farmers that the approach was sound. Over the next few months, the commission's flow modelers found calculations of benefits much easier to perform on specific sites than on the whole river. They predicted that 400 to 600 gigaliters would produce enough flow to show demonstrable improvements at the sites.

Thus, on November 14, 2003, the Ministerial Council announced its First Step decision on environmental flows: the six governments committed themselves to return to the river 500 gigaliters a year to "maximize environmental benefit" for six significant localities along the river. The announcement was both historic and anticlimactic, pioneering and timid, a half-empty glass and a half-full one. Instead of Blackmore's 1,000 gigaliters, the five participating governments pledged to return to the river only half as much—according to the Expert Reference Panel report, a number with a low probability of producing a healthy river. Indeed, the council made no pretense of arguing that 500 additional gigaliters a year would return the Murray to health. For that matter, there was no guarantee that *any* water would be available for environmental flows. Given the continuing shortage of rainfall and the increasing evidence of climate change, scientists were beginning to consider the possibility that what the basin was experiencing was not, technically speaking, a drought, but was rather its new climate. If the drought continued, the commission would have to cut back environmental flows to meet its other obligations.

On the other hand, the simple achievement of an environmental flows accord was stunning. Out of a tunnel-visioned, engineer-heavy, obscure bureaucratic outpost, Blackmore had fashioned a scientifically sophisticated, multidimensional instrument of change. He'd introduced the basin to the reality of water scarcity, and he'd persuaded distrustful states to act for the common good. He'd shown that the river could not be considered apart from the land surrounding it and devised programs to take land use into account. He'd developed a sophisticated set of river-related data and grounded the commission's operations in its findings. When necessary, he'd developed new technologies—state-of-the-art salt-disposal basins and fish ladders able to accommodate phlegmatic Australian fish—and put in place the tools (in the form of new policies) to wield them rationally. As recently as a decade earlier, the likelihood of generating a serious debate about environmental flows in the basin was remote, yet Blackmore had navigated through the debate all the way to the accomplishment. He'd been a supple tactician, who'd made the drought his ally, and he managed to deliver all the major elements of his vision.

He spoke about the council's action with the pride and delicacy of a politician. "It was an extraordinarily courageous decision," he said. "It's a very important outcome that reflects well on our stable democracy."

But given his previously expressed hope for 1,000 gigaliters, I asked him, didn't the amount disappoint him? No, he said, cutting off my question midsentence. Given other sources of water that had been pledged to the basin—as much as 140 gigaliters from a New South Wales flow management plan, 100 gigaliters already provided annually by Victoria and New South Wales for the environmental benefit of the Barmah Forest, and about 75 gigaliters from the Snowy River environmental flow plan—the real number the river had acquired was 750 or 850, nearly all that he'd hoped for. That, he believed, was "about as big a bite, initially, as rural communities could take." Besides, the river's managers at the commission and state departments weren't experienced enough to handle a bigger number; they needed to work with this amount to maximize the environmental benefit. This wasn't the end of the journey, he said, but it was an adequate number for now.

Embedded in the Ministerial Council's decision were pledges by the five governments whose territories embrace the Murray to provide a total of more than $300 million to compensate farmers for their lost water rights. This enabled the council to avoid inflicting far graver pain on farmers by simple appropriation of their rights. Most farmers found the decision palatable. The twenty-one-thousand-member Victorian Farmers Federation, the largest Australian farmer lobbying group, called the decision "a victory for common sense over the radical environmental movement." Even Murray Irrigation Limited, the fractious farmer-owned New South Wales irrigation cooperative to which David May belongs, said the decision was "welcome," though it later sponsored the research that cast aspersions on the quality of the science underpinning environmental flows.

Among the commission's key constituents, only environmentalists were unenthusiastic. They argue that since the Expert Reference Panel report specifies that at least 1,500 gigaliters a year are required for a "moderate" chance of delivering a healthy, working Murray, any lower number is scientifically indefensible. Paul Sinclair, author of *The Murray* and the lead river campaigner for Environment Victoria, the most important environmental group in the state, said that pegging the level of environmental flows at 500 gigaliters reflected the commission's failure to communicate the dire implications of its scientific research to the basin community. Farmers demanded certainty that the sacrifices they'd make in the form of relinquished water would result in a rejuvenated Murray—"which is impossible," Sinclair said, "because the science isn't based on certainties, it's based on probabilities. Whether we're dealing with ecosystems or social systems, we've got multiple causes and effects, yet many people want a single cause and a clear effect. It's a replication of the engineering frame of mind."

Tim Fisher, a river campaigner at the Australian Conservation Foundation, was more sanguine. Since the ruling coalition in the Commonwealth government included the rural-based, farmer-influenced National Party, it was unlikely that the government would take a bolder first step, he said. Even so, the power of Australian environmentalists continues to expand. "A decade ago, I was the only person in the environmental movement paid to work on rivers. Now there are many more, and the interest of volunteers has grown enormously. I can only see awareness on river health continuing to grow, so it's really just a matter of time before these issues are again in the spotlight.

"The question is"—and here he sounded ominous—"whether we've got enough time for the Murray-Darling Basin."

23 THE COORONG

To understand the Murray and its dismantling, go at last to the bottom of
the drain. The river arrives at its mouth as a whisper, nearly ethereal, an
incantation of its sixteen-hundred-mile length but lacking its elemental
substance. Even before the European era, the river occasionally may
have ceased flowing short of its mouth. Now, with nearly three-quarters
of the Murray's flow diverted for human use and a drought on top of that,
only dredging prevents the mouth from closing more or less permanently. Instead of the European or American idea of a river, gaining mass
and breadth as it hastens to the sea, the last reaches of the Murray resemble a child's maze, with numerous hairpin curves, shifting sandbars, and
dead ends, one of which forms the remarkable ninety-mile-long lagoon
called the Coorong—the Aboriginal "place of the long neck." The
Murray's flow rushes nowhere, but rather leaves the basin reluctantly, as
if expelled, a few drops spilled from the saucer. In the predam era, the
river paused as it opened to broad, shallow Lake Alexandrina and its
smaller, equally shallow offshoot, Lake Albert; then, its waters having
languidly eased through the lakes, the Murray slid to the sea through five
separate channels, skirting Hindmarsh Island and other sandy aits before
reuniting, nearly depleted, at the mouth. Even here, some water eluded
the ocean, as it was pushed away from the mouth into the long pencil of
the lagoon. Near the lagoon's open end, the river concluded in a di-
aphanous smear of sand and water that found a single opening in a hundred miles of peninsular sand hill and seeped feebly into the Southern
Ocean. That's where all the energy resides, in the raucous ocean, expended in never-ending row upon row of lavish waves against the stark
Ninety Mile Beach. Along that uninhabited, majestic strand of sand the
color of yellowed parchment paper, all that energy is turned into foam,
which is hurled against the shore, then disappears. Inevitably, the waves
have found the Murray's weakness. The lakes, the lagoon, and the channels to the mouth once constituted a magnificent estuary, thick with fish,
where hundreds of thousands of birds dined. Within it, freshwater con-

veyed by the Murray mixed with salt water that penetrated the mouth and created a variety of brews: fresh, brackish, as salty as the ocean, saltier than the ocean. The water's salinity varied with location, depth, and season, thereby sustaining a diverse array of birds, fish, and invertebrates. The Murray brought freshwater and nutrients to the northern, estuarine half of the Coorong, and high tides and winds washed them over its mudflats; wading birds consumed the resulting effusion of aquatic worms and other invertebrates as the birds promenaded along the edges of receding tides. In the predam era, the Murray almost certainly possessed enough force to maintain a rough parity between sand swept into and out of the mouth, but that stopped being true decades ago. Now a sand blanket covers most mudflats, eliminating most avian feeding grounds, and bird and fish species have drastically declined. All this is a consequence of the Murray's dams, weirs, barrages, and water diversions, whose destructiveness is now accelerated by drought.

Even at the outset of the European era, the Murray mouth was nearly invisible: the entrance is so inconspicuous that oceanic explorers consistently missed it. In 1802, geographers Matthew Flinders and Nicolas Baudin were on mapping missions for their respective countries, Britain and France, when they spotted each other's ships across the bay outside the Murray mouth. The two countries were distrustful rivals, having concluded a war only two weeks earlier. Flinders readied weapons as his boat approached Baudin's, but upon meeting, the two men were sociable, if somewhat reserved. Flinders had covered far more Australian coastline than Baudin, and wrote in his journal after their conversation, "I did not apprehend that my being there at this time, so far along the coast, gave him any great pleasure."* Indeed, each man thought he was conducting Australia's first circumnavigation, and was disappointed to discover that the distinction he'd taken for granted was contested. The

---

*Flinders was another brave but hapless Australian explorer. He completed the continent's first circumnavigation ahead of Baudin and was the first European to map much of the Australian coast. But on his way back to England, he stopped in Mauritius to repair his leaking vessel, whereupon the French colonial authorities there jailed him for six years. By the time he got home, in 1810, Baudin had published his expedition results, in which he claimed for France "discoveries" made by Flinders. (Of course, the true discoverers of all these places were Aboriginals, who arrived in the Coorong region about six thousand years earlier.) Physically spent, Flinders rushed to prepare an account of his travels, but the first published copy was placed in his hands just after he fell mortally unconscious. He died the next day.

meeting gave the inlet its name—Encounter Bay—but the mouth, perhaps then at low flow, eluded the explorers. Flinders saw only the long dune now known as Younghusband Peninsula, with its hills of up to a hundred feet of sand, and had no notion of the extravagant lagoon on its other side. "Low sandy shore, topped with hummocks of almost bare sand," is how he described it. Nearly three decades passed before Charles Sturt finally located the mouth in 1830.

Because the Coorong is shallow and delicate and made largely of sand, its landforms shift constantly. The mouth has moved along the shoreline four or five miles over the last three thousand years, and has been observed moving nearly fifty feet in twelve hours. Regulation of the Murray has slowed the mouth's movement, and the river's reduced flow has caused it to become clogged. Aerial photographs of the Murray mouth taken over the last twenty-five years illustrate the process. No two photographs look alike. They look lurid, like X-rays rendered in finger paint, full of ominous shadows, fleshy curves, and curling, inky liquids, unidentifiable but disturbingly familiar, as if depicting bodily functions. In form, they resemble Kandinsky paintings, with their agglomerations of straight lines (the parallel rows of waves facing the beach, the shoreline) and curved ones (the channels, the islands, the lakes). Twice in the first half of the twentieth century, the Murray mouth became shallow enough to walk across, but dating from the arrival of Europeans, it never quite sealed up. Then, in 1981, a new era arrived: the mouth closed for seventeen weeks. With its feathery deposits of green-streaked tan sand filling the water channels, the 1981 photograph, considered abstractly, is beautiful. It's only when you focus on its subject—the collapse of the Murray mouth—that you feel an Antarctic chill. In the 1978 image, the channels appear open, but three years later, after a drought, they're clogged, and the mouth has become a sandy, desiccated delta. In 1983, one of the old channels still looks blocked, while narrow rivulets have formed across the surface of the others. By 2002, sandbars have formed on the ocean side of the mouth, presaging another, more lasting closure, warded off so far only by continuous dredging.

The Coorong is only a few thousand years old. Until perhaps twenty thousand years ago, the land it now occupies consisted of vast rows of dunes, overlooking an ocean several hundred feet lower than it is now. As the ocean rose, it deposited debris on top of the dunes, which led to the formation of long barrier islands. Two or three thousand years ago, the islands fused, and the southern outlet to the sea was closed. The islands turned into the Younghusband Peninsula, the barrier between the newly formed lagoon and the sea. Most estuaries are saltiest at their mouths, but the Coorong, a "reverse estuary," is saltiest farthest away, ninety miles to the southeast, where the water is four times as saline as the ocean—that's because water there has no outlet, so its salt can't be washed away. Even before river regulation, the southern end—the so-called South Lagoon—was much saltier than the sea. It received its freshwater only every decade or two, when the Murray produced a large flood; at other times, evaporation and the inflow of salt intensified the water's salinity. There, the only fish that thrived was the two-inch-long small-mouth hardyhead, a staple for pelicans and cormorants. Even close to the mouth, where freshwater was more abundant, the Coorong was not exceptionally hospitable to humans. In the late eighteenth century, as few as six hundred Aboriginals may have lived around the lagoon, and the Europeans never mastered habitation there at all. The water was too salty, the wind too harsh, and the flies too persistent.

Birds, on the other hand, thrived. Sturt writes at the Murray mouth of "thousands of wild fowl that rose before us, and made a noise as of a multitude clapping hands in their clumsy efforts to rise from the waters." George French Angas,* an English naturalist who explored the Coorong in 1844, found himself in "a fairy scene of birds and solitude," where "red-legged gulls, plovers, and sandpipers were for ever busy in search of marine insects, or paddling in gentle ripples of the mimic waves, in undisturbed enjoyment." Looking down from a promontory at the Coorong's "Narrows," where the lagoon in summer is only a foot or two deep and 140 feet wide, Angas reports that "myriads of ducks, swans, pelicans, and every variety of sea fowl darkened the water beneath us . . . Humming bees wandered over the odouriferous plants in the scrub, and bustards, ground-parroquets, bronze-winged pigeons, robins, and a variety of birds of brilliant plumage, constantly appeared." Even now, some eighty-five species of waterbirds feed in the Coorong, and it remains the most important haven for migratory waders in temperate Australia. Most

---

*Angas, an enthusiastic and meticulous naturalist who began his career in South Australia at the age of twenty-one, is the rare early European who saw Australia clearly. Unlike other European artists in Australia who filled their drawings with vaguely English-looking plants, Angas depicted identifiable Australian flora.

of the Coorong is only two or three yards deep, and a seasonal variation of a foot is enough to expose a third of its entire surface. The mudflats are therefore ideal feeding grounds for the twenty species of small, short-legged migratory waders that feed there; among them is the six-inch-long red-necked stint, which in its perpetual pursuit of summer fattens itself for its annual eight-thousand-mile flight to breeding grounds in Siberia and Alaska. In good years, some three or four thousand Australian pelicans, possessors of the world's largest avian beaks, breed in the Coorong. Gulls, terns, and grebes dive for fish in its waters, and land birds inhabit the dunes. For all these reasons, a large portion of the Coorong was declared a national park in 1966, and it was listed under the Ramsar Convention as a wetland of international importance in 1985, but neither action has prevented its continuing decay. The number of the Coorong's migratory waders probably had already declined by the early 1980s, when counts ranged as high as 235,000; in 2004, the number was 59,000.

The crucial event in the modern history of the region was the insertion of five barrages into the channels just upstream from the Murray mouth in the late 1930s. Together with the channel islands to which they are linked, the barrages form a thirteen-mile-long barrier that cuts the estuary in two: on one side are Lakes Alexandrina and Albert, and on the other are the lagoon and the mouth. Barrages are the lowest form of dam. The simplest of them neither allow passage of boats nor hold back floodwater nor possess gates to release what water they do contain. The Murray barrages range in length from 800 feet to 2.3 miles. Some of them possess gates, but they're opened only in times of high flow; because of the drought, they were recently kept continuously closed for nineteen months, a record. The gates I saw look more like fence than wall—they resemble highway barriers affixed to the riverbed. In the Murray, their role is to separate fresh and salt water, and in this, they have succeeded all too well. Even before the dam era, the Murray in low flow produced saltwater intrusion into the lakes and as far as 150 miles upstream. When upstream diversions extended seawater's reach, the River Murray Commission built the barrages. In response to one set of unintended consequences caused by the diversions, the barrages created another set. The barrages solved a few problems at once. By lifting the lakes' surface by a couple of feet, they promoted navigation and gravity-powered irrigation, and they turned the lakes into freshwater storages for use by farmers and city dwellers. But they cut the estuary to a tenth its former size, and probably reduced fish and bird populations by a similar proportion. Migratory fish, which move during their life cycles between fresh and salt water, found their paths obstructed; eight fish species once commonly seen in the Coorong disappeared. With the Murray's flow to its mouth restricted to times of bountiful water, when the barrage gates were opened or water spilled over them, no force resisted the sand thrust through the Murray mouth by the ocean's waves—sandbars began building up inside the mouth, and the mouth grew narrower. The influence of tides inside the mouth decreased, until sand covered many of the Coorong mudflats where waders once dined. Now, despite the barrages, the lakes are again becoming saltier, thanks to the Murray's heightened salinity and low flow. The lakes' water is still potable, but it is close to becoming useless to wineries that rely on it, and without a substantial flow it will grow worse. The ecosystems downstream from the barrages, in the Coorong, the channels, and the mouth, are at risk of degrading into what a commission publication delicately calls "a uniform condition," which is not much different from a terminal one.

Like Chowilla, the Coorong region is not heavily touted in tourist brochures—one refers to it as "unspectacular"—but this is blindness, the old tendency to look at Australia and wish for Europe. Looking toward the mouth from Richard Owen's house on Hindmarsh Island, I was easily beguiled. The house is an expanded shack, and from its deck, looking toward Sir Richard Peninsula north of the mouth, I could see no other sign of human habitation. I saw a strand of amber grassland, mottled with scrub; then a strip of sparkling estuary; the low hump of the peninsula; the pale, cloudless sky. The ocean beyond the peninsula wasn't visible, but the breeze carried its prickly scent. It was April 2004, when the dredging operation a few miles down the peninsula had been running twenty-four hours a day for eighteen months, but from Owen's porch, the machinery was inaudible. In fact, the dearth of sound comprised the only jarring element in the scene: the few birdcalls I heard seemed to reverberate in the silence, as if the Coorong's grace was nullified by its faunal emptiness. "Provided you're not at the mouth," Owen said, "there's no sense of the tragedy that's going on here." Though casual tourists fly over the Coorong in World War II warplanes and cruise through it on tour boats, water skis, and jet skis, most miss the tragedy entirely.

Until six years earlier, Owen had been a librarian writ large, who managed South Australia's community college library network. Then he retired, and devoted his life to environmental restoration of the Coorong. Built in the '50s, his house is one of about three hundred vacation homes on Hindmarsh Island, at roughly twenty square miles the largest of the islands near the Murray mouth. Hindmarsh's houses are humble: they're the sort of places, in Owen's words, "where it doesn't matter too much what it looks like." Owen added a deck to his house with recycled plantation pine ferried from the mainland, and he installed a gutter on the roof to collect rainwater, the only water he uses. (Rainwater here is sufficiently pure that Owen drinks it unfiltered.) Fifty-eight years old, he sported a beard and graying ponytail, and his frayed and torn sweatshirt covered a Falstaffian belly. Having made a conscious choice to sacrifice a measure of comfort to live on the edge of a wilderness, he looked comfortable in his frugality, wearing it without appearing impoverished.

The trouble with Owen's life in Elysium is that the Coorong is dying, and the wilderness is becoming less wild. During the 1980s, developers built a small marina at the western end of Hindmarsh Island and made plans for a vast expansion including nine hundred houses, a one-hundred-bed motel, a heliport, and the largest marina in the Southern Hemisphere. To realize the plan, the developers needed a bridge between the mainland and Hindmarsh Island, and they began trying to persuade the Commonwealth and South Australian governments to build it. The ensuing imbroglio, pitting environmentalists and local Aboriginals against the developers, became known as the Hindmarsh Island Bridge Affair. It provoked so much bitterness that in towns near the Murray Mouth, bridge supporters and opponents refused to patronize each other's shops, and some merchants banned bridge discussions on their premises to keep peace. After Owen formed a community group to oppose the bridge, the developers obtained court injunctions to force the group to cease activity. When the injunctions lapsed, the developers began filing lawsuits, eventually charging not just Owen and other local bridge opponents but also the Australian Broadcasting Company, Adelaide University, the South Australian and Commonwealth governments, even the Commonwealth minister for Aboriginal affairs. Of the eighteen counts cited in the initial defamation suit against Owen, a colleague, and the environmental group Conservation Council of South Australia, seventeen were dismissed after trial or appeal, but the eighteenth, requiring the defendants to pay $55,000 in damages to the developers, was upheld on a two-to-one vote of the South Australia Supreme Court. Yet dismissal of even the last claim would have amounted to no more than a Pyrrhic victory for Owen, since the bridge was completed in 2001, the marina is expanding, and about three hundred new houses have been built or are under construction. "Huge palatial ostentatious bullshit" is how he describes the new homes.

Farmers cleared Hindmarsh and the other channel islands of native vegetation more than a century ago. As a result, Owen estimates, only 1 to 3 percent of native vegetation still grows in the Coorong region. Given that fact alone, he says, "Is it any wonder the river is in trouble?" To counter this, he and his allies have planted more than thirty thousand trees and plants on the island. Behind his house, he has started a seed orchard for ground cover he hopes to plant throughout the Coorong to anchor its dunes and ward off introduced weeds. Most of Owen's two hundred–plus specimens of native salt-tolerant ground cover known as saltbush were two or three feet in diameter. Splayed in geometric rows across a vast polypropylene tarp called a weed mat, the plants looked like Brobdingnagian green buttons on a factory assembly line. Though Owen started the orchard only seven months earlier, he had already reaped his first harvest of seeds. To show me how, he lifted some stems and shook them, causing hundreds of seeds to fall to the mat; eventually, he'll sweep them up and plant them.

From the back of the house, I could see the remains of a small wharf that had functioned until 1940, the year the barrages were completed. Now it was nothing but a few poles sticking out of dry land. The process that stranded the wharf has not stopped. The mouth has moved half a mile in the last ten years, and its near closure has accelerated sand accumulation inside it. The wharf's fate constitutes a warning that the marina developers have chosen not to heed.

Together with Owen's son Peter, an environmental activist for the Conservation Council of South Australia, we climbed in Owen's motorboat and headed toward the lagoon. On the way, we passed some of the old Hindmarsh Island vacation houses. They were as eccentric and ramshackle as Owen had suggested. An alien fir tree grew through an opening in the roof of one of them, while many were linked to the lagoon by rickety private piers. In place of Angas's "myriads," we saw birds only occasionally: a cormorant; a few geese; a sacred ibis with black, pointed beak; half a dozen whistling kites circling over carrion they'd spotted on the peninsula. Near the mouth, we heard the rumble of the dredging operation and saw its smoke, and inspected the sinuous, moss-encrusted pipe that conveyed dredged sand a few hundred yards up the coast—not far enough, Owen thinks, to prevent the sand's eventual return to the mouth. The Murray-Darling Basin Commission introduced dredging as an eight- to ten-month emergency project, but because of the drought, eighteen months have passed and the dredging has never stopped—if it did, the mouth probably would seal up within a few weeks. "If the channel closes, the Coorong is gone," Owen said. This isn't hyperbole: the mouth's closure would eliminate the Coorong estuary. In winter, its water level would be higher, covering the mudflats; in summer, the water level would drop because of evaporation, but the water would become more saline than the ocean. Fish that travel between the Coorong and the ocean would be stranded, and migratory waders would lose most of their feeding grounds. With the Coorong already at a historical low point, two-thirds dead by some estimates, the mouth's sealing would finish off the job.

We stopped at Godfrey's Landing, near the peninsula's narrowest point, where the Coorong National Parks and Wildlife Service has installed a yard-wide boardwalk over the ridge to the ocean. Godfrey's Landing is indistinguishable from the rest of the lagoon's sandy shoreline except for a scant park outpost there: a sunshade made of corrugated tin, two bathrooms, a sheltered sign. The sign is wrong from the outset. "The Murray Mouth Discovery," it says and proceeds to describe the exploits of Flinders and Sturt. Someone has scratched an "X" through the word "Discovery" and has written next to it the truer word, "Conquest." As we proceeded along the boardwalk, we found evidence of the conquered: an expanse over hundreds of yards of sandy hill and swale that glistened with shards of shells—we were standing on an Aboriginal midden. The Coorong was once so bountiful that as recently as two centuries ago, Aboriginals lived semipermanently here. From the lagoon on one side and the ocean on the other, they extracted cockles, oysters, and crabs, and gathered just beneath the peninsula's sandy ridge, on its lee side. Here, they ate and left their shells on the spot; at my feet, shell remnants formed a mosaic in the sand. "This is one of the biggest open-air cafés you'll ever see," Pete said, and I tried to imagine Aboriginals perched on the hill, picnicking.

We heard the surf before we reached the dune's crest. At the top, the wind stiffened, and we found ourselves looking out over the turquoise ocean, the white rows of breakers, the alabaster sand. On one side of the crest, headlong decay; on the other, eternity.

As understanding of the destructiveness of the barriers across the River Murray has spread over the last couple of decades, environmentalists have cautiously advanced proposals to remove the barrages and some of the weirs. The dams, Hume and Dartmouth, are invulnerable, for no one can imagine the basin without them. No one, that is, except the basin's Aboriginals, who for fifty or sixty millennia did not just imagine the unobstructed river but lived it. Nevertheless, in debates about the Murray's future such as the deliberations over environmental flows, Aboriginals' voices have been barely audible. To be sure, Blackmore has worked hard to consult Aboriginals in disputes that directly involve them, as at Lake Victoria, but Aboriginal concerns almost certainly didn't figure in the Ministerial Council calculations that arrived at 500 gigaliters. This is both unjust and comprehensible, for dam mitigation is not a process bound to resonate among Aboriginals. The river's harnessing, after all, enabled European domination of the basin and furthered Aboriginal displacement. Some Aboriginals are therefore less interested in softening the effects of the barriers than in eliminating them.

I met Tom Trevorrow, a Coorong Aboriginal leader, on the day after I visited Owen. Trevorrow is a member of the Ngarrindjeri (roughly, nah-RIN-doo-ree) community, which once flourished in the Coorong and along 250 miles of the Murray corridor—the Ngarrindjeri comprised one of the largest of Australia's Aboriginal nations. Trevorrow belongs to the last generation of Ngarrindjeri to grow up around the Coorong. Until he was eight years old, he and his extended family lived on government-owned land in huts made of corrugated tin, kerosene drums, farmers' wheat bags, uprooted vegetation, and whatever else they could find. As an adult, he became a Coorong National Park ranger. Now, at forty-nine, he runs Camp Coorong, a 250-acre outpost of Aboriginal pedagogy, where the remnants of Ngarrindjeri history and culture

are taught to four thousand outsiders a year. The camp sits on scrubland
a few miles from the Coorong. The main building consists of some
sparsely furnished offices, a reception room, and a museum displaying
Ngarrindjeri boats, medicines, and shelters. Trevorrow is six feet tall, but
sitting behind a large table in his capacious, bare-bones office, he looked
dwarfed, and wore the pained expression of a man not entirely reconciled to his surroundings. Even his body did not quite match up with his
clothes. His tight pants enclosed slim hips, but his stomach hung over
his belt, and the two pocket flaps of his purple canvas shirt pointed up
and down respectively. His shoulder-length brown hair was tucked behind his ears.

He spoke in a grave, deep voice that seemed at odds with his informal
appearance. "The river is a living thing," he said. "When you start inter-
fering with it, everything dies." Environmental flows are a "Band-Aid" —
what's needed is a full-blown river restoration. "Restore the life of the
river, then you'll restore the lakes and the Coorong and everything that
revolves around them."

He spoke softly and did not smile; his theme, after all, was loss. The
Coorong was not the same Coorong that he knew as a child, he said.
"Then it was alive with fish and birds and grasses, you could smell that it
was alive. Today, there's not much there: the fishermen can't even catch
fish. And no fish are going to come anywhere near the Murray mouth
until the dredging machinery is removed. One problem after another.

"When we lived along the Coorong, and camped in places from one
end to the other, we had no problems getting a feed of ducks, a feed of
fish, whatever we wanted, but we were taught the old way, to take only
what we need. You could walk around and see the shape of a flounder in
the sand, and you could spear it. You can't spear fish anymore because
there are no more fish. You could go to an inland soak and dig a meter
down to a little well, and get as much freshwater as you needed. Those
soaks have dried up now. The Coorong was a mixture of salt and fresh
water, but there was a distinct line between fresh and salt, and you knew
where you'd catch saltwater fish and where you'd catch freshwater fish.
Now the Coorong is very salty, and there's no mixing of the waters anymore.

"The pelicans breed on the islands in the southern part of the
Coorong. They used to hop off their islands and swim around and get a
good feed of fish, and they fed their young. My uncles and aunts would

say, 'Look at all those fish out there,' and I'd say 'Where?' They'd say, 'Where all those pelicans are—see them?' You can't read the pelicans anymore because there's no fish for them."

We went on a drive in my rented car along the lagoon's long north-eastern edge. Here, halfway down the Coorong, past the still-functioning part of the estuary, the lagoon looked inert, a watery equivalent of the moon. Here were no birds, no fish, no trees, not even as much water as I'd expected. In some places, the Coorong's seasonal contraction left exposed broad swaths of salt-whitened lagoon bed so smooth they looked polished. It was a bestiary without beasts, alluring but inanimate, cold and hard.

Our last stop was at a crest overlooking the lagoon. Trevorrow pointed in the distance to an area abutting the shoreline, where stones had been piled on top of a reef to form a nearly perfect circle: it was an ingenious, effortless fish trap. When the tide rose over the stones, fish swam inside the circle, only to be snared as the water ebbed; a single tidal cycle could yield hundreds of fish. The site was near one of the Ngarrindjeris' seasonal camps, which they visited for a few months out of the year; when they broke camp, they breached the wall so that no fish would be trapped in their absence. Now the trap contained nothing but a puddle. Piles of stones lay on either side of the gap in the wall, right where the Ngarrindjeri had considerately placed them, perhaps not knowing it was for the last time. In the seven or eight decades since then, the Coorong has lost nearly everything, but the trap remains intact, as if awaiting the return of fish, the return of the Murray, the return of deep water.

EPILOGUE

In early September 2002, Medha Patkar tried to drown herself for the last time. A successful monsoon had lifted the Sardar Sarovar reservoir to an unprecedented height, and all of Domkhedi, including the huts in which we'd talked and slept a year earlier, was inundated. Medha was in a "sacrifice squad" of six women protesting the official indifference to resettlement. The women stood in the water until it reached their necks. Medha was, in her own words, "ready for the end." Instead, about twenty village women who'd been watching from the shore jumped into the water and dragged the protesters to safety. They couldn't allow Medha to die, they told her; they wanted her in the struggle. Medha felt the act as a "sudden blow" that left her, for once, speechless. "We could understand their feelings," she wrote in an e-mail, "yet didn't know how to respond." She eventually declared that in view of this "decision of the people," she would no longer try to drown herself.

She did, however, continue to go on fasts for other causes. In February 2005, after a protest against government clearance of inhabited Bombay slums resulted in her arrest, she launched an indefinite fast in prison. Six days later, she showed signs of failing, and was hospitalized and forcibly injected with a saline solution. Soon afterward, a government offical revoked the arrests of her and other activists, and agreed to allow slum inhabitants to remain temporarily. Medha then ended her fast.

Sardar Sarovar had been built to 363 feet, four-fifths its planned height, when heavy monsoon rains struck in August 2004. The first casualty was the canal, the "lifeline to Kutch and Saurashtra." The Heran River is one of numerous channels that the canal bisects; its floodwater is designed to flow through a tunnel beneath the canal. But when the Heran River flooded, the tunnel was clogged with sediment. As a result, floodwater broke through the canal wall, destroying a fifty-yard length of it and submerging nearby villages. A week later, the swollen Narmada filled to the dam's brim, engulfing hundreds of farms and houses for the first time.

Government resettlement plans continued to be carried out apathetically, if at all. In an excruciatingly belated acknowledgment of the resettlement debacle, in March 2005 the Indian Supreme Court responded to an Andolan petition by suspending dam construction until the resettlement program is expanded and improved. The ruling is likely to delay Sardar Sarovar's completion by at least a year.

Throughout 2002 and into 2003, Tonga resettlers in Lusitu suffered from hunger but not starvation, largely thanks to relief efforts conducted by the Zambian government, nonprofits, and religious groups. A reasonable 2003 crop and a better one in 2004 have relieved hunger and have enabled some resettlers to replenish dwindling cattle herds. At the same time, the resettlers are being plagued by increasingly fearless elephants that cross the Zambezi River from Zimbabwe and devastate agricultural fields and knock down huts in search of food.

Filling of the Mohale reservoir began in November 2002. Thayer Scudder's prediction that the World Bank would lose what leverage it had over Lesotho and South Africa once construction was completed proved accurate. Though the treaty that established the project specified that no villagers should be worse off because of it, the joint Lesotho–South Africa commission that oversees the project declined to pay advance compensation for predicted losses to villagers in reaches far downstream from the dam. As a result, a Bank aide-mémoire declared, "24,000 of the poorest households in Lesotho must see their precarious standard of living deteriorate further." The Development Authority also skipped its first two scheduled environmental flow releases, violating its own policies. And the lodge whose proposed construction on an island in the Mohale reservoir had infuriated Scudder nevertheless was built, in violation of area conservation and development standards. Indeed, the Bank discovered that landscaping on the site included thirty-four plants exotic to the highlands, whose presence "makes a mockery of the zoning designation of the island and ignores the importance of the island's existing flora." As of September 2004, three global companies had been found guilty of bribery in a Lesotho court, and nine other companies awaited trial. After declaring that Acres International, the first convicted company, could continue to participate in new Bank projects, the Bank announced in April 2004 that Acres would be barred from Bank projects for three years. Debarment of other convicted companies is considered likely.

Don Blackmore resigned from the Murray-Darling Basin Commission in March 2004 and became a consultant on national and international water issues. He was made a Member of the Order of Australia and was the honoree at a farewell dinner at Parliament House. The 350 guests were read a letter from Prime Minister John Howard citing Blackmore's "significant contribution to Australia's federation, society and govern-ment." Blackmore's fourteen years as chief executive, Howard said, were "testament to the dedication you have shown in discharging your duties and the respect which you have earned from all governments in the Murray-Darling Basin."

Dredging of the Murray mouth continued into 2005.

In February 2003, in apparent repudiation of the World Commission on Dams report, the World Bank announced its new "water resources" pol-icy. It advocates again taking up "high-reward/high-risk investments"— which is to say, large dams. John Briscoe, the Bank's senior water adviser, said in late 2003 that because dams require so much planning, the change would not be apparent at first, but that in three to four years, "major increases" in Bank support for dams would occur. The first evi-dence of the change was the Bank's approval in March 2005 of financing for Nam Theun 2 in Laos.

We erect dams assuming that they are eternal, as if they'll never topple over or be dismantled or fill with sediment or lose their financial ration-ale. Yet all dams will die. All that remains unknown is the manner of their passing: whether they will be drained and safely put out of opera-tion—"decommissioned," in the lingo—or will, out of indifference or misjudgment or an act of war, collapse, to the enormous peril of all be-ings downstream. Dams are, of course, loaded weapons aimed down rivers, pointed at ourselves; they're proof of the gambling nature of the societies that build them. A typhoon toppled a couple of dams in China's Henan Province in 1975, and two hundred thousand people died in the ensuing flood, epidemics, and famine. So oblivious are we to dams' inevitable fate, so smitten with the idea of their power, that feasibility studies don't consider the cost of decommissioning; the omission amounts to a form of subsidy and inflates dams' value. Dams get old at radically different rates: Hoover may last a thousand years, but other dams have cracked or leaked or filled with silt within a decade. There have been only fifty or so recorded dam failures that caused ten or more people to die (and an unknown but certainly substantial number of unrecorded ones), but the numbers will grow as the planet's quotient of dams approaches old age. Around the world, five thousand large dams are at least fifty years old; the average U.S. dam is in its forties. And the whiff of mortality has reached the dam industry, driving some multinationals out of the business. A chronological graph of international large dam construction resembles an elongated, inverted U: in the 1950s, dam building exploded, then flattened during the '70s, and dropped precipitously after that. In the United States, still the trend-setting dam-building country, six hundred thousand miles of river—a sixth of the nation's total—have been drowned behind dams, but dams in this country are now being decommissioned more quickly than new ones are being built.

As dams age, the cost of repairing them increases until it overtakes profits, and the owners decide they're not worth operating anymore. Over the last century and a half, owners have often abandoned unwanted dams: they became the ultimate litter. Out of the American inventory of seventy-five thousand dams over six feet high, some 15 percent are of "indeterminate ownership." States supported by federal funds often end up paying for decommissioning—we leave it to government to bear the cost of our profligacy, and then we castigate government for wasting our money. Decommissioning is itself a tricky business. Some dams don't qualify because no one can figure out how to dispose of the sediment that has accumulated behind them. There's usually too much of it to pump out, and no place to put it, besides. If the dam is downstream from agricultural or mining operations, its sediment is laced with pesticides, fertilizer, or tailings: release that stuff, and watch the river wither. Even without toxins, the sediment can overwhelm the downstream river's plants and animals if released in a sudden flood.

The larger the dam, the more expensive its dismantling becomes, until, in the case of a hydroelectric megadam, the expense may surpass the cost of erecting it. Imagine these dams in five hundred or a thousand years, after their useful life has ended, when an earthquake from the fault line beneath Sardar Sarovar fractures it, or the bankruptcy of Zambia and Zimbabwe leads to Kariba's fatal neglect, or Three Gorges fills with sediment, or Katse gets too expensive to maintain—and when every one of them suffers from an altered hydrological regime as a result of climate change. Take your pick of mortal scenario: cumulatively, they're more plausible than the assumption that megadams will be successfully financed, adroitly managed, and properly maintained into perpetuity. Some dams will crumble into the basins from which they rose, while others may still be intact but no longer storing water, which instead runs over or through or around them. They'll be relics of the twentieth century, like Stalinism and gasoline-powered cars, symbols of the allure of technology and its transience, of the top-down, growth-at-all-costs era of development and international banks, of the delusion that humans are exempt from nature's dominion, of greed and indifference to suffering. If there are tourists then, they'll circulate through the ruins as they now inspect the Pyramids, awed by the structures' technological prowess and unimaginable cost. The people who lived around the dams will have been dispersed, of course, and their cultures shattered, and the rivers and their valleys may still be depleted. But the dams' ephemerality, not the rivers', will then be on display. They'll be reminders of an ancient time when humans believed they could vanquish nature, and found themselves vanquished instead.

say, 'Look at all those fish out there,' and I'd say 'Where?' They'd say, 'Where all those pelicans are—see them?' You can't read the pelicans anymore because there's no fish for them."

We went on a drive in my rented car along the lagoon's long north-eastern edge. Here, halfway down the Coorong, past the still-functioning part of the estuary, the lagoon looked inert, a watery equivalent of the moon. Here were no birds, no fish, no trees, not even as much water as I'd expected. In some places, the Coorong's seasonal contraction left exposed broad swaths of salt-whitened lagoon bed so smooth they looked polished. It was a bestiary without beasts, alluring but inanimate, cold and hard.

Our last stop was at a crest overlooking the lagoon. Trevorrow pointed in the distance to an area abutting the shoreline, where stones had been piled on top of a reef to form a nearly perfect circle: it was an ingenious, effortless fish trap. When the tide rose over the stones, fish swam inside the circle, only to be snared as the water ebbed; a single tidal cycle could yield hundreds of fish. The site was near one of the Ngarrindjeris' seasonal camps, which they visited for a few months out of the year; when they broke camp, they breached the wall so that no fish would be trapped in their absence. Now the trap contained nothing but a puddle. Piles of stones lay on either side of the gap in the wall, right where the Ngarrindjeri had considerately placed them, perhaps not knowing it was for the last time. In the seven or eight decades since then, the Coorong has lost nearly everything, but the trap remains intact, as if awaiting the return of fish, the return of the Murray, the return of deep water.

ACKNOWLEDGMENTS

I have felt privileged to write this book because of both the subject's gravity and the extraordinary support shown it by so many people. Indeed, since a report on global water commodification called *Blue Gold* floated into my consciousness in mid-1999, I have felt borne along a gentle current to sagacious editors, large-hearted funders, perspicacious experts, and richly complicated, enthusiastic principals. Bob Dawson, an environmental photographer who later accompanied me to India, gave me the report—probably the only time in our considerable friendship that he passed me such a document. Written by Maude Barlow and published by the International Forum on Globalization, it alerted me to the water shortages overtaking dozens of regions throughout the world. The magazine proposal I wrote as a result reached Colin Harrison, then *Harper's Magazine*'s deputy editor, just as he'd decided to look for a writer on water. Colin assigned the piece at six thousand words but encouraged me to go longer: I wrote twenty thousand, of which he published fifteen thousand. The piece, called "Running Dry," landed on the desk of Paul Elie, an editor at Farrar, Straus and Giroux, and that convergence led to this book. I am fortunate to have been guided by Colin and Paul, the two best editors of my professional life. Paul's vision, experience, and willingness to toil for many hours in the grime of words have enormously benefited this book.

In an act of extraordinary friendship, my classmate Art Kern not only bestowed upon me a remarkable grant but also persuaded the William and Flora Hewlett Foundation to match it. The Marin, California, Arts Council and the Fred Gellert Family Foundation also provided generous grants. On top of this, the book won the J. Anthony Lukas Work-in-Progress Award, a kind of writer's-fondest-fantasy prize, consisting of a celebratory evening at Harvard and a substantial stipend. All these grants provided sustenance for my family and me during the three-plus years of book writing, and enabled the extensive traveling the book required. I am deeply grateful to Michael Fischer, then head of the Hewlett Foundation environmental program; Linda Healey and Marion Lynton, guiding spirits of the Lukas Prize Project; Lance Walker, grants program director of the Marin Arts Council; Fred Gellert, Annette Gellert, and Peggy Lauer of the Gellert Family Foundation; and, of course, Art.

When I first considered writing a book on dams, Achim Steiner, secretary general of the World Commission on Dams, heartened me with his enthusiasm and facilitated a month-long research stint at commission headquarters in Cape Town, South Africa. There and later, Commissioners Deborah Moore, Jan Veltrop, and Judy Henderson provided important guidance. Commission consultant Mark Halle offered insightful counsel. And I benefited greatly from the friendly ministrations of advice and support from Jamie Workman, the commission's media liaison. Throughout the project, Juliette Majot and Patrick McCully of the International Rivers Network provided valuable support and information.

At least 150 people answered questions I posed during and after my trips in interviews, phone conversations, and e-mail. They are too numerous to name, but I am grateful to all of them. On each trip, I had the good fortune to encounter an array of good-hearted people who provided all sorts of assistance. In Baroda, Dr. Dilip Desai provided a room in his house where we spent our first night in Gujarat. In Domkhedi, Ashwini Deo interpreted youth camp discourse; Vikram and M. K. Sukumar of the Narmada Bachao Andolan played the same role in interviews with resettled and threatened villagers. Joe Athialy arranged our Domkhedi trip and later patiently answered scores of questions I e-mailed him from California. So did Himanshu Thakkar of the South Asia Network on Dams, Rivers, and People, who is a remarkable repository of information about the interrelationship of water and politics in India. In Bombay, journalists Kalpana Sharma, Sameera Khan, and Meena Menon offered invaluable advice, and filmmakers Sanjay Kak and Anand Patwardhan showed me their movies about the Narmada conflict. After the trip, Bob Dawson gave me proofs of hundreds of his exceptional photographs, and Malavika Vartak of the International Rivers Network translated the signs and slogans that Bob photographed. In San Francisco, Raj Desai helped me decipher Medha Patkar's taped comments.

In Lesotho, Sefeane Keketso of the Lesotho Highlands Development Authority graciously gave up a weekend with his family to conduct a World Commission on Dams official and me through the mountains. In Botswana, Lars Ramberg, director of the Harry Oppenheimer Okavango

Research Centre, provided invaluable logistical advice. In Zambia, Ben Clark proved a superlative guide. Tom Savory graciously allowed me to stay overnight at his farmhouse in Monze while en route to Mazulu. Richard and Emmy stalwartly interpreted the comments of fellow Mazuluans, and Delly ably watched over our campsite. (Richard, by the way, is a pseudonym, as are Zaka and Jeffra; in all other respects, the book aspires to factuality.) Anthropologists Lynn Morgan of Mount Holyoke College and Dan Aronson of the World Bank gave me generous dollops of time to help me understand their discipline.

In Australia, Judy Andrews looked after me with nearly familial concern, and Anita Hancock of Corporate Traveller Canberra Central provided logistical support. Jeff Parish, Henry Jones, Richard Owen, Peter Owen, Tom Trevorrow, Michael Harper, Jack Seekamp, Peter Forward, George Warne, and Mac Paton all provided revealing glimpses of their respective corners of the Murray Basin. I am particularly grateful to David and Robin May, who not only showed me their farm but provided a hearty dinner and comfortable bed; Denis Flett, who provided both a long interview and a night's lodging, and Peter Dower, who gave up a day off to show me Dartmouth Dam. Kevin Goss and David Dole of the Murray-Darling Basin Commission and Jane Roots of the South Australia Department for Water, Land and Biodiversity Conservation graciously devoted many hours to answering my questions.

In all three regions, many people read portions of the manuscript and offered valuable comments. Among them are Roxanne Hakim, the late Monua Janah, Dilip D'Souza, Jackie King, Karen Ross, Lynn Morgan, Lars Ramberg, David Dole, Leith Boully, Paul Sinclair, Tony Sharley, Jane Roots, Jacqueline Koopman, and Don Gastwirth. From every one of them, I received useful comments that broadened my perspective and strengthened the text. Nancy Daniels, Martin Woollacott, Marilyn Kriegel, and my estimable agent Kathy Anderson all heroically read a full draft and provided insights and encouragement.

My debt to the three principals of *Deep Water*—Medha Patkar, Thayer Scudder, and Don Blackmore—is, of course, immense. All endured my investigations into their lives and professional activities with grace, patience, and extraordinary openness. In Domkhedi, Altadena, and Tabourie Lake, respectively, I was the recipient of extraordinary hospitality offered by the principals and their spouses, Molly Scudder and Sharron Blackmore.

From the beginning, I have been guided and informed by the work of other writers on water and dams, who help comprise the field's intellectual lineage. I owe a hefty debt to the late Marc Reisner, Patrick Mc-Cully, Sandra Postel, and Peter Gleick.

My daughter, Sarah, has endured my frequent absences from home with characteristic grace and kindness. And if not for my wife, the gloriously monikered Leslie Leslie, this book would not exist. It was her passionate environmental concern and outrage at injustice that deepened my interest in the earth's biological fate, and her involvement in environmental causes that led me to Huey Johnson, California's environmental elder statesman and precious mentor. Leslie cajoled, pondered, critiqued, and transcribed; she read every page of every draft, often on the day I wrote them. Her enthusiasm, acuity, and emotional awareness sustained me from the book's inception to the writing of these words.

November 2004
Mill Valley